TESTING AND EVALUATION OF INFRARED IMAGING SYSTEMS

Gerald C. Holst

JCD Publishing Co
2932 Cove Trail
Maitland, Florida 32751

Notice:
The Publisher and Author have taken great care in preparing the information and guidelines in this book. However, the guidelines and other material within are of a general nature only. The Publisher and Author take no responsibility with respect to the use of the information, guidelines and material furnished and assume no responsibility for any damages or costs sustained while using the guidelines.

The Publisher and Author further disclaim any and all liability for any errors, omissions or inaccuracies in the information, guidelines and material given in this book whether attributable to inadvertence or otherwise, and for any consequence arising therefrom.

Copyright © 1993 Gerald C. Holst

All rights reserved. No part of this book may be reproduced in any form by any means without permission in writing from the copyright owner.

This book is dedicated to the people who have touched my life:

> Cathy
> Mom H.
> Dad H.
> Mother P.
> Dad P.
> Liz
> Hedy
> AJ
> Kathryne-laura
> Eugene

PREFACE

Testing and evaluating infrared imaging systems is the laboratory assessment of system performance using standard tests and standard targets. This book describes the characterization of all infrared imaging systems that includes infrared (thermal) imaging systems, infrared search and track systems, IRST, machine vision systems, and line scanners. These systems have applications in aerospace, airborne reconnaissance, astronomy, medical imaging, remote sensing, robotics, and spectroscopy. Although this book emphasizes infrared imaging systems, the test methodologies apply to all imaging systems. These include electronic still cameras, image intensifiers, night-vision goggles, solid state cameras, and TVs.

Because of different test equipment, test methodologies and data analysis techniques, results from different laboratories have been difficult to compare. The objective of this book is to start the standardization process of measurement techniques. If all laboratories use the same methods then the test results should be repeatable and therefore *believable*.

This book consists of three major sections: (1) an introduction that sets the stage for system characterization [Chapters 1 through 4], (2) the tests in which the data is obtained with laboratory test equipment [Chapters 5 through 9], and (3) tests where an observer makes decisions about image quality [Chapter 10].

Chapter 1 introduces the concept of image quality and the standard metrics used to evaluate infrared imaging systems. Overall system operation is discussed in Chapter 2 with emphasis on those characteristics that affect data analysis and interpretation. Artifacts are image modifiers that are not present in the real world but are created by the imaging system. They include sampling effects, shading, AC coupling, fixed pattern noise, and image processing. Some of these effects are only objectionable when viewing specific targets. For example, many test targets accentuate sampling effects whereas sampling effects are far less obvious when viewing the real world that has texture and irregularly shaped objects. Chapter 3 discusses radiometry, the ΔT concept, and normalization. Chapter 4 describes generic test methodology with emphasis on data collection and statistical data analysis.

Resolution and focussing are presented in Chapter 5 because, typically, the same targets and test techniques are used for both tests. The only difference is that the targets are *calibrated* for resolution testing. Chapter 6 presents system responsivity which is simply an output/input transformation. It consists of two different measures depending upon the variable. With fixed target size, the output becomes a function of target intensity yielding the traditional responsivity function. The slope of the linear portion of the responsivity function is the signal transfer function (SiTF). With fixed target intensity, the output becomes a function of target size to provide the aperiodic transfer function (ATF) or the slit response function (SRF).

Noise affects system performance. Chapter 7 introduces the three-dimensional noise model. Methods of measuring high frequency and low frequency noise both temporally and spatially are presented. This separation yields fixed pattern noise and nonuniformity measures. Referring the noise to the input via the SiTF provides the noise equivalent differential temperature (NEDT). The system response to square waves and sinusoids - the contrast transfer function and modulation transfer function respectively - is described in Chapter 8. Chapter 9 discusses the geometric mapping of the object to the image. This includes field-of-view measurements and distortion. Faithful object-to-image mapping is important to machine vision systems.

In Chapter 10, the observer's interpretation of image quality as specified by the minimum resolvable temperature (MRT) and minimum detectable temperature (MDT) is described. MRT is a measure of the ability to resolve detail and is inversely related to the MTF whereas the MDT is a measure to detect something. Very small targets cannot be resolved and therefore MDT is inversely proportional to the ATF. The MTF and ATF are the system's response to high contrast noiseless targets whereas the MRT and MDT deal with an observer's ability to perceive low contrast targets embedded in noise.

Test configurations, test procedures, and data analyses are explained in detail. The theoretical basis for each test is given. However, minimal math is required to perform the test or analyze the test results. The test engineer can bypass the detailed math and go on with the tests. Many examples illustrate test configurations, dimensional analysis and data analyses.

This book does not address the issues of spectral response. They are omitted because these tests tend to be performed on the optical/detector subsystem. Methods to measure the spectral response can be found in "Detectors", D. G. Crowe, P. R. Norton, T. Limperis, and J. Mudar, in <u>Electro-Optical Components</u>, W. D. Rogatto, ed., which is Volume 4 of <u>The Infrared and Electro-Optical Systems Handbook</u>, Environmental Research Institute of Michigan, Ann Arbor Mich. (1993), and <u>Fundamentals of Infrared Detector Testing</u>, J. D. Vincent, John Wiley and Sons, New York, (1990). Characterization of the optical, detector or electronic subsystems is not specifically addressed but the system test methodology described here also applies to those subsystems (see also <u>Radiometric Calibration: Theory and Methods</u>, C. L. Wyatt, Academic

Press, Orlando (1978) and <u>Radiometric System Design</u>, C. L. Wyatt, Macmillan Publishing Co., New York (1987)).

The personnel who perform the tests are usually called the test team. However, for success, the team needs the support of design engineers, analysts, specification writers, and managers. All these people as well as the customer must understand the purpose and complexity of the measurements and the success criteria. In a global sense, all these individuals are part of the test team. This book is for the entire test team.

The author extends his deepest gratitude to all his co-workers and students who have contributed to the ideas in this book. They are too many to mention by name. The author especially thanks all those who read draft copies of the manuscript: Curtiss Webb, US Army NVESD; W. Michael Farmer, Bionetics; David A. Bloom, James D. High, Frank Link, George Spencer, Alan Taylor, C. Gilbert Young, Martin Marietta; Herb Huey, Northrop; James Sterritt, Space Optics Research Labs; Jonathan Mooney, Rome Laboratory, Hanscomb AFB; Brian Rich, Stephen W. McHugh, Santa Barbara Infrared; Tom Stapleton, Stapleton Communications; Steven Park, College of William and Mary; David Gallinger, McDonnell Douglas, and Marshall Weathersby, Nichols Research Corporation. Don Davison produced all the graphics.

Gerald C. Holst

September 1993

TABLE OF CONTENTS

CHAPTER 1: INTRODUCTION 1
   1.1. INFRARED IMAGING SYSTEMS 2
   1.2. IMAGE QUALITY 5
      1.2.1. PHYSICAL MEASURES 8
      1.2.2. SUBJECTIVE EVALUATION 9
   1.3. TEST PHILOSOPHY 12
      1.3.1. TEST PLAN 13
      1.3.2. TEST EQUIPMENT 13
      1.3.3. DATA ANALYSIS 14
      1.3.4. DOCUMENTATION 15
   1.4. FIELD TESTING 16
   1.5. SUMMARY 17
   1.6. REFERENCES 19
   EXERCISES 20

CHAPTER 2: INFRARED IMAGING SYSTEM OPERATION 21
   2.1. OPTICS 23
      2.1.1. OPTICS 23
      2.1.2. SCANNERS 26
   2.2. DETECTORS and DETECTOR ELECTRONICS 30
      2.2.1. SCANNING SYSTEMS 32
      2.2.2. STARING SYSTEMS 36
   2.3. DIGITIZATION 36
   2.4. IMAGE PROCESSING 44
      2.4.1. GAIN/LEVEL NORMALIZATION 45
      2.4.2. IMAGE FORMATTING 48
      2.4.3. GAMMA CORRECTION 48
   2.5. RECONSTRUCTION 51
   2.6. MONITORS 51
   2.7. SUMMARY 51
   2.8. REFERENCES 54
   EXERCISES 56

CHAPTER 3: BASIC CONCEPTS IN IR TECHNOLOGY . 57
   3.1. RADIOMETRY 58
      3.1.1. PLANCK'S BLACKBODY LAW 59
      3.1.2. EXTENDED SOURCE, DIRECT VIEW 61

       3.1.3. EXTENDED SOURCE IN COLLIMATOR ... 63
       3.1.4. POINT SOURCE 66
       3.1.5. ΔT CONCEPT 73
  3.2. NORMALIZATION 79
  3.3. SPATIAL FREQUENCY 82
  3.4. SUMMARY 85
  3.5. REFERENCES 88
  EXERCISES 88

CHAPTER 4: GENERAL MEASURING TECHNIQUES .. 90
  4.1. BLACKBODIES 90
  4.2. TARGETS 93
       4.2.1. STANDARD EMISSIVE TARGETS 94
       4.2.2. NOVEL EMISSIVE TARGETS 98
       4.2.3. REFLECTIVE TARGETS 99
       4.2.4. TARGET WHEELS 101
       4.2.5. PASSIVE TARGETS 101
       4.2.6. SOURCES AS TARGETS 102
       4.2.7. SPECIAL CONSIDERATIONS 103
  4.3. COLLIMATORS 105
  4.4. ATMOSPHERIC TRANSMITTANCE and
      TURBULENCE 112
  4.5. MOUNTING FIXTURE 115
  4.6. DATA ACQUISITION 115
  4.7. STATISTICAL ANALYSIS 124
  4.8. SUMMARY 128
  4.9. REFERENCES 131
  EXERCISES 132

CHAPTER 5: FOCUS and SYSTEM RESOLUTION 134
  5.1. TEST METHODOLOGY 135
  5.2. FOCUS TESTS 137
       5.2.1. VISUAL METHOD 137
       5.2.2. ANALOG VIDEO AMPLITUDE METHOD .. 141
       5.2.3. MTF METHOD 142
       5.2.4. EDGE DETECTION ALGORITHMS 144
  5.3. SYSTEM RESOLUTION 144
       5.3.1. DEFINITIONS 144
       5.3.2. RESOLUTION TARGETS 153
  5.4. SUMMARY 155
  5.5. REFERENCES 159
  EXERCISES 159

CHAPTER 6: SYSTEM RESPONSIVITY 160
  6.1. SIGNAL TRANSFER FUNCTION 160
    6.1.1. SYSTEM RESPONSE 160
    6.1.2. RESPONSIVITY UNIFORMITY 167
    6.1.3. SiTF TEST PROCEDURE 169
  6.2. APERIODIC TRANSFER FUNCTION AND
       SLIT RESPONSE FUNCTION 176
  6.3. DYNAMIC RANGE and LINEARITY 184
  6.4. SUMMARY 188
  6.5. REFERENCES 190
  EXERCISES 191

CHAPTER 7: SYSTEM NOISE 192
  7.1. NOISE STATISTICS 193
  7.2. THREE-DIMENSIONAL NOISE MODEL 195
  7.3. NOISE MEASUREMENTS 200
    7.3.1. NOISE EQUIVALENT DIFFERENTIAL
           TEMPERATURE 205
    7.3.2. FIXED PATTERN NOISE 211
    7.3.3. NONUNIFORMITY 214
    7.3.4. NOISE EQUIVALENT FLUX DENSITY ... 219
    7.3.5. NOISE POWER SPECTRAL DENSITY 222
    7.3.6. MEAN-VARIANCE TECHNIQUE 225
  7.4. SUMMARY 229
  7.5. REFERENCES 232
  EXERCISES 233

CHAPTER 8: CONTRAST, MODULATION and
PHASE TRANSFER FUNCTIONS 234
  8.1. CONTRAST TRANSFER FUNCTION 238
  8.2. MODULATION TRANSFER FUNCTION 247
    8.2.1. INTRODUCTION 247
    8.2.2. ISOPLANATISM 250
    8.2.3. SPATIAL SAMPLING 250
    8.2.4. SYSTEM LINEARITY 254
    8.2.5. TEST EQUIPMENT DIGITIZATION 256
    8.2.6. BACKGROUND REMOVAL 260
    8.2.7. JITTER 261
    8.2.8. NOISE 264
    8.2.9. LSF SYMMETRY 265
    8.2.10. FOURIER TRANSFORM 266
    8.2.11. AMPLITUDE NORMALIZATION 270

xii *TESTING & EVALUATION OF IR IMAGING SYSTEMS*

       8.2.12. FREQUENCY SCALING 271
       8.2.13. TEST CONFIGURATION MTF 275
       8.2.14. MTF TEST PROCEDURES 275
  8.3. PHASE TRANSFER FUNCTION 279
  8.4. SUMMARY 282
  8.5. REFERENCES 285
  EXERCISES 287

CHAPTER 9: GEOMETRIC TRANSFER FUNCTION 288
  9.1. FIELD-OF-VIEW 289
  9.2. GEOMETRIC DISTORTION 294
  9.3. SCAN NONLINEARITY 298
  9.4. MACHINE VISION PERFORMANCE 302
       9.4.1. MEASURING LOCATION AND
               COUNTING OBJECTS 302
       9.4.2. ALGORITHM EFFICIENCY 305
  9.5. SUMMARY 305
  9.6. REFERENCES 307
  EXERCISES 307

**CHAPTER 10: OBSERVER INTERPRETATION
OF IMAGE QUALITY** 308
  10.1. OBSERVER VARIABILITY 309
       10.1.1. FREQUENCY OF SEEING RESPONSE 310
       10.1.2. VISUAL ANGLE 312
       10.1.3. NOISY IMAGES 316
       10.1.4. OBSERVER QUALIFICATION 317
       10.1.5. MEASURED VALUES VERSUS
                SPECIFICATIONS 319
  10.2. MRT and MDT TESTS 323
       10.2.1. SUBJECTIVE TEST METHODOLOGY 328
       10.2.2. SEMIAUTOMATIC TEST
                METHODOLOGY 334
       10.2.3. OBJECTIVE TEST METHODOLOGY 337
  10.3. SUMMARY 340
  10.4. REFERENCES 342
  EXERCISES 344

INDEX 345

TESTING AND EVALUATION OF INFRARED IMAGING SYSTEMS

1

INTRODUCTION

Infrared imaging system characterization is the laboratory assessment of image quality using both subjective and quantitative methods. In the most general form, system characterization is the measurement of various input-to-output transformations. The particular test specifies the input and it is well-defined both in shape and intensity. The sensor converts the input into a measurable output which is the *image*. It may be in many different forms such as that displayed on a monitor, the analog video signal or digital data stored in a memory. The output may be measured in volts, monitor brightness, analog to digital converter units, ADUs, or may be an observer's impression of image quality (Figure 1-1).

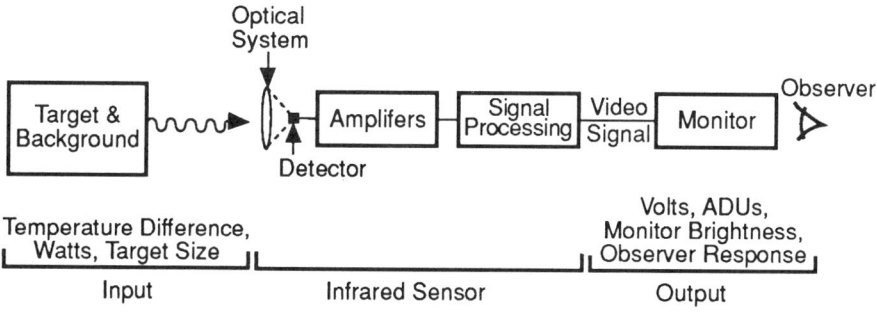

Figure 1-1: Generic sensor operation that applies to all electronic imaging systems.

Good image quality must first be defined before any measurement program is undertaken. As image quality definitions are refined, test methodologies become more precise. Since *good* image quality is so elusive, many parameters are used to define the characteristics of an infrared imaging system. With complete unambiguous definitions, it becomes possible to write clear test procedures, identify appropriate test equipment, identify data collection methodologies and select appropriate statistical methods for data analysis. A thorough test plan permits reproducible results of both the physical evaluation and subjective evaluation of infrared imaging systems.

2 TESTING & EVALUATION OF IR IMAGING SYSTEMS

Image evaluation, the physical measure of image quality, has been and continues to be a powerful tool for predictive modeling, system design, performance assessment, and quality control. It is used not only to verify the final design but the results are used by analysts to validate their various models. Predictive modeling drives future system design, system requirements and quality assurance specifications. Connecting specifications to well-understood physical parameters makes the designer, manufacturer and customer more confident that the design objectives have been achieved. Image evaluation includes resolution, responsivity, noise, modulation transfer function, contrast transfer function, and distortion measurements.

A diverse set of parameters, both internal and external to the observer, governs perceived image quality. It includes a whole set of psychophysical aspects that cannot be easily quantified. It is a subjective measure that cannot be placed upon an absolute scale. There exist large variations in observers' judgment as to the rank ordering of images according to *quality* or from *best* to *worst*. As a result, there can be considerable variability in any test involving observer interpretation. However, the data analysis methodology presented in Chapter 10 treats the observer variability statistically. This provides reproducible minimum resolvable temperature (MRT) and minimum detectable temperature (MDT) results.

1.1. INFRARED IMAGING SYSTEMS

In 1969, Hudson[1] listed over 100 separate applications for thermal imaging systems. He divided the list into 4 major categories: military, industrial, medical and scientific. Each category was then subdivided into (1) search, track and range, (2) radiometry, (3) spectroradiometry, (4) thermal imaging, (5) reflected flux, and (6) cooperative source. The list is surprisingly complete and only a few new applications have been added since then. Today, two broad categories are in usage: military and commercial. Table 1-1 highlights a few applications in each category. Military and commercial systems are similar in basic design; but each system is built for a specific purpose. As a result, military and commercial systems tend to be described by different performance parameters. Some generic differences are listed in Table 1-2. For measurement purposes, these systems are combined into two broad categories: imaging and machine vision. Imaging systems have an observer as the image interpreter whereas machine vision imagery is assessed by hardware and/or software. An important subset of machine vision is the infrared search and track type systems (IRST). These systems are designed to detect point sources. The specific system design depends upon the application, the atmospheric transmittance and availability of optics and detectors.

Table 1-1
THERMAL IMAGING APPLICATIONS

COMMUNITY	APPLICATIONS	
MILITARY	Reconnaissance Target acquisition Fire control Navigation	
COMMERCIAL	CIVIL	Law enforcement Fire fighting Border patrol
	ENVIRONMENTAL	Earth resources Pollution control Energy conservation
	INDUSTRIAL	Maintenance Manufacturing Non-destructive testing
	MEDICAL	Mammography Soft tissue injury Arterial constriction

Table 1-2
TYPICAL DESIGN REQUIREMENTS

DESIGN AREAS	MILITARY	COMMERCIAL
Vibration stabilized	Yes	Usually not required
Image processing algorithms	Application specific (e.g., target detection or automatic target recognition)	Menu-driven multiple options
Resolution	High resolution (resolve targets at long distances)	Typically not an issue since the image can magnified by moving closer
Image processing time	Real time	Real time not usually required
Target signature (sensitivity requirements)	Usually just perceptible (low NEDT)	Usually high contrast target (NEDT not necessarily a dominant design factor)

4 TESTING & EVALUATION OF IR IMAGING SYSTEMS

Due to the atmospheric spectral transmittance, electronic imaging system design is partitioned into six generic spectral regions (Figure 1-2) of which four are associated with infrared imaging systems. The ultraviolet (UV) region ranges in wavelength from 0.2 to 0.4 μm. The visible spectral region ranges in wavelength from 0.4 to 0.7 μm. TVs, electronic still cameras and most solid state cameras* operate in this region. The near infrared imaging spectral region (NIR) spans approximately 0.7 to 1.1 μm. Low light level TVs (LLLTV), image intensifiers, star light scopes and night vision goggles operate in this region. For historical reasons, the UV, visible and NIR technologies have developed their own terminologies. For this book, the first infrared imaging band is the short wavelength infrared imaging band (SWIR) which approximately covers 1.1 to 2.5 μm. The second infrared band is the mid-wavelength infrared (MWIR) spectral region that covers approximately 2.5 to 7.0 μm. The third infrared band is the long wavelength infrared (LWIR) spectral band. It covers the spectral region from approximately 7 to 15 μm. The fourth infrared band is the far infrared (FIR) or very long wave infrared (VLWIR) region. It applies to all systems whose spectral response extends past 15 μm. The MWIR and LWIR regions are sometimes called the first and second thermal bands respectively. The terminology is author dependent. Typical detectors used in the MWIR are PtSi, PbS, PbSe, and InSb. HgCdTe detectors are the most popular for the LWIR.

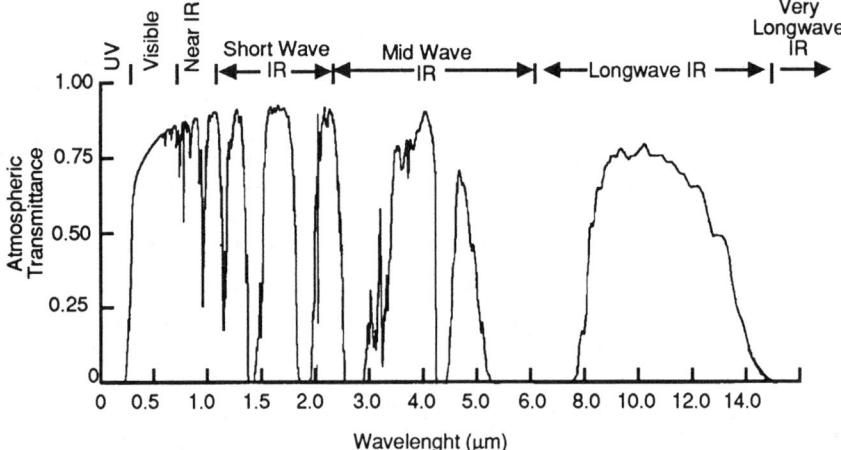

Figure 1-2: Atmospheric transmittance over a 1 km path length calculated by LOWTRAN7 for a mid-latitude location, 23 km visibility, and a rural aerosol.

Solid state cameras are popularly called CCD cameras because most contain charge-coupled device detector arrays.

INTRODUCTION 5

The precise spectral response for any system depends upon the design. To say a system is a LWIR system only means its response is somewhere in the LWIR region. For example, a LWIR system may have a spectral response from 7.7 to 11 μm or 8 to 12 μm. The test engineer must know the precise spectral response so that he may select the appropriate source, targets, and collimator to correctly assess system performance.

1.2. IMAGE QUALITY

System characteristics, observer experience, scene content, atmospheric transmittance, monitor settings and a variety of miscellaneous factors affect the perceived image quality (Figure 1-3). The cause of poor imagery cannot be discerned simply by examining the image. The purpose of laboratory testing is to be able to quantify the most important variables that support good image quality (Figure 1-4). Optimal image quality cannot be determined by simply viewing an image. To determine if optimum has been achieved, it is necessary to verify focus, adjust gain, adjust level, tune the monitor, measure the response to different sized targets, and measure the response to different target intensities (i.e., quantify the various input-to-output transformations shown in Figure 1-4).

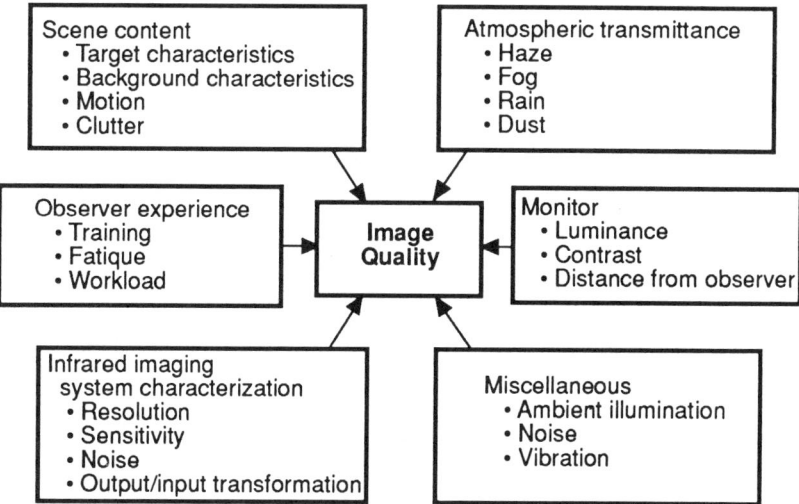

Figure 1-3: Image quality contributors. All these factors affect the perceived quality.

6 *TESTING & EVALUATION OF IR IMAGING SYSTEMS*

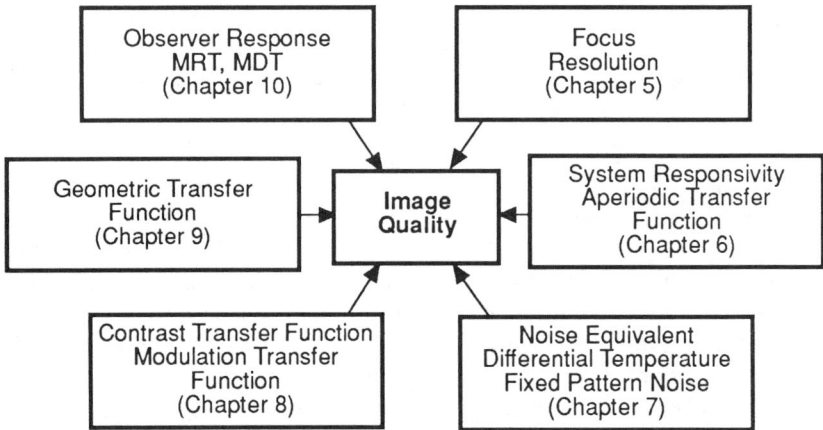

Figure 1-4: Laboratory evaluation of image quality. Measurement techniques for each image quality contributor are provided in Chapters 5 to 10.

An overwhelming majority of imaging quality discussions center on resolution or sensitivity. Resolution has been in use so long that it is thought to be something fundamental which uniquely determines image quality. It is the smallest detail that can be perceived. It may be specified by a variety of sometimes unrelated metrics such as the Rayleigh criterion, the spatial frequency at which the modulation transfer function drops to 2%, or the instantaneous-field-of-view. The appropriate metric depends upon the intended system application. Resolution does not include the effects of target contrast nor system noise. There is a clear distinction between resolution (the ability to see detail) and detection (the ability to see something).

Sensitivity deals with the smallest signal that can be detected. It is that signal which produces a signal-to-noise ratio of unity at the system output. Sensitivity is dependent upon the light-gathering properties of the optical system, the responsivity of the detector and the noise of the system. It is independent of resolution.

Overall system response depends on both sensitivity and resolution. As shown in Figure 1-5, the MRT is bounded by sensitivity and resolution considerations. Different systems (Figure 1-6) may have different MRTs. System A has a better sensitivity. It has a lower MRT at low spatial frequencies. At mid-range spatial frequencies, the systems are approximately equivalent and

INTRODUCTION 7

it can be said they provide equivalent performance. System B has better resolution and can display finer detail than System A. Figure 1-6 illustrates that neither sensitivity, resolution nor any other single parameter can be used to compare systems. Rather, all the quantities illustrated in Figure 1-4 must be specified for complete system-to-system comparisons.

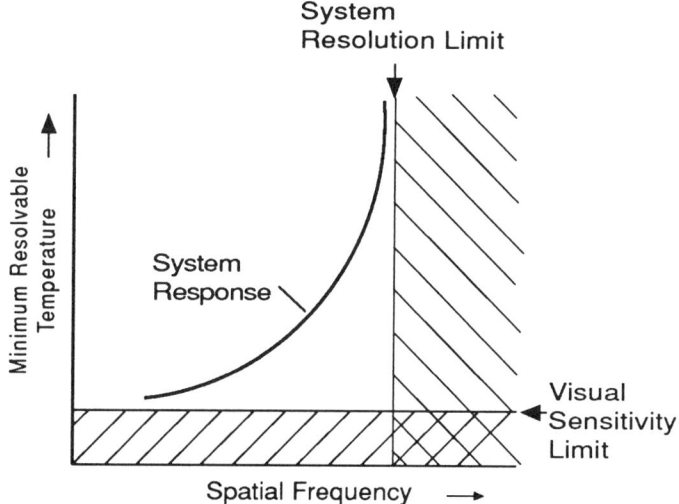

Figure 1-5: MRT is bounded by the system's resolution and the visual sensitivity limit.

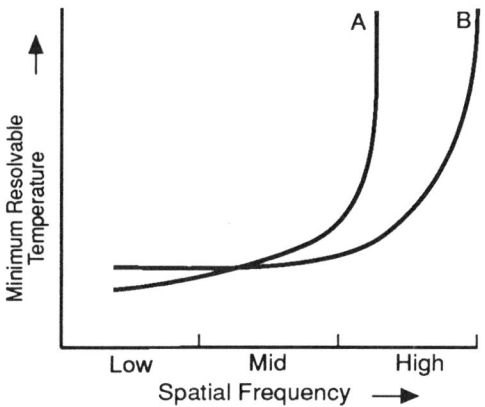

Figure 1-6: Two systems with different MRTs. Whether system A is better than B depends upon the specific application.

8 *TESTING & EVALUATION OF IR IMAGING SYSTEMS*

1.2.1. PHYSICAL MEASURES

Physical measures of image quality define or describe various image statistics compared with a base line or ideal image. In the laboratory, the test environment is usually well controlled. The atmospheric transmittance, monitor, scene content and miscellaneous effects shown in Figure 1-3 (page 5) are usually well characterized. Laboratory measures of image evaluation are given in Table 1-3. The test procedures, data collection techniques and data analysis methodologies are discussed in the chapters listed.

Table 1-3
IMAGE EVALUATION PARAMETERS

CHAPTER	TEST
CHAPTER 5	Focus Resolution
CHAPTER 6	Signal transfer function Responsivity uniformity Aperiodic transfer function Slit response function Dynamic range
CHAPTER 7	Noise equivalent differential temperature Fixed pattern noise Non-uniformity Noise equivalent flux density Noise power spectral density
CHAPTER 8	Contrast transfer function Modulation transfer function Phase transfer function
CHAPTER 9	Field-of-view Geometric distortion Measuring location and counting objects Algorithm efficiency

1.2.2. SUBJECTIVE EVALUATION

The MRT and MDT are laboratory measures of image quality and have become standardized in the infrared imaging community. The MRT or MDT is sometimes obtained only once. However, every time the system is turned on the observer subconsciously evaluates the image quality according to his internal rating scale.

Many arguments between system designers and the user are on what represents good imagery. These disagreements could have been averted if the test objectives were clearly defined and rating scales were accepted by all. Before creating a subjective evaluation scale, the following questions must be answered: What is the intended use of the system? What is the task? Does the rating system answer the right question? What is the user being asked to rate? How will the experimental results be used?

Why do we need a rating scale? To reiterate, every time the system is turned on, all observers subconsciously evaluate the image quality according to their internal rating scale and these scales should be standardized. The Cooper-Harper scale[2] has become a standard in the US Air Force and is used extensively for evaluation of aircraft handling qualities and the associated mental workload. This test has been modified[3] for assessing workload situations where perceptual and communication activity is present. It provides a reliable assessment of operational effectiveness on a relative basis. This scale combines usability with perception, monitoring and evaluation. If properly structured, all observers should assign approximately the same rating to a system.

Following the Cooper-Harper methodology, a subjective rating scale for infrared imaging systems can be established. The observer first determines if adequate performance is obtainable. Then he redefines the rating by selecting a more detailed category. This results in a rating of one to ten that represents a global impression of performance. The rating system presented in Figure 1-7 is only a guideline. A rating system should not rate something as excellent or optimum for this implies that no further improvement is necessary. The scale should be a consensus between the manufacturer and the customer about what is excellent, acceptable, and unacceptable. Examples of each rating level must be clearly defined and, if possible, actual recorded imagery of these examples should be available.

10 *TESTING & EVALUATION OF IR IMAGING SYSTEMS*

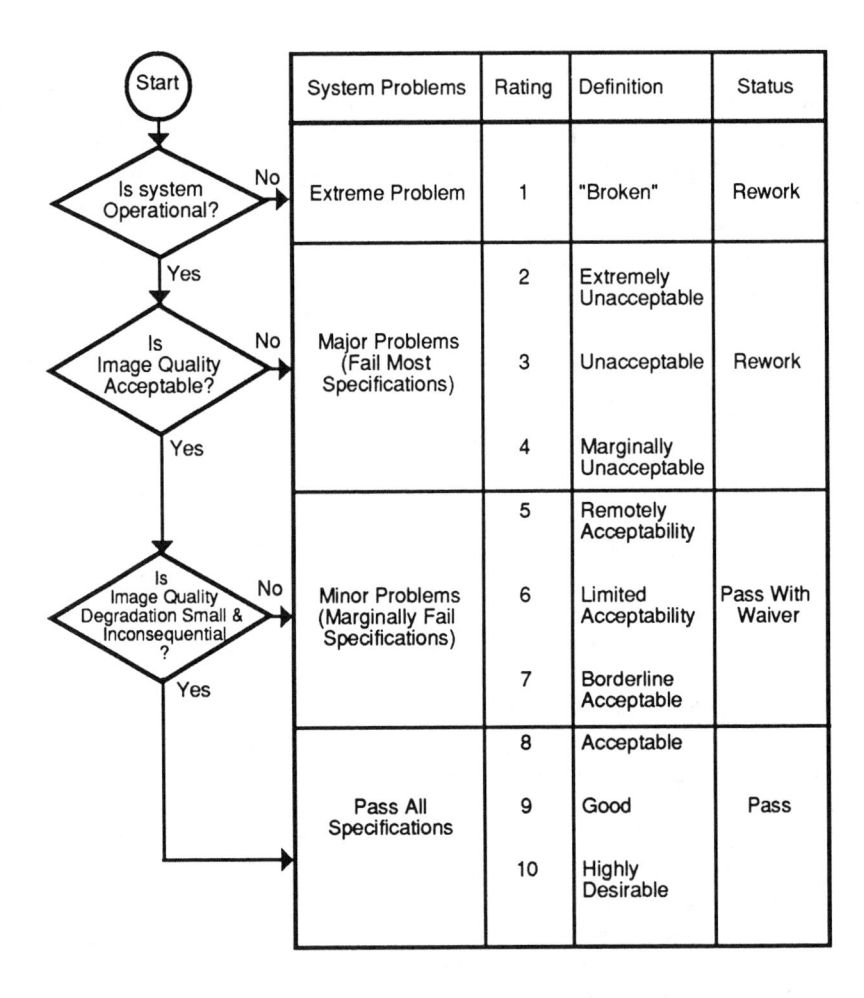

System Problems	Rating	Definition	Status
Extreme Problem	1	"Broken"	Rework
Major Problems (Fail Most Specifications)	2	Extremely Unacceptable	
	3	Unacceptable	Rework
	4	Marginally Unacceptable	
Minor Problems (Marginally Fail Specifications)	5	Remotely Acceptability	
	6	Limited Acceptability	Pass With Waiver
	7	Borderline Acceptable	
Pass All Specifications	8	Acceptable	
	9	Good	Pass
	10	Highly Desirable	

Figure 1-7: An example of a subjective rating scale. A standardized rating scale leads to consistent definitions of image quality.

INTRODUCTION 11

Rating	Examples	System Impact
1	No Image	"Fix It"
2	Barely Recognize Large Objects Double Images (Forward/Interlace Errors) Dead Detectors In Center of Field of View	Redesign is Mandatory
3	Flashing Lines in Center of Field of View Extremely noisy Excessive Microphonics	Redesign is Recommended
4	Cannot be Adequately Focussed Wrong Dynamic Range: Too Low (No Contrast) or Too High (Scene Saturation)	Redesign is Desirable
5	Noise Level is 2 Times Specification	Design Change is Recomended
6	Flashing Lines in Periphery Noticeable Microphonics Streaking/Blooming	Design Change is Very Desirable
7	Slight Defocusing During Environmental Extremes Perceptible Fixed Pattern Noise Perceptible Shading or Narcissus	Design Change Would Enhance System
8	Just Pass Specifications Just Detect/Recognize Targets at Required Ranges	Design is Adequate
9	Passes All Specifications with Reasonable Margin	Design is Good
10	Greatly Exceeds All Specifications Noise is 1/2 Specification Detect/Recognize Targets at Extremely Long Ranges	Design Appears Optimized

Figure 1-7: Continued.

1.3. TEST PHILOSOPHY

Standardized test methodologies, test equipment requirements, and data analyses will provide consistent results. Clear performance requirements is the starting point for a comprehensive test program. Test plans should be written and followed to the letter. Data analysis methodology similarly should be carefully followed. Proper testing includes the acquisition, analysis and storage of volumes of data. With some forethought, appropriate equipment and procedures can be set up to handle the data flow.

Test procedures can be developed only when the operation of the system is fully understood (discussed in Chapter 2) and the test objectives are fully understood. An intimate understanding of how the system operates and the expected results are a fundamental necessity. Test early and in all phases of system buildup. It is false economy to skimp on testing during the early phases of buildup. Finding problems well after substantial buildup represents an impact on cost and delivery[4]. It is important that any problem be quickly identified during each phase of build up. For example, if a discrepancy is found, further diagnostic testing should be in the test plan for backup. This saves valuable time in test set and personnel utilization. This requires that personnel knowledgeable in system design, performance, and testing be always available.

Complete characterization may take hundreds of hours. While this may be viable for the prototype, it does not lend itself to production testing. Thus a decision that trades off adequate testing consistent with delivery time and cost must be made. Carefully select the appropriate tests and the detail of the testing for each level of buildup. Hence a dilemma: reduced testing implies that the statistics of quality control have been established and the statistical variations associated with all measured parameters are known. However, these variations are not known until exhaustive testing has been performed on perhaps a hundred units to establish the data variance. The most often asked question is "What is the minimum number of tests required?" The answer lies in the statistical variations associated with all the parameters and the intended use of the system. Prototypes should be fully tested. As system development progresses, the tests can be reduced to those that provide a snapshot of system performance. At a minimum these include the NEDT and the MRT for imaging systems.

Too much emphasis is often placed upon meeting a delivery date at the exclusion of proper testing and test result documentation. However, proper testing cannot be done overnight and the testing activity cannot wait until the night before delivery to decide how or what to test. Often the actual test takes a short amount of time whereas the setup procedure and data analysis can be time consuming.

Before the initial test, estimate expected signals and noise voltages. Analyze representative data to verify that the data analysis techniques and plotting/printing routines are appropriate. Sources of errors should be identified and addressed. For those areas where anomalous data are possibilities, identify multiple test methods. It is important to reduce as much data as possible during the test. This provides confidence in the data. Errors found can usually be corrected immediately. If the errors are found a week later, the unit under test, test equipment or test personnel may not be readily available to repeat the test.

1.3.1. TEST PLAN

A written test plan states the objectives, requirements, data analysis methodology with accepted error ranges (error bars) and the success criteria of the test. The test configuration and measuring conditions (discussed in Chapter 4) must be clearly stated. The plan also specifies how the data should be presented (graphical and/or tabular). A good plan will have the attributes listed in Table 1-4.

Table 1-4
TEST PLAN ATTRIBUTES

Simple
Easy to read
Clearly defines test equipment
Clearly define test procedures
Clearly define data analyses

1.3.2. TEST EQUIPMENT

Because of the complexities of modern infrared imaging systems, it is essential to employ state-of-the-art measuring equipment. Invest in good test equipment. Buying marginal equipment because it is less expensive is a false economy. If the test procedure must be changed, it is highly probable that the marginally acceptable equipment becomes useless. However, locating adequate test equipment becomes a challenge as infrared imaging system technology

14 *TESTING & EVALUATION OF IR IMAGING SYSTEMS*

advances. With the emerging requirements for more sensitive and higher resolution infrared imaging systems, the test equipment capabilities become a concern. Chapter 4 describes test equipment requirements.

1.3.3. DATA ANALYSIS

Statistics provide the uncertainty levels in the estimated population averages and standard deviations (discussed in Chapter 4). Sampling statistics are used to determine the number of data points for adequate analysis. Understanding the causes of MRT variability allows the separation of observer variability from system response (the desired result). Only with a sound statistical approach can data be presented with confidence.

If a system fails a test, rely on the test engineer to alert engineering/manufacturing personnel that a problem exists. Although rarely stated, the test engineer is relied upon to identify the most likely cause of the problem. Only the test engineer, who has seen hundreds of similar units, can rapidly identify subtle changes based upon his internal rating scale. Rely on the test team to identify faults, not to simply pass or fail the equipment. If the test team fails the unit with no comment, then the test must be repeated to identify the problem areas. One way of documenting anomalies is to record the image on videotape with complete annotation for later playback and interpretation. If a system fails a test, evaluate both the system and the test procedure: Is the system substandard (the purpose of the test) or did the system fail because of an artifact created by the test equipment, data analysis, test procedure or test engineer?

The pass/fail criterion simply suggests that the system will perform adequately. For most tests, the value is first obtained and then compared to the specification to determine if the system has passed. For example, to determine if the NEDT is less than the specification, it is necessary to measure the NEDT. The exact value is important to the design engineer and quality control engineers. Since the exact value is obtained, it is prudent to record and save the value for trend analysis. Table 1-5 illustrates a typical data presentation scheme that provides the specification, test result, and the required pass/fail.

INTRODUCTION 15

Table 1-5
TYPICAL DATA PRESENTATION

TEST	SPECIFICATION	TEST RESULT	PASS/FAIL
NEDT	0.1° C	0.08° C	PASS
MRT @ 1 cy/mrad	0.2° C	0.12° C	PASS
Non-uniformity	3%	3.5%	FAIL

1.3.4. DOCUMENTATION

It is extremely important to archive data results for trend analyses and fatigue analyses. In Figure 1-8, there is a gradual trend in increasing noise level. It is reasonable to predict that the manufacturing process is out of control and that sometime in the future the system will fail. With this trend data, time is available to alter manufacturing techniques without ever failing the specification. In Figure 1-9, several components were changed which adversely affected the noise level. The possibility of failing the noise specification was only a matter of time after the change. With trend analysis, it is possible to determine exactly why the noise increased. This determination would not be possible with a simple pass/fail criterion.

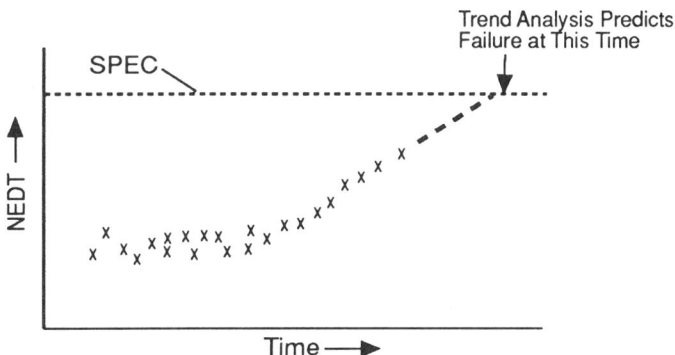

Figure 1-8: Trend analysis can predict anticipated failure.

16 TESTING & EVALUATION OF IR IMAGING SYSTEMS

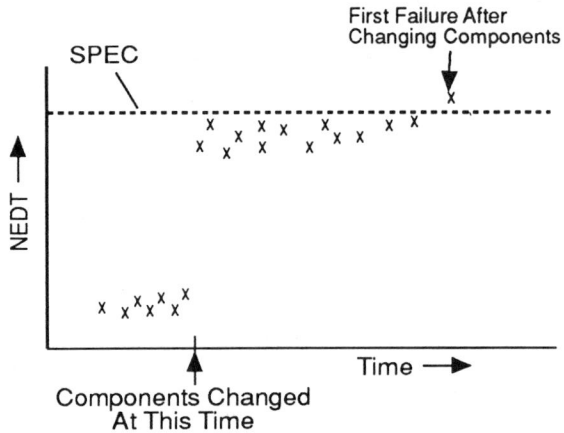

Figure 1-9: Potential failure caused by redesign.

The process of recording, analyzing, interpreting, presenting, and archiving data must be done rigorously and faithfully. Data should be in a format useful to multiple users and multiple analysts. With the widespread use of personal computers, it is prudent to store all data in a format consistent with the more common spreadsheets. The test team has achieved all its goals when the final report is written.

1.4. FIELD TESTING

Nothing provides a more conclusive demonstration of adequate performance than actual field testing. The main objective of field testing is to prove that the system will operate under a variety of field conditions. The field test should also prove that the user, who often lacks the technical skill of an engineer, can use the system safely and efficiently.

Field tests are difficult to control. As such they are demonstrations and are not usually intended for complete system characterization. Field test results often show a large variation in results due to the variation in environmental conditions (Figure 1-3: page 5). In poor weather conditions, it may not be possible to perceive the target due to signal-to-noise limitations. This is an effect of the environment and not the infrared imaging system. To appropriately analyze field

INTRODUCTION 15

Table 1-5
TYPICAL DATA PRESENTATION

TEST	SPECIFICATION	TEST RESULT	PASS/FAIL
NEDT	0.1° C	0.08° C	PASS
MRT @ 1 cy/mrad	0.2° C	0.12° C	PASS
Non-uniformity	3%	3.5%	FAIL

1.3.4. DOCUMENTATION

It is extremely important to archive data results for trend analyses and fatigue analyses. In Figure 1-8, there is a gradual trend in increasing noise level. It is reasonable to predict that the manufacturing process is out of control and that sometime in the future the system will fail. With this trend data, time is available to alter manufacturing techniques without ever failing the specification. In Figure 1-9, several components were changed which adversely affected the noise level. The possibility of failing the noise specification was only a matter of time after the change. With trend analysis, it is possible to determine exactly why the noise increased. This determination would not be possible with a simple pass/fail criterion.

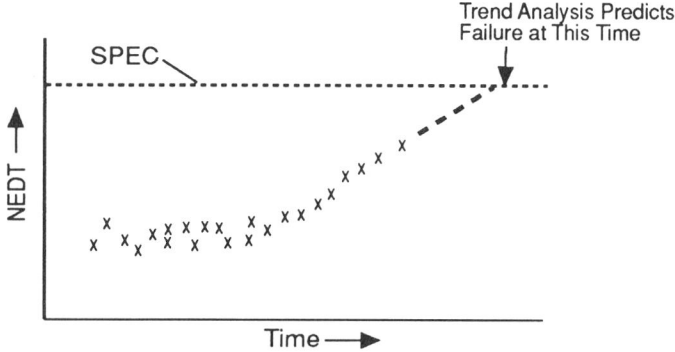

Figure 1-8: Trend analysis can predict anticipated failure.

16 TESTING & EVALUATION OF IR IMAGING SYSTEMS

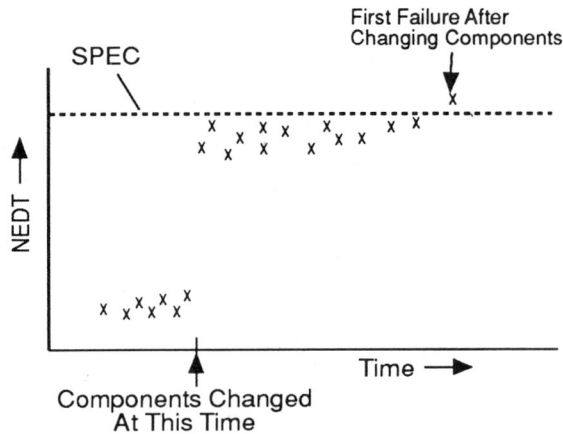

Figure 1-9: Potential failure caused by redesign.

The process of recording, analyzing, interpreting, presenting, and archiving data must be done rigorously and faithfully. Data should be in a format useful to multiple users and multiple analysts. With the widespread use of personal computers, it is prudent to store all data in a format consistent with the more common spreadsheets. The test team has achieved all its goals when the final report is written.

1.4. FIELD TESTING

Nothing provides a more conclusive demonstration of adequate performance than actual field testing. The main objective of field testing is to prove that the system will operate under a variety of field conditions. The field test should also prove that the user, who often lacks the technical skill of an engineer, can use the system safely and efficiently.

Field tests are difficult to control. As such they are demonstrations and are not usually intended for complete system characterization. Field test results often show a large variation in results due to the variation in environmental conditions (Figure 1-3: page 5). In poor weather conditions, it may not be possible to perceive the target due to signal-to-noise limitations. This is an effect of the environment and not the infrared imaging system. To appropriately analyze field data, the field test engineer must be fully aware of variations in target signature[5,6] and the atmospheric transmittance[7]. Analysis of field data is difficult since it is nearly impossible to standardize the test conditions. Since the weather and scene contrast change rapidly, the test results often cannot be reproduced enough times to provide good statistical analyses. Field calibration[8] is more difficult than laboratory calibration and requires extra care.

A word of caution: The system may pass all test specifications in the laboratory but not perform well in the field. This may happen if the laboratory environment does not simulate field conditions. For example, the system may perform poorly in a hot desert environment but well in an air-conditioned laboratory. This may be caused by the cooler's inability to handle the increased heat load. As a detector heats up, its noise and responsivity change. Field power may not be well regulated. Auxiliary field equipment may produce power surges as it cycles on and off. Grounding may also be different. It takes the concentrated effort of design engineers, laboratory test engineers and field test engineers to determine why performance may be different in the laboratory and field.

1.5. SUMMARY

Reviewing the specifications is the starting point for any test activity. The test team must understand the system operation and the intended use of the infrared imaging system. The test philosophy includes a simple, easy-to-read test plan, clear test procedures, and definitive data analysis methodologies and documentation procedures. A thorough test plan addresses the following: What is the intended use of the system? What are the appropriate test parameters? Are the specifications clearly written? How will the experimental results be used? Documentation includes archiving of data and the final report.

Analysts, designers, test engineers, specification writers, managers and the customer must all participate in testing and evaluating of infrared imaging systems. Each, however, will make decisions based upon needs. If the test plan is clearly and concisely written and the tests are performed correctly, then each user will glean the information needed from the test results (Table 1-6). All test team members should review the test plan to insure that it is comprehensive and that the tests provide the results needed[9]. This may introduce some additional tests that are not called out in the specifications.

Test requirements vary according to the system application; use Table 1-7 as a guide. For example, subjective tests (MRT and MDT) are not appropriate for machine vision systems where the image is evaluated by hardware and/or software. As the system moves into production, the number of tests may be reduced. Specific test requirements are determined jointly by the customer and the manufacturer.

Table 1-6
NEEDS OF VARIOUS USERS

USER	NEED
Analyst	Validation of performance models
Designer	Validation of subsystem performance in terms that are specifically related to subsystem parameters (e.g., electronic bandwidth, optical MTF, etc.)
Manager	Pass/fail criteria and the impact on cost and delivery if the unit fails
Customer	System will perform as intended

Every time the system is turned on, the observer subconsciously evaluates the image quality according to his internal rating scale. With agreed-upon rating scale, every observer should rank-order the imagery approximately the same. A rating scale is also useful to describe transient events such as flashing lines and focus problems. The eye can see many subtle effects that are often missed during the traditional data collection activity.

If difficulty is encountered in determining a subjective rating scale or if MRT or MDT is not the appropriate measure for the specific application, then it may be necessary to create a new figure of merit. Biberman[10], Lloyd[11], and Farrell and Booth[12] present many data describing the physical aspects of images that are considered acceptable to the eye in terms of good image quality.

Finally, tests were originally developed to characterize US common module systems. As appropriate, the test procedures have been modified to characterize common module derivatives and systems containing staring arrays. These systems tend to employ similar optical designs, detectors, and electronics. As new technologies emerge such as uncooled detectors[13] and quantum well detectors, test methodologies may change.

Table 1-7
TYPICAL TESTS FOR PRODUCTION TESTING

TEST	General Imaging	Machine Vision
MRT	x	
MDT	x	
NEDT	x	x
Non-Uniformity	x	x
MTF	x	x
ATF		x
Field-of-view	x	x
Distortion	x	x
Resolution	x	
Focus	x	x
Dynamic range	x	
Responsivity	x	

1.6. REFERENCES

1. R. D. Hudson Jr., Infrared System Engineering, Chapters 16 to 19: John Wiley and Sons, New York (1969).
2. G. E. Cooper and R. P. Harper, "The Use of Pilot Rating in the Evaluation of Aircraft Handling Qualities", NASA Ames Research Center, Moffett Field: NASA Technical Note TN-D-5153 (1969).
3. W. W. Wierwille and J. G. Casili, "A Validated Rating Scale for Global Mental Workload Measurement Applications", in Proceedings of the 27th annual Human Factors Society Meeting, pp. 129-132 (1983).
4. T. F. Greene, "Infrared Sensor Test Requirements and Critical Issues", in Infrared Scene Simulation: Systems, Requirements, Calibration, Devices and Modeling, R. B. Johnson and M. J. Triplett, eds.: SPIE Proceedings Vol. 940, pp. 18-25 (1988).
5. M. V. Mansi and I. A. Walls, "Prediction of Temperature Differences in Background Scenery", in Passive Infrared Systems and Technology, H. M. Lamberton, ed.: SPIE Proceedings Vol. 807, pp. 61-68 (1987).
6. T. M. Lillesand and R. W. Kiefer, Remote Sensing and Image Interpretation, pp. 402-414: John Wiley and Sons, New York (1979).

7. R. Richter, "Infrared Simulation Model SENSAT-2", Applied Optics, Vol. 26(12), pp. 2376-2382, (1987).
8. P. Chevrette and D. St-Germain, "Field Calibration Software for Thermal Imagers and Validation Experiments", in Characterization, Propagation and Simulation of Infrared Sources, W. P. Watkins, F. H. Zegel and M. J. Triplett, eds.: SPIE Proceedings Vol. 1311, pp. 2-15 (1990).
9. C. Coles, W. Phillips and J. D. Vincent, "Reporting Data for Arrays with Many Elements", in Infrared Imaging Systems: Design, Analysis, Modeling and Testing II, G. C. Holst, ed.: SPIE Proceeding Vol. 1488, pp. 327-333 (1991).
10. L. M. Biberman, ed., Perception of Displayed Information: Plenum Press, New York, (1973).
11. J. M. Lloyd, Thermal Imaging Systems: Plenum Press, New York (1975).
12. R. J. Farrell and J. M. Booth, Design Handbook for Imagery Interpretation Equipment, Reprinted with corrections: Boeing Aerospace Company, Seattle, Wash (February 1984).
13. R. E. Flannery and J. E. Miller, "Status of Uncooled Infrared Imagers", in Infrared Imaging Systems: Design, Analysis, Modeling and Testing III, G. C. Holst, ed.: SPIE Proceedings Vol. 1689, pp. 379-395 (1992).

EXERCISES

1. Explain why observer age, IQ, education, training, motivation, personality and fatigue may affect subjective evaluation of image quality.
2. Explain how vibration and extraneous noise may affect subjective evaluation of image quality. List three other extraneous factors.
3. In Figure 1-6 (page 7), system A performs better at low spatial frequencies than system B. Describe a scenario in which low spatial frequency response is important. (Hint: Very large targets are associated with low spatial frequency.)
4. In Figure 1-6 (page 7), system B performs better at high spatial frequencies than systems A. Describe a scenario in which high spatial frequency response is important. (Hint: Very small targets (or target detail) are associated with high spatial frequency.)
5. List the advantages and disadvantages of the pass/fail criteria.
6. Describe two environmental factors that affect field testing.
7. List three applications for a MWIR system. (Hint: Aircraft engines which are approximately 800 K emit significant radiation in the MWIR region.)
8. List three applications for a LWIR system. (Hint: Most terrestrial objects which are at 300 K emit significant radiation in the LWIR region.)
9. Figure 1-2 (page 4) illustrates a typical terrestrial atmospheric transmittance. On the other hand, deep space has nearly unity transmittance at all wavelengths. A "star wars" satellite has a detector system which has a spectral response of 4.2 to 4.3 μm. Why? (Hint: Draw a picture depicting the earth, atmospheric layer and a missile.)

2
INFRARED IMAGING SYSTEM OPERATION

An infrared imaging system consists of many subsystems. Each subsystem processes information differently and may create artifacts or variations in the processed image that were not present in the original scene. If not understood, these artifacts may be construed as data anomalies or, in an extreme, as a malfunctioning system. These artifacts include, but are not limited to, cosine$^N\theta$ variation[*], AC coupling effects, phasing effects, incomplete gain/level normalization, line-to-line interpolation, and gamma correction. Through appropriate test design, the test engineer can minimize artifacts and interpret the data results correctly when the artifacts are present.

The functional electro-optical block diagram shown in Figure 2-1 illustrates five major subsystems: optics and scanner, detector and detector electronics, digitization, image processing, and image reconstruction. These subsystems were simply called the infrared sensor in Figure 1-1 (page 1). Figure 2-1 applies to scanning and staring systems. The specific design depends upon the number of detector elements and the required output format. The optics image the radiation onto the detector(s). Scanners optically move the detector's instantaneous-field-of-view (IFOV) across the field-of-view to produce an output voltage proportional to the local scene intensity. In a scanning system, the output of a single detector represents the scene intensity across a line. With a staring array there is no scanner and adjoining detector outputs provide scene variations.

The detector is the heart of the infrared system because it converts infrared radiation into a measurable electrical signal and target spatial information into electrical temporal information. Amplification and signal processing creates an *image* in which voltage differences represent scene intensity differences. The detector electronics is matched to the detector characteristics and required output. Many systems digitize signals because of the relative ease to create an image. In addition, many digital image enhancement algorithms are available. Machine vision systems typically operate in the digital domain.

It may be necessary to format the *image* into signals and timing that is consistent with monitor requirements. To produce a linear input-to-output system, a gamma correction algorithm removes the monitor's nonlinear response. The monitor may or may not be an integral part of the infrared imaging system.

[*]*Text books illustrate cosine$^4\theta$ shading. Depending upon the optical design, shading follows cosine$^N\theta$ where $2 \leq N \leq 4$.*

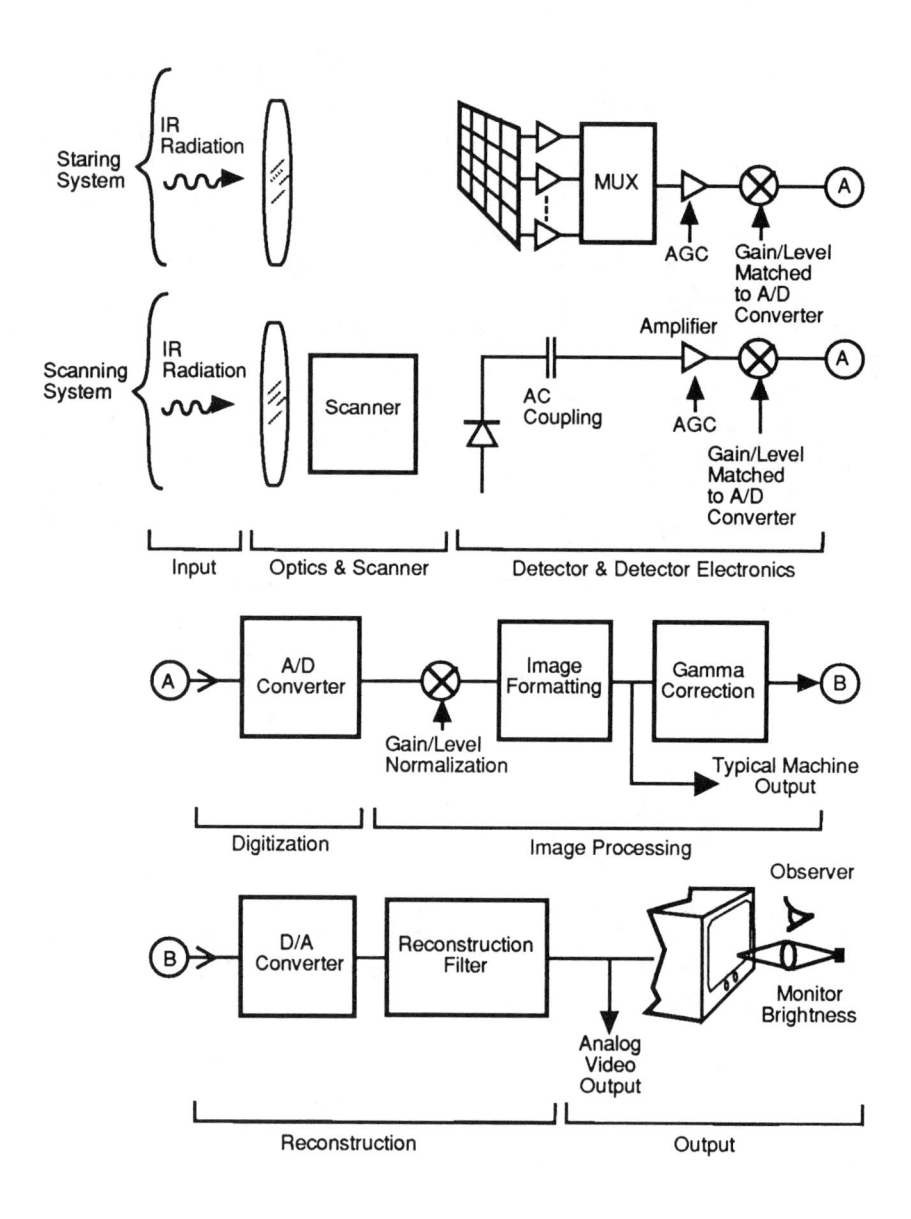

Figure 2-1: Typical electro-optical functional block diagram.

IR IMAGING SYSTEM OPERATION 23

Figure 2-1 is a representative block diagram. Each detector will have its own amplifier. The amplifier outputs are multiplexed together and then digitized. The number of channels multiplexed together depends upon the specific design. Systems may have several multiplexers and several A/D converters operating in parallel.

This chapter highlights the operation of the five major subsystems with specific attention given to the artifacts created by those subsystems. The artifacts appear as *apparent* data anomalies and they affect specific test results. At the end of each test procedure in Chapters 5 though 10, there is a table listing possible causes for poor test results. Those tables list the artifacts described in this chapter.

2.1. OPTICS and SCANNER

2.1.1. OPTICS

Broad spectral response systems may have considerable chromatic aberrations. Whether the chromatic aberrations are noticeable depends upon the spectral content of the source. Optical subsystems are usually "color-corrected" which means that the chromatic aberrations have been minimized at certain wavelengths but will be aberrated at other wavelengths. As test targets move across the field-of-view, aberrations can cause variations in all input-to-output transformations in which target detail is important (e.g., resolution, MTF, CTF, ATF, MRT, and MDT).

$Cosine^N\theta$ shading effect is a geometrical phenomenon that reduces the intensity reaching off-axis detectors. It depends upon the optical design and the physical location of apertures and the detectors. With a single detector in a scanning system (Figure 2-2a), the detector is always on-axis and there is no $cosine^N\theta$ shading. It is the scanner that allows the detector to sense off-axis radiation. With a linear array of detectors (Figure 2-2b), the $cosine^N\theta$ variation is in the direction of linear array (cross scan direction). Staring arrays exhibit radially symmetrical $cosine^N\theta$ roll-off as measured from the center of the field-of-view (Figure 2-2c). Figure 2-3a illustrates the signals from three different lines for the linear scanning array depicted in Figure 2-2b. Figure 2-3b illustrates the signals for the staring array shown in Figure 2-2c. Gain/level normalization minimizes this effect. Figure 2-3c depicts the normalized signal as a function of line number for a system suffering from extreme $cosine^N\theta$ effects.

24 *TESTING & EVALUATION OF IR IMAGING SYSTEMS*

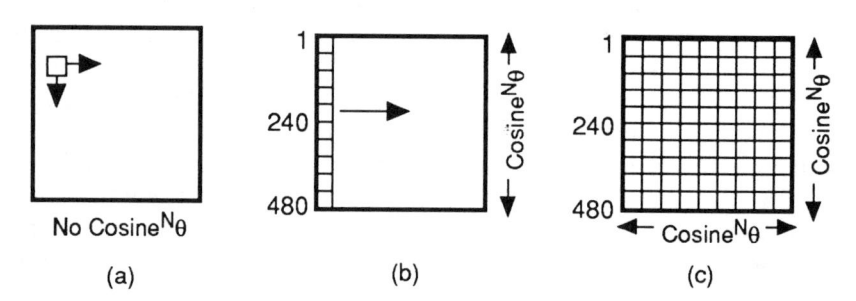

Figure 2-2: Cosine$^N\theta$ Effect. (a) Single detector, (b) scanning linear array consisting of 480 x 1, and (c) staring array consisting of 480 x 480 detectors.

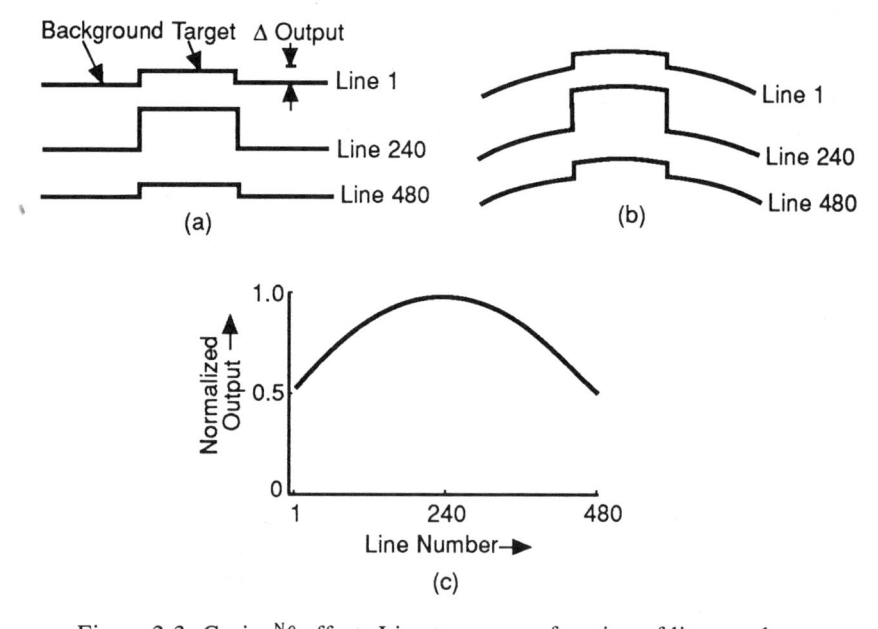

Figure 2-3: Cosine$^N\theta$ effect. Line traces as a function of line number for (a) a linear scanning array and (b) a staring array. (c) The normalized output as a function of line number for a linear scanning array. Gain/level normalization minimizes cosine$^N\theta$ for staring arrays.

Vignetting appears in a very similar way (i.e., reduction in output across the field-of-view) but does not obey cosine$^N\theta$. Vignetting tends to be rotationally symmetrical and is independent of the number of detectors and the scanning scheme. Both vignetting and cosine$^N\theta$ cause location sensitive variations in all input-to-output transformations in which the source intensity is the input (e.g., responsivity, NEDT, uniformity, MRT, and MDT). Electronic boost can compensate for the cosine$^N\theta$ roll-off. Here, the image will appear cosmetically appropriate. However, any boost circuitry also will increase the noise so that the signal-to-noise ratio remains constant.

Narcissus[1,2] occurs when there is a reflection from a lens or window such that the detector can *see* itself (Figure 2-4). Narcissus becomes more pronounced and annoying as the reflected image becomes more in focus. The result is a dark spot (cold spike) on the monitor. A system with a single detector produces a single dark area. For a detector array, a long rectangular dark area, consistent with the array dimensions, is seen. The narcissus signal may be an image of the cold shield. The gain/level normalization algorithm usually removes the narcissus signal in staring arrays. For scanning systems, reflected energy may vary as a function of scan angle. The resultant variation in the output is called scan noise. Efficient anti-reflection coatings, designing lenses such that the surfaces are not confocal with the detector and tilting flat windows minimizes narcissus and scan noise.

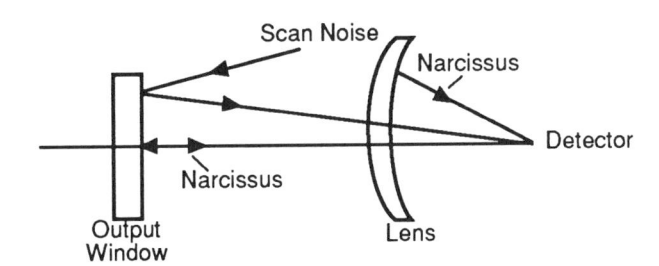

Figure 2-4: Narcissus and scan noise are present in all infrared imaging systems. Appropriate optical design can reduce these effects below perceptibility.

Glare is not normally a concern during laboratory evaluation of infrared systems. It becomes apparent when high intensity sources are near the field-of-view. Lens imperfection or multiple reflections produce glare (Figure 2-5). Efficient anti-reflection coatings and appropriate baffling reduce multiple reflections. Glare superimposes extraneous signals onto the image thereby making data analysis more difficult.

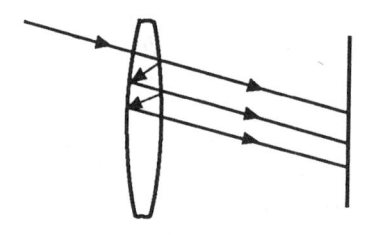

Figure 2-5: Multiple reflections produce glare.

2.1.2. SCANNERS

The function of the scanner is to dissect the image sequentially and completely. That is, the scanner moves the detector IFOV around the system FOV in a way that is consistent with the monitor requirements. A large variety of scanning schemes are available. There is nothing inherently good or bad with any particular scanning scheme: each has its own advantages and disadvantages. The most common scanners are rotating drums, polygons, refractive prisms and oscillating mirrors. With scanning systems, the detector output creates the image during the active scan time and during the inactive scan time, the detector output is ignored. The inactive time provides the time necessary for the scanner to come into the appropriate position for the next frame or scan line. Detector characteristics may dictate the scan direction. Figure 2-6 illustrates a system with a single detector with unidirectional (raster) scan. SPRITE (Signal processing in the element) detectors require an unidirectional scan. Figure 2-7 illustrates a linear array that uses bidirectional scanning (parallel scan). US common module systems employ bidirectional scanners. Figure 2-8a illustrates a series of detectors whose outputs are summed together to provide time-delay and integration, TDI. Systems with TDI usually require unidirectional scan. The scan velocity must be matched to the time delay in the integrating element (Figure 2-8b). Scan linearity is important when using TDI to avoid geometric distortion and MTF degradation. A full staring array (Figure 2-2c: page 24) does not have a scanner.

IR IMAGING SYSTEM OPERATION 27

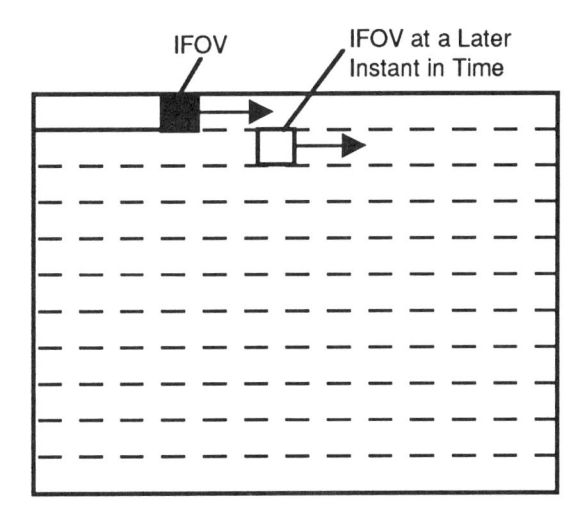

Figure 2-6: Single detector scan pattern employing unidirectional (raster) scanning.

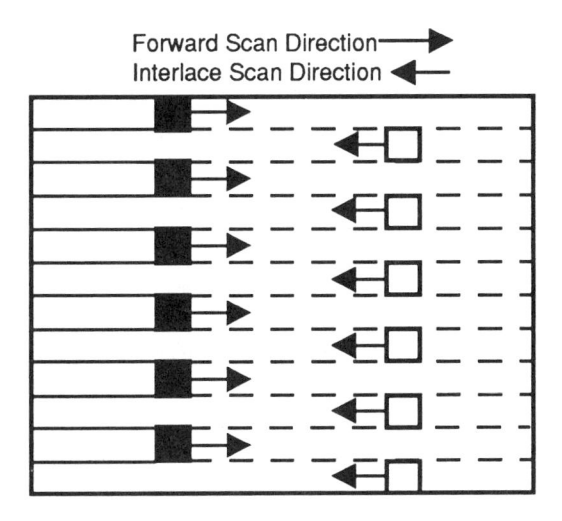

Figure 2-7: Linear array employing bidirectional scan and 2:1 interlace.

28 TESTING & EVALUATION OF IR IMAGING SYSTEMS

Figure 2-8a: Multiple detectors operating in a pure serial scan mode.

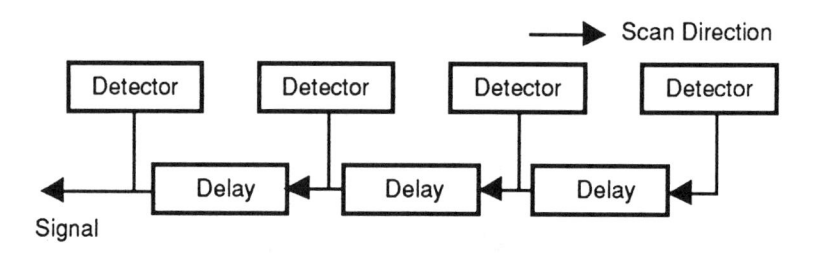

Figure 2-8b: Multiple detectors operating in a pure serial scan mode require TDI delay elements.

Since all objects emit infrared radiation, if the detector senses radiation from any object other than the scene, an artifact (scan noise) is introduced. During the inactive scan time the detector may sense radiation from the housing or from other objects. Narcissus and scan noise (Figure 2-4: page 25), which are nonexistent in the visible, or at most noncritical, pose special design problems for infrared systems. These anomalies can appear in the image in a variety of ways ranging from slight brightness gradations over the entire FOV to a sharp bright or dark area. The exact category for a myriad of image defects is not standardized and is author dependent. Other terms include shading and ghosting. Although the mechanisms differ, they result from the detector sensing variations in the infrared radiation and are usually dependent upon scan angle.

IR IMAGING SYSTEM OPERATION 29

Uniformity (or nonuniformity) is a measure of infrared radiation variations. Since scanning systems are often not rotationally symmetrical, images may be affected differently in the vertical direction than in the horizontal direction.

☞

Example 2-1
SCAN EFFICIENCY

What is the scan efficiency for a LWIR common module system?

The US common module system contains 180 detectors that are interlaced by an oscillating mirror to create 360 infrared scene lines (Figure 2-7: page 27). As illustrated in Figure 2-9 the detector senses radiation from the FOV when the scanner is between -5° and +5°. This is the active scan time. The scan mirror requires time to decelerate, stop, reverse direction and accelerate to a linear velocity as shown in Figure 2-10. The common module system operates at a frame rate of 30 Hz. Each frame lasts 1/30 sec and each field lasts 1/60 sec. If the active scan time is 12 msec for each field, then the scan efficiency is 12/16.7 = 72%. During the inactive time ($\theta < $ -5° and $\theta > $ +5°), the detector senses radiation from parts of the scene and the housing (Figure 2-9c). This radiation introduces scan noise if it reaches the detector during the active scan time.

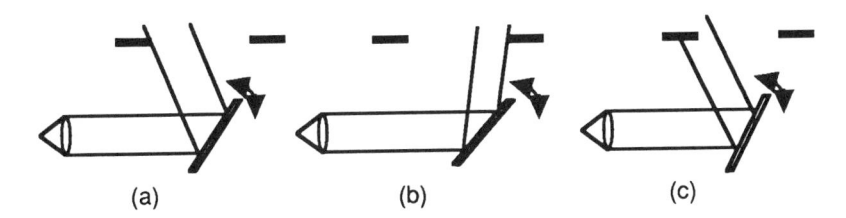

(a)   (b)   (c)

Figure 2-9: Beam location as a function of scan angle when employing an oscillating mirror. (a) Scan angle = +5°, (b) scan angle = -5° and (c) mirror looking outside the field-of-view.

30 *TESTING & EVALUATION OF IR IMAGING SYSTEMS*

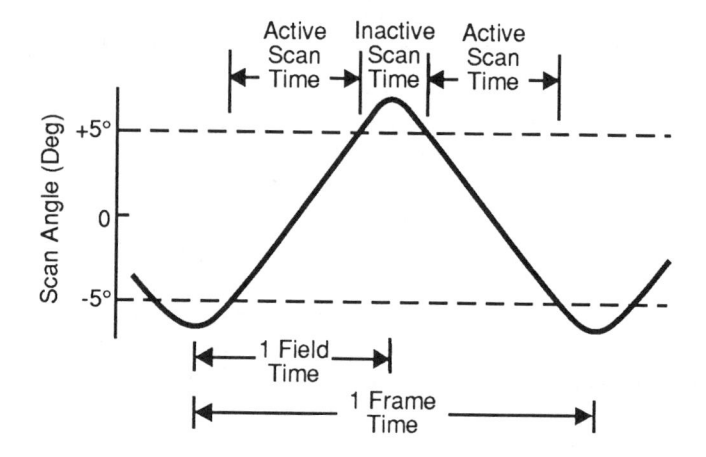

Figure 2-10: Scan angle as a function of time for an oscillating mirror. The ratio of active scan time to one field time is the scan efficiency.

2.2. DETECTORS and DETECTOR ELECTRONICS

For most infrared detectors, the responsivity and noise are functions of the detector temperature. Figure 2-11 illustrates a typical temperature profile for a US common module LWIR HgCdTe detector. The detector temperature depends upon the cooling capacity of the cooler, the ambient temperature and the heat load induced by the electronics near the detector. As temperatures change, the detector characteristics change and all input-to-output transformations change. This change could take place during an experiment or from day-to-day.

Each detector will have its own amplifier. Each detector/amplifier combination will have a different gain and level offset (Figure 2-12). These variations produce fixed pattern noise (FPN). For a linear array of elements (Figure 2-2b: page 24), each line may have a different gain and offset. This produces FPN in the vertical direction only and will appear as horizontal streaks. For staring arrays (Figure 2-2c), each detector will have a different response resulting in a two-dimensional FPN. Electronic gain/level normalization removes FPN. Incomplete removal results in residual FPN noise. Single element systems (Figure 2-6: page 27) and pure serial systems (Figure 2-8: page 28) do not have FPN. FPN affects all input-to-output transformations that are functions of noise (e.g., NEDT, MRT, and MDT).

IR IMAGING SYSTEM OPERATION 31

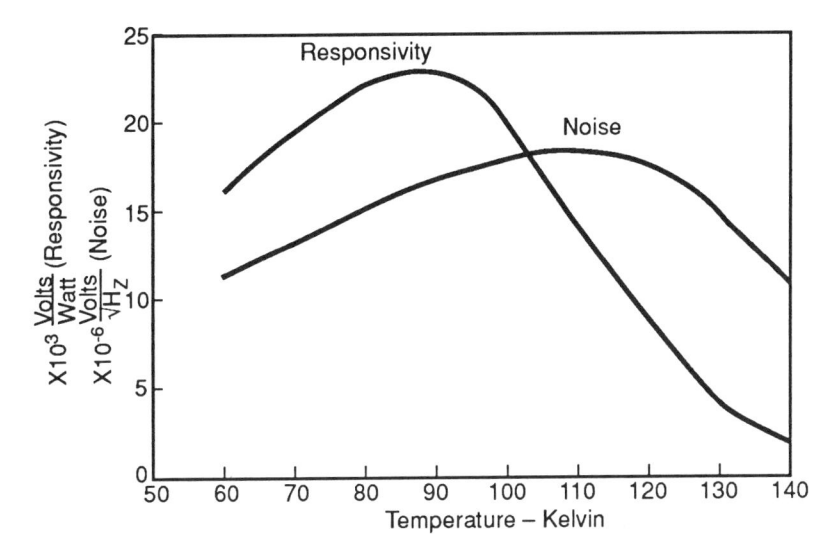

Figure 2-11: Typical US common module HgCdTe detector temperature characteristics. Most detectors are optimized to operate at liquid nitrogen temperatures (77 K).

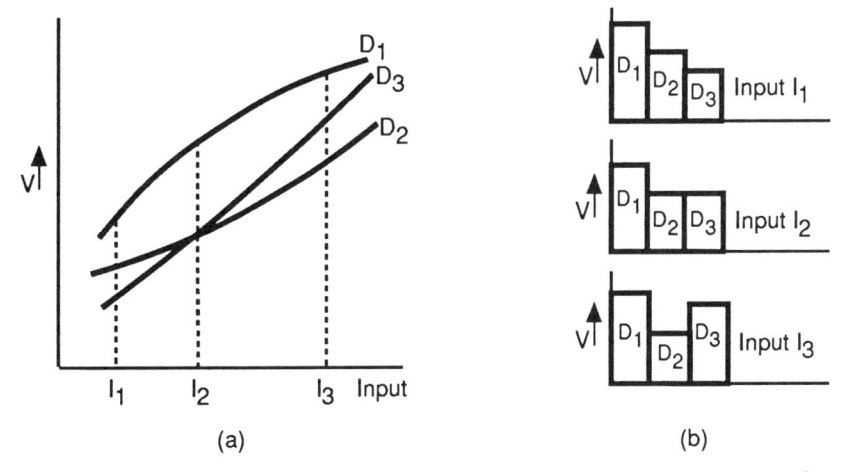

Figure 2-12: Responsivity for three differences detectors (D_1, D_2, and D_3). (a) Typical responsivities and (b) detector outputs for three different input intensities. The responsivity variations cause fixed pattern noise.

32 TESTING & EVALUATION OF IR IMAGING SYSTEMS

When detector elements are defective, the output of an adjoining detector often replaces the output of the defective element. This is called strapping the detectors together. This results in two output lines (for a scanning system) with the same signal and noise characteristics.

The responsivity function provides information about linearity, dynamic range and saturation. The entire responsivity function is typically S-shaped due to the dark current at low input irradiances and due to saturation at high irradiances. For staring systems, saturation occurs when the charge well is filled. For many imaging systems, the average ambient value is a high irradiance level. To maximize the A/D converter dynamic range, the detector output produced by this high irradiance level is scaled into the center of the A/D converter range. Saturation in the positive and negative directions about this average value is then limited electronically by the dynamic range of the A/D converter.

There also may be an automatic gain control (AGC) circuit. The AGC insures that all signals are within the dynamic range of the subsequent electronics. Background intensity, target intensity, target size or target location may affect the AGC. Some systems have an AGC window. Only targets within this window affect the AGC. Ideally, the gain should be fixed for all measurements. Otherwise the target intensity or size will affect the gain and confound the results. The gain may be fixed either electronically (e.g., turned off) or optically by using a specially designed AGC target (described in Chapter 4).

2.2.1. SCANNING SYSTEMS

AC coupling is used in scanning systems to suppress a large uninformative background and so that the small variations about ambient temperature can be amplified. The average level of each detector output is held constant such that the displayed image no longer is radiometrically correct. The same object can appear either as light gray or dark gray depending upon the immediate surroundings. Figure 2-13a illustrates a temperature controlled board. The radiometric levels are shown in Figure 2-13b but after AC coupling, the video lines appear as that shown in Figure 2-13c. Figure 2-13d illustrates the resultant image. For AC coupled systems, *hot* and *cold* are relative to the local background and not the absolute temperature. For example, a 300 K target will appear *hot* on a 299 K background and *cold* against a 301 K background. Very large uniform areas will appear as the same intensity on the monitor independent of the absolute temperature. AC coupling affects only the line that is AC

IR IMAGING SYSTEM OPERATION 33

coupled. Any measurement of signal difference must be made on the same AC coupled line (same detector output). DC restoration is a process that adds signal to an AC coupled image so that it appears radiometrically correct.

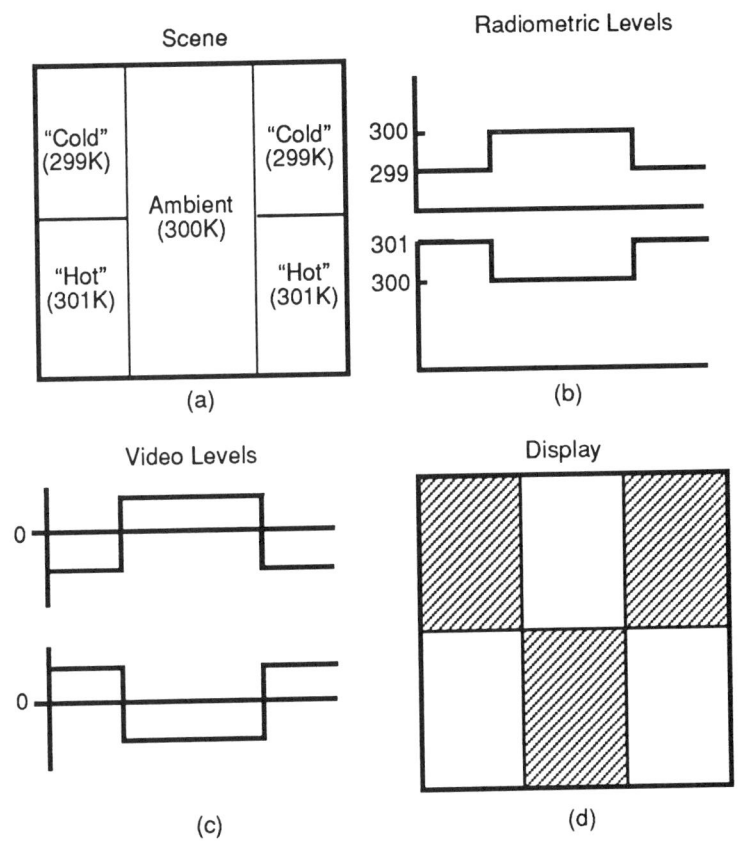

Figure 2-13: The AC coupling problem. (a) A calibrated target board with three different temperatures, (b) the radiometric values, (c) the output after AC coupling, and (d) the appearance on a monitor.

AC coupling causes droop. The amount seen depends upon the target size and circuit design. Small targets exhibit minimal droop whereas large targets exhibit the most. With bidirectional scanning, the interlace scan is achieved with the scanner moving in the opposite direction from the forward scan. AC coupling appears reversed from the forward to interlace scan and time appears "reversed" (Figure 2-14). When displayed on a monitor, droop appears reversed from one scan to the next. To minimize this alternating and often distracting effect, sometimes two lines are summed together. Summing lines together reduces vertical resolution and reduces the noise level. The averaging of lines also means that adjacent lines are no longer independent and any test requiring statistically independent lines must be carefully considered. DC restoration cannot remove droop.

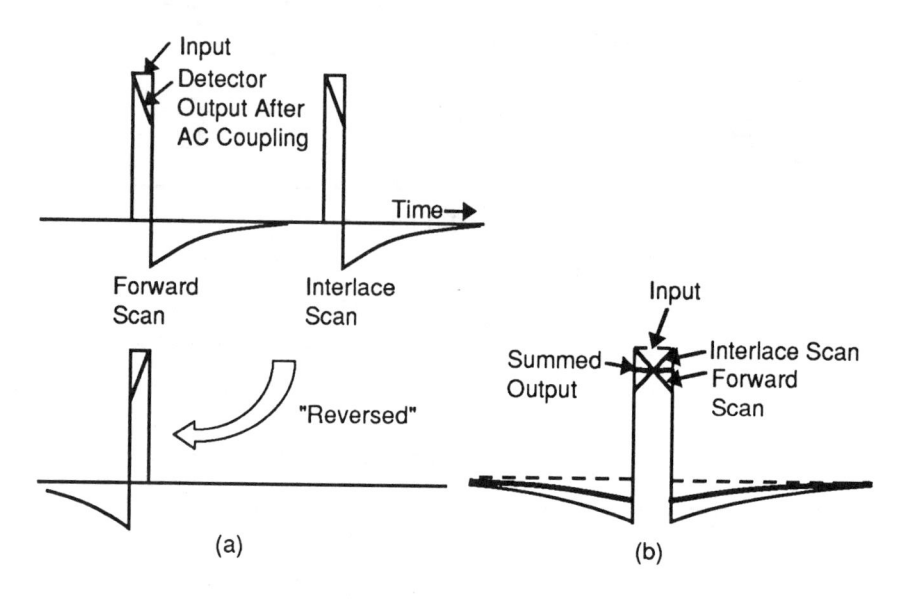

Figure 2-14: The displayed image of a target with bidirectional scanning and significant droop. (a) Output of one detector for forward and interlace scans. The interlace scan information is "reversed" in time. (b) Summing 2 lines to minimize droop.

With AC coupling, the output of the detector circuitry depends upon both the scene content and what the detector *sees* during the inactive time. For example, the relative difference between the detector housing temperature (seen during the inactive time) and the scene temperature (seen during the active scan time) affects the level of each line (Figure 2-15). With severe droop, the displayed intensity of each line will vary across the line depending upon the relative difference between the housing temperature, T_H, and the scene temperature, T_S. It is this variation that will be measured during the nonuniformity tests. By adjusting the scene temperature, nonuniformity can be minimized[3]. However, with some systems, due to internal motors and electronics, the internal housing may not stay at a constant temperature and the nonuniformity results may constantly change.

IR IMAGING SYSTEM OPERATION 35

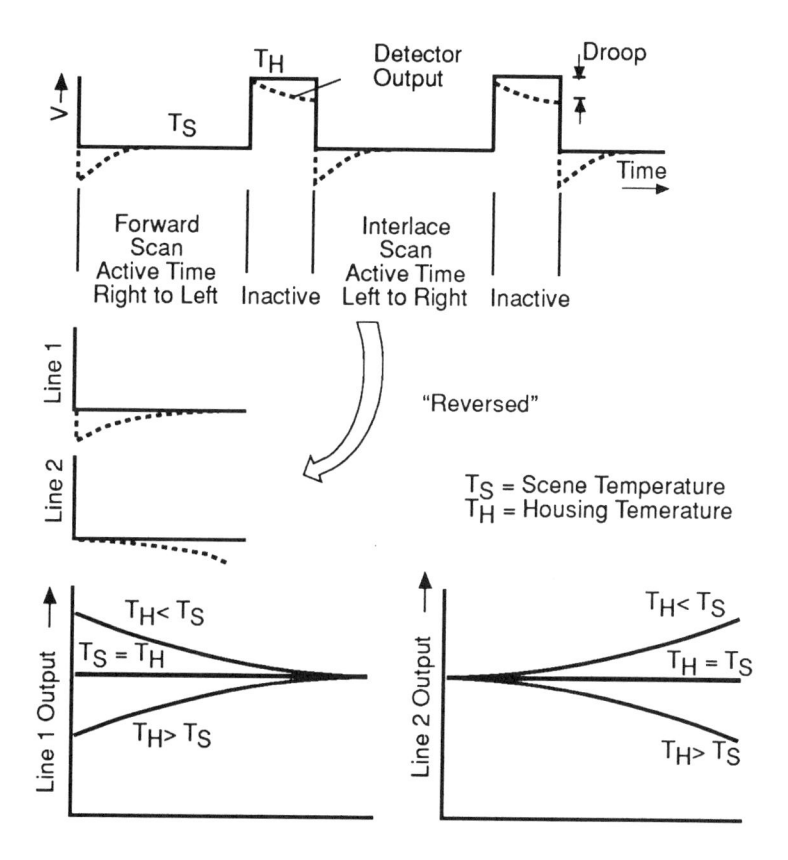

Figure 2-15. The displayed output with bidirectional scanning. If the detector *sees* a large variation in radiation from the inactive scan time to the active scan time, the displayed image will appear nonuniform in intensity. Because time is "reversed" for the interlace field, the nonuniformity is different for the two lines.

2.2.2. STARING SYSTEMS

Figure 2-16a illustrates a full focal plane array (no scanner) where the detectors are contiguous. Many detectors do not completely fill the cell area (Figure 2-16b). The ratio of active detector element area to the cell area is the fill factor. If a small object is imaged onto one of these dead areas between the detectors, there will be no output. As the small object moves, its image will move on and off the active detector element resulting in a twinkling of the object. Finite fill factors affect all input-to-output transformations in which detail is important (e.g., resolution, MTF, CTF, ATF, MDT, and MRT).

Fill Factor = 100%
Full Focal Plane Array
(a)

Fill Factor = $\dfrac{A_{Detector}}{A_{Cell}}$
Low Fill Factor Focal Plane Array
(b)

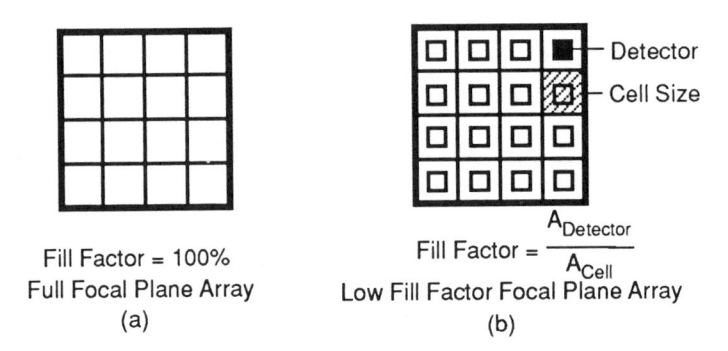

Figure 2-16: Fill Factor definition. (a) 100% fill factor and (b) finite fill factor.

2.3. DIGITIZATION

Infrared systems spatially sample the scene because of the discrete location of the detectors. Staring systems, because of detector location symmetry, tend to have equal sampling rates in both the horizontal and vertical directions. With scanning systems, the detector output in the scan direction can be electronically digitized at any rate whereas in the cross scan direction, the detector locations define the sampling rate. Therefore, in a scanning system, the sampling rate may be different in the horizontal and vertical directions.

Signals can be undersampled, appropriately sampled or oversampled. The highest frequency that can be faithfully reproduced by a digital system is one-half the sampling rate. Any input signal above the Nyquist frequency, f_N, (which is defined as one half the sampling frequency, f_s) will be aliased down to a lower frequency. That is, an undersampled signal will appear as a lower

IR IMAGING SYSTEM OPERATION 37

frequency (Figure 2-17). Any input frequency, f_x above f_N will appear as $2f_N - f_x$. After aliasing, the original signal can never be recovered. Aliasing is obvious when viewing periodic targets such as those use for system characterization. Undersampling creates Moire patterns (Figure 2-18). Diagonal lines appear to have jagged edges or "jaggies". Aliasing is not obvious when viewing complex scenery and, as such, is rarely reported during actual system usage although it is always present.

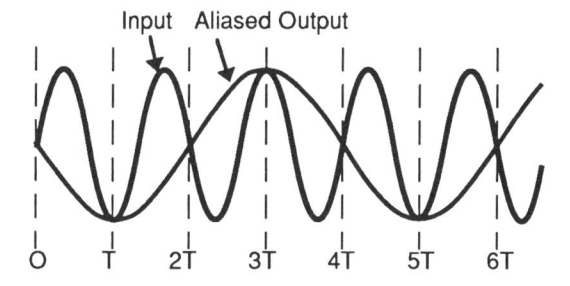

Figure 2-17: Aliasing. T is the sample-to-sample time. The sampling frequency is $f_s = 1/T$. An undersampled sinusoid will appear as a lower frequency after sampling.

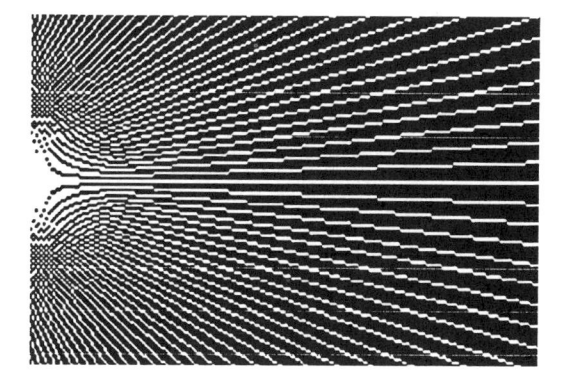

Figure 2-18: Moire pattern. A raster scan system creates Moire patterns when viewing wedges or star bursts.

Sampling theory states that the *frequency* can be unambiguously recovered for all input frequencies below Nyquist frequency. It states nothing about the intensity nor the appearance of the signal. Furthermore, sampling theory deals with sinusoids whereas infrared characterization tests tend to employ bar (square wave) patterns. The expansion of square waves into a Fourier series clearly shows that square waves consist of an infinite number of sinusoidal frequencies. The fundamental frequency may be below the Nyquist frequency but the higher order frequencies may not. During digitization, the higher order frequencies will be aliased down to lower frequencies and the square wave will change its appearance. There will be intensity variations from bar-to-bar and the bar width will not remain constant.

When a signal of frequency f_x is sampled at a frequency of f_s, side bands occur that replicate the frequency at $\pm nf_s \pm f_x$. Amplitude and pulse width variations are caused by sum and difference frequencies (beat frequencies). The first (n = 1) beat frequency is $f_B = (f_s - f_x) - f_x = 2(f_N - f_x)$. Lomheim et. al.[4] created these beat frequencies when viewing multiple bar patterns with a CCD camera. The beat frequency period lasts for N input frequency cycles:

$$N = \frac{f_x}{2(f_N - f_x)} \qquad \text{2-1}$$

N is plotted in Figure 2-19 as a function of f_x/f_N. As f_x/f_N approaches 1, the beat frequency becomes obvious. For example, if $f_x/f_N = 0.968$, the beat frequency is equal to 15.1 cycles of the input frequency (Figure 2-20). Here, the target must contain at least 16 bars (cycles) to see the entire beat pattern. When f_x/f_N is less than about 0.6 (Figure 2-21), the beat frequency is not obvious. Now the output nearly replicates the input but there is some slight variation in pulse width and amplitude. In the region where f_x/f_N is approximately between 0.6 and 0.9, adjacent bar amplitudes are always less than the input amplitude (Figure 2-22).

Figures 2-20 to 2-22 illustrate the output of an ideal staring system when viewing an infinite periodic target. Standard characterization targets, however, consist of a finite number of bars. Therefore the beat pattern may not be seen when viewing the 4-bar MRT target. The 4-bar pattern would have to be moved $\pm \frac{1}{2}$ IFOV to change the output from a maximum value (in-phase) to a minimum value (out-of-phase). This can be proved by selecting just 4 bars in Figure 2-20. When f_x/f_N is less than about 0.6, a 4-bar pattern will always be seen (select any 4 adjoining bars in Figure 2-21). When $f_x/f_N < 0.6$, phasing effects are minimal and a phase adjustment of $\pm \frac{1}{2}$ IFOV will have little effect on an observer's

IR IMAGING SYSTEM OPERATION 39

ability to resolve a four-bar target. In the region where f_x/f_N is approximately between 0.6 and 0.9, 4-bar targets will never *look* correct (Figure 2-22). One or two bars may be either much wider than the others or one or two bars may be of lower intensity than the others. As a result, the MRT will appear worse in the $0.6 f_N$ to $0.9 f_N$ spatial frequency region[5,6].

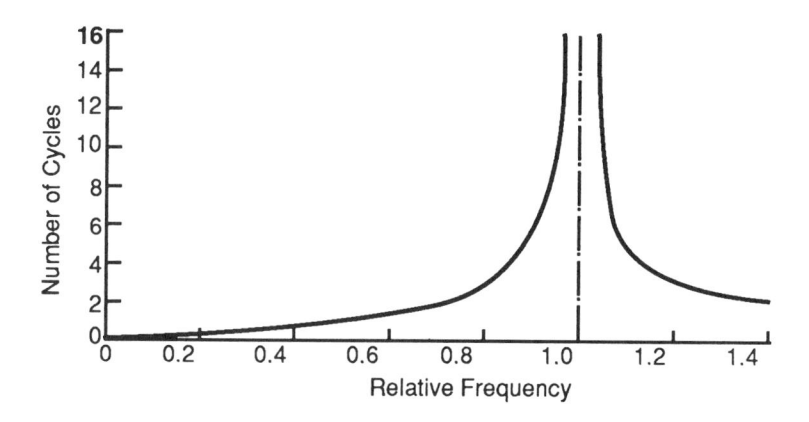

Figure 2-19: Number of input frequency cycles required to see one complete beat frequency cycle as a function of f_x/f_N.

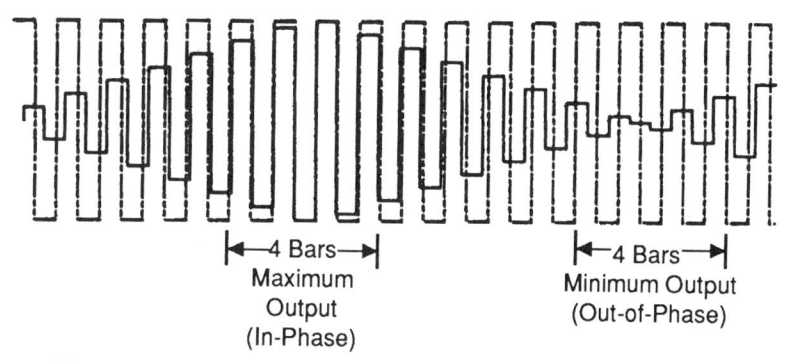

Figure 2-20: Beat frequency produced by an ideal staring system when $f_x/f_N = 0.968$. N = 15.1. The dashed line is the input and the solid line is the detector output. A 4-bar pattern may either be replicated or provide a negligible output depending upon the phase.

40 TESTING & EVALUATION OF IR IMAGING SYSTEMS

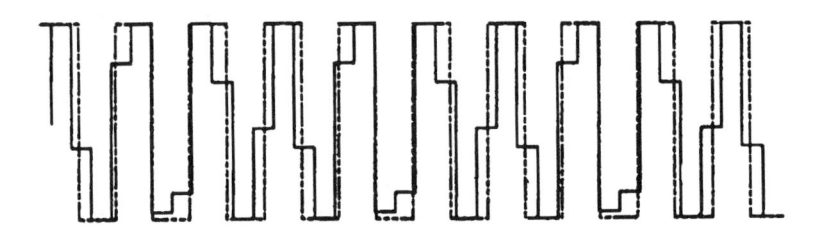

Figure 2-21: Ideal staring system output when $f_x/f_N = 0.545$. N = 0.599. The dashed line is the input and the solid line is the detector output. The output nearly replicates the input when $f_x < 0.6$.

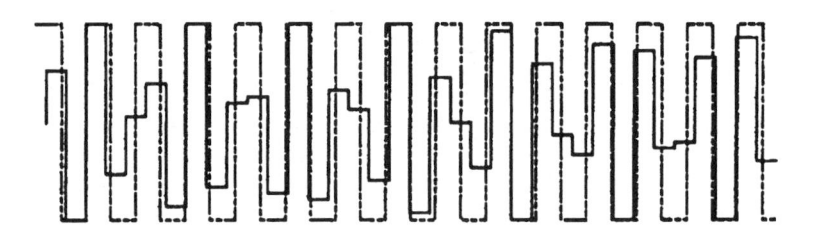

Figure 2-22: Ideal staring system output when $f_x/f_N = 0.811$. N = 2.14. The dashed line is the input and the solid line is the detector output. The output never *looks* quite right when f_x/f_N is between 0.6 and 0.9.

Input frequencies of $f_x = f_N/k$, where k is an integer, are faithfully reproduced[7] (i.e., no beat frequencies). When k = 1, as the target moves from in-phase to out-of-phase, the output will vary from a maximum to a zero. Selection of f_N/k targets avoids the beat frequency problem but significantly limits the number of spatial frequencies selected.

Signals whose frequencies are above Nyquist frequency will be aliased down to lower frequencies. This would be evident if an infinitely long periodic target was viewed. However, when f_x is less than about 1.1 f_N, it is possible to select a phase such that 4 adjoining bars *appear* to be faithfully reproduced (Figure 2-23). These targets can be *resolved* although the underlying fundamental frequency has been aliased to a lower frequency. The highest frequency that can be faithfully reproduced is the called the system cutoff frequency. System cutoff is the smallest of the optical cutoff, detector cutoff or the Nyquist frequency.

|← 4 Bar →|
Pattern
(In-Phase)

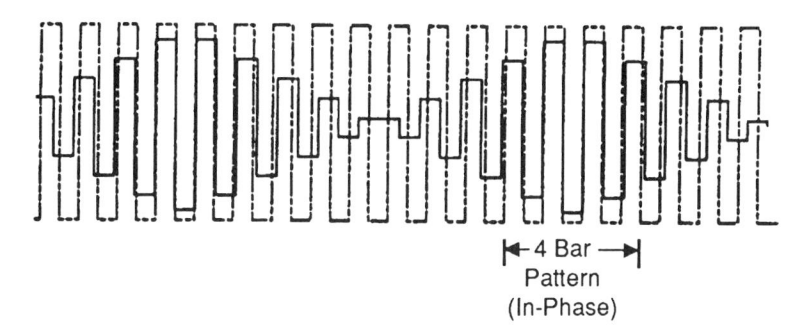

Figure 2-23: Ideal staring system output when $f_x/f_N = 1.053$. The dashed line is the input and the solid line is the detector output. By selecting the appropriate phase, the output *appears* to replicate an input 4-bar pattern.

Example 2-2
NYQUIST FREQUENCY

What are the horizontal and vertical Nyquist frequencies for a LWIR common module system? The IFOV is 0.2 mrad horizontally by 0.2 mrad vertically. The field-of-view is a square format with 72 mrad on a side. The horizontal line (scan direction) is electronically digitized at 2 samples per IFOV.

A LWIR common module system consists of 180 detectors that are interlaced to produce 360 infrared scene lines. In the horizontal direction, there are $72/0.2 = 360$ independent IFOVs. With 2 digital samples per IFOV there are 720 digital samples in the horizontal direction. The Nyquist frequency is one-half the sample frequency. Equivalently there are 360 cycles in the horizontal direction and the Nyquist frequency is $720/(2·72) = 5$ cy/mrad. This is identical with 1/IFOV. In the vertical direction, there are only 360 independent samples (the number of detectors) representing 180 cycles. Thus the vertical Nyquist frequency is $360/(2·72) = 2.5$ cy/mrad. Line interpolation may further degrade this value.

Example 2-3
SYSTEM CUTOFF

The LWIR system described in Example 2-2 is electronically digitized at 4 samples per IFOV. What is the horizontal system cutoff?

With 4 digital samples per IFOV, there are 1440 digital samples in the horizontal direction and the A/D converter Nyquist frequency is 10 cy/mrad. However, the highest spatial frequency that can be faithfully reproduced by the detector occurs at 2 samples per IFOV or 5 cy/mrad. Oversampling minimizes the phasing effects illustrated in Figures 2-20 though 2-22 (pages 39-40).

Example 2-4
STARING ARRAY CUTOFF

A staring array consists of detectors that are 40 μm x 40 μm in size. The detector pitch (the center-to-center spacing) is 60 μm. The effective focal length is 30 cm. What is the system cutoff?

The IFOV is $40 \times 10^{-6}/0.3 = 0.133$ mrad. The detector cutoff is 1/IFOV = 7.5 cy/mrad. The detector pitch provides sampling every $60 \times 10^{-6}/.3 = 0.2$ mrad for an effective sampling rate at 5 cy/mrad. Since the Nyquist frequency is 1/2 of the sampling frequency, the system cutoff is 2.5 cy/mrad. For staring arrays, the detector pitch defines the system cutoff.

HISTORICAL NOTE: Early thermal imaging systems were designed for a single observer. He viewed the output of light emitting diodes (LED) whose intensities were directly proportional to the scene irradiance levels. For multiple observers, the LED outputs were converted into a TV format with a vidicon and were called EO multiplexed or EO mux systems. These were truly analog systems in the horizontal direction and therefore did not suffer from phasing effects. The vertical direction had a raster pattern and was sampled in the vertical direction.

IR IMAGING SYSTEM OPERATION 43

Example 2-5
STAGGERED ARRAY CUTOFF

The staggered elements, as illustrated in Figure 2-24, increase the vertical spatial sampling rate. If the effective focal length is 40 cm, what is the system cutoff?

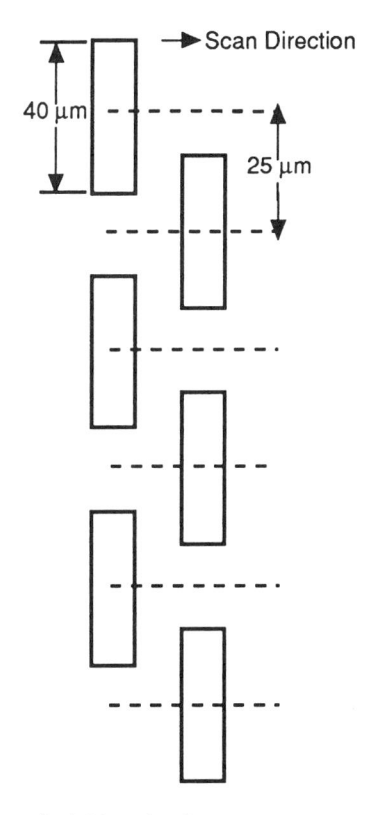

Figure 2-24: The vertical Nyquist frequency can be increased with a staggered array.

The IFOV is $40 \times 10^{-6}/0.4 = 0.1$ mrad representing a 10 cy/mrad capability. The sampling is every $25 \times 10^{-6}/.4 = 0.0625$ mrad or at a rate of 16 cy/mrad. The highest spatial frequency that can be faithfully reproduced is 8 cy/mrad. This represents $16/10 = 1.6$ samples per IFOV. The system is still undersampled since the Nyquist criterion requires 2 samples per IFOV.

44 TESTING & EVALUATION OF IR IMAGING SYSTEMS

Electronic digitization quantizes the analog signal. Quantization places a lower limit on the detectability of targets. If an 8 bit digitizer is used (256 gray levels), the minimum detectable signal is about one gray level, one least significant bit (LSB), or 1/256 of the total input range. Figure 2-25 illustrates the LSB for an 8, 10 and 12 bit system. For example, if the maximum signal range is 25 degrees and an 8 bit A/D converter is used, the smallest signal that can be detected is about $25/256 \approx 0.1$ degrees. The smallest noise that can be measured with confidence[8] is approximately an LSB.

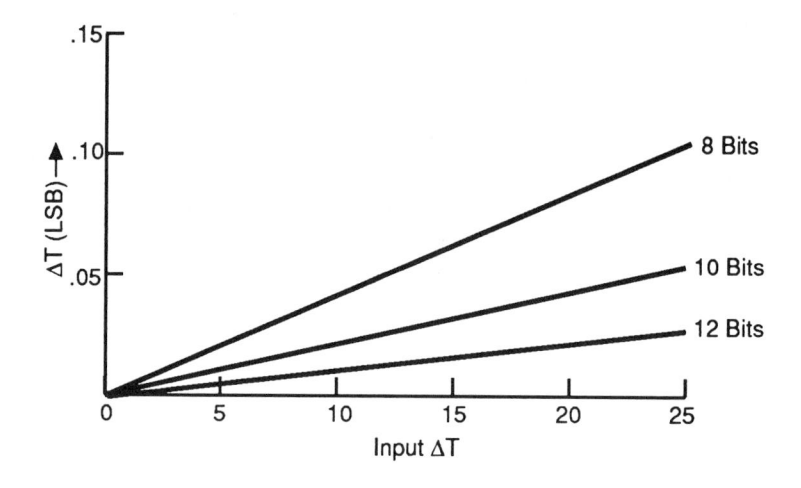

Figure 2-25: Least significant bit as a function of input signal range and A/D converter dynamic range. ΔT is the maximum input.

2.4. IMAGE PROCESSING

Image processing algorithms can be used to enhance images, suppress noise and put the image data into a format consistent with monitor requirements. The algorithms also may minimize the effects described in previous sections. Digital algorithms can provide boost and interpolation. Boost increases the signal amplitude at specific spatial frequencies but it does not affect the signal-to-noise ratio. For systems that are not noise limited, boost may improve image quality. However, for noisy images, the advantages of boost are less obvious.

2.4.1. GAIN/LEVEL NORMALIZATION

Every detector/amplifier combination will have a different gain (responsivity) and offset. These variations result in fixed pattern noise or spatial noise. If large deviations in responsivity exist, the image may be unrecognizable. As a result, systems employing more than one detector may require gain/level normalization or nonuniformity correction (NUC) to produce an acceptable image. Although most literature discusses NUC for staring systems, it applies to scanning systems that have more than one detector in the cross scan direction (Figure 2-2b: page 24). For good imagery, the individual detector outputs are normalized (made equal) for several discrete input intensities. These normalization intensities are also called calibration points, temperature references or simply points.

Figure 2-12a (page 31) illustrated the responsivity of three different detectors before gain/level normalization. The outputs for input intensities I_1, I_2, and I_3 were shown in Figure 2-12b. Figure 2-26 illustrates the normalized output after correction at two points. If all the detectors had linear responsivities, then all the curves would coincide. As individual detectors deviate from linearity, the responsivity differences become more noticeable[9]. It is this variation in responsivity that creates the fixed pattern noise after gain/level correction.

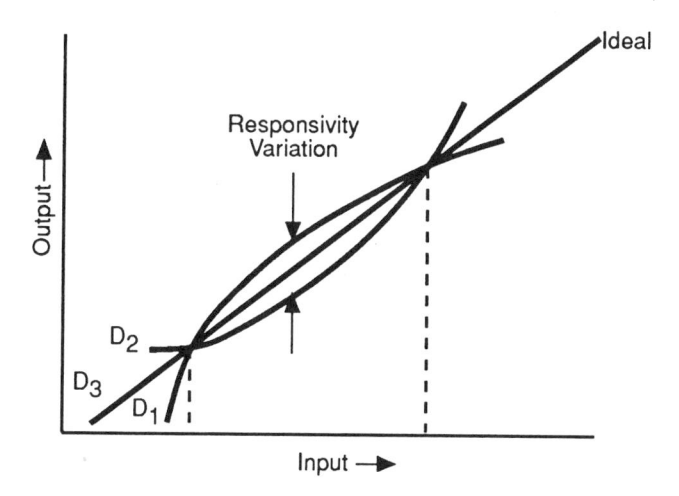

Figure 2-26: Responsivity curves after 2-point correction. Responsivity deviations from linearity produce fixed pattern noise.

46 TESTING & EVALUATION OF IR IMAGING SYSTEMS

With single point correction, the noise (fixed pattern plus random) will be a minimum at the reference intensity. With perfect correction there will be no fixed pattern noise at the reference intensity (Figure 2-27). Truncation errors in the normalization algorithm and different spectral response detectors[10] produce residual FPN. As the background temperature deviates from the reference calibration temperatures fixed pattern noise will increase. The amount depends upon how far the detector responsivity curves deviate from linearity. For two-point correction, the spatial noise will be minimum at two reference intensities. For any other intensity, the spatial noise increases. But in the region between the two references[11] (Figure 2-28), the spatial noise is less compared to the spatial noise outside the two references. Figure 2-29 illustrates the effect of changing the reference temperature[12]. As the reference temperatures are brought together, the fixed pattern noise between the reference inputs decreases and the fixed pattern noise outside the region increases. As depicted in Figures 2-27 through 2-29, all input-to-output transformations affected by noise will be a function of the reference temperature and the background test temperature (e.g., NEDT, uniformity, MRT, and MDT).

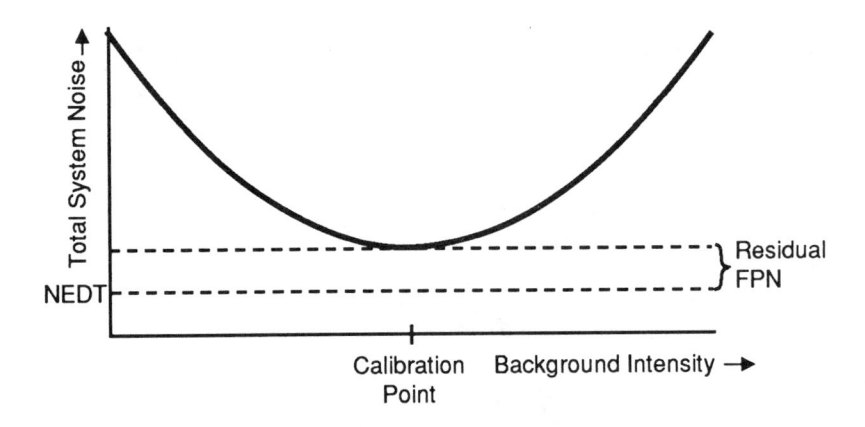

Figure 2-27: System noise after single point correction. Noise will be a minimum at the correction temperature (calibration point).

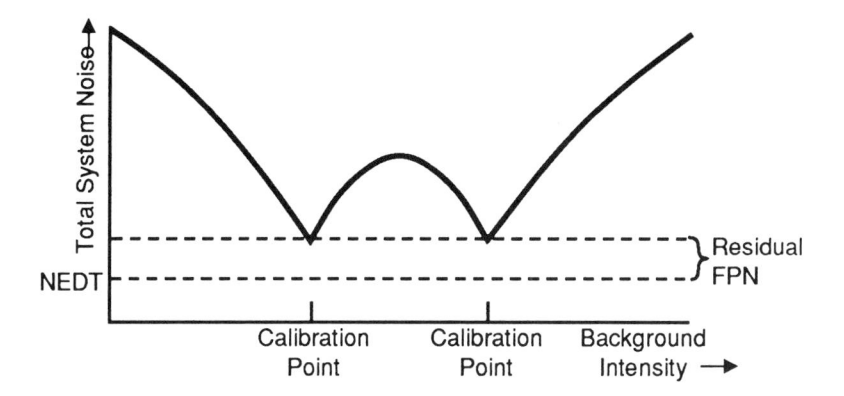

Figure 2-28: System noise after 2-point correction. The amount of noise between the 2-correction points depends upon the system spectral response, detector responsivities and the correction algorithm.

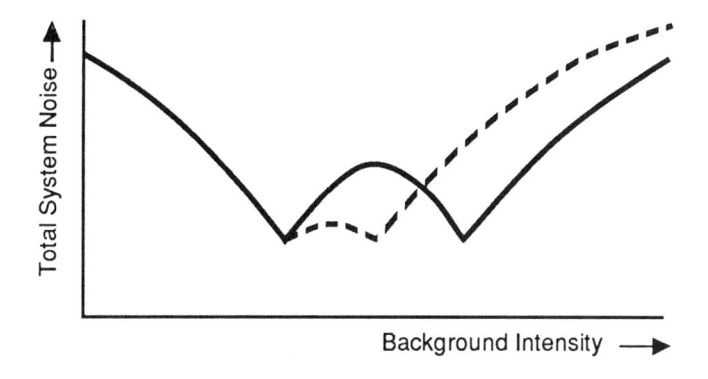

Figure 2-29: Effect of calibration intensities on system noise. The noise between the calibration points decreases as the calibration points are brought together.

Although the fixed noise pattern is labeled as *fixed*, it typically changes at a very slow rate. 1/f noise produces a slowly varying output signal that if uncorrected, appears as fixed pattern noise. The amount of noise depends on the 1/f characteristics and the time the data are collected after the last correction has been made[12,13]. If a system is corrected only once when it is turned on, the noise

48 *TESTING & EVALUATION OF IR IMAGING SYSTEMS*

may slowly increase with time. Here, the measured noise will be a function of test time. 1/f noise may not be dominant in some arrays and the effect illustrated in Figure 2-30 may not be significant.

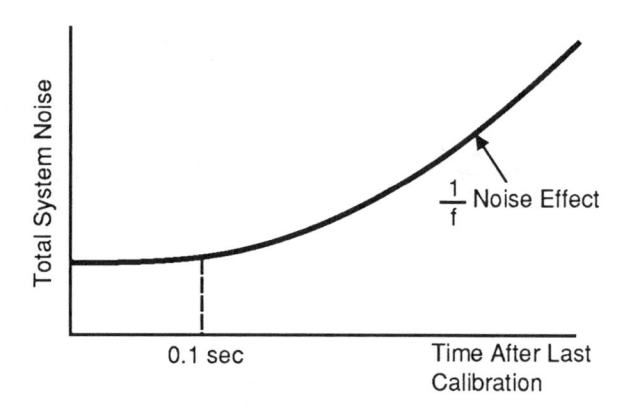

Figure 2-30: 1/f noise manifests itself as fixed pattern noise.

2.4.2. IMAGE FORMATTING

For systems that have detector arrays that are not consistent with analog video formats requires image formatting. Monochrome video formats may either be the US standard[14] RS 170 that displays 485 TV lines or may be the European CCIR standard that displays 577 lines. Generally, infrared imaging systems output 480 active lines with 5 blank lines to be consistent with RS 170. A LWIR common module system, which creates infrared 360 lines, requires vertical interpolation. The interpolator increases the line rate and may change the video sample rate to provide the number of TV lines required by the monitor. Interpolation can be simply achieved by the duplication of lines or may be a more complex algorithm. The interpolator can dramatically affect vertical resolution and therefore can greatly affect the ability to resolve vertical detail. Adjacent lines are no longer independent with interpolation. As a result, it will affect any statistical analysis that requires independent outputs.

2.4.3. GAMMA CORRECTION

It is desirable to have a linear system such that if the signal intensity doubles, then the monitor luminance is double. However, monitors tend to be nonlinear in terms of output luminance versus input voltage. The slope of this relationship on a logarithmic scale is the monitor gamma (Figure 2-31). To provide a linear relationship, the inverse gamma is inserted as an image

IR IMAGING SYSTEM OPERATION 49

processing algorithm (gamma correction). When present, there is no longer a linear relationship between the analog output video voltage and the input signal intensity. This nonlinear relationship affects all data collected from the analog video signal. Its effect is precisely quantifiable and can be mathematically removed from the data. If a gamma correction algorithm is present, linearity is only assured when viewing the monitor.

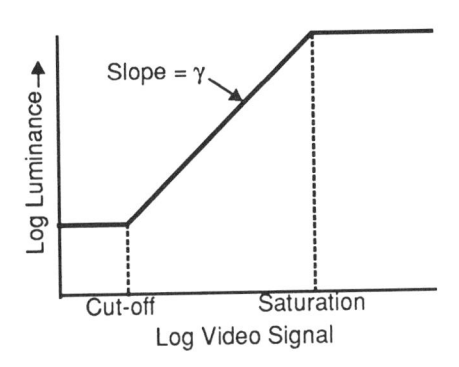

Figure 2-31: Definition of monitor gamma. An ideal system will have a gamma of unity.

☞ ──

Example 2-6
GAMMA CORRECTION

What effect does gamma correction have on the measured MTF?

$$MTF = \frac{V_{MAX} - V_{MIN}}{V_{MAX} - V_{MIN}} = \frac{\frac{V_{MAX}}{V_{MIN}} - 1}{\frac{V_{MAX}}{V_{MIN}} + 1} \qquad 2\text{-}2$$

where V_{MAX} and V_{MIN} are the maximum and minimum voltage levels produced by a sinusoidal target just before the gamma correction circuitry. The analog output voltage with gamma correction is

$$V_{analog\ video} = V_{IN}^{\left(\frac{1}{\gamma}\right)} \qquad 2\text{-}3$$

50 *TESTING & EVALUATION OF IR IMAGING SYSTEMS*

Substituting yields:

$$MTF_{analog\ video} = \frac{A^{\frac{1}{\gamma}} - 1}{A^{\frac{1}{\gamma}} + 1} \qquad 2\text{-}4$$

where

$$A = \frac{1 + MTF_{IN}}{1 - MTF_{IN}} \qquad 2\text{-}5$$

The results are plotted in Figure 2-32 for various gammas. The monitor luminance is

$$L = K(V_{analog\ video})^{\gamma} = K\left(V_{in}^{\frac{1}{\gamma}}\right)^{\gamma} = K V_{IN} \qquad 2\text{-}6$$

so that the monitor luminance is proportional to the detector output and target intensity.

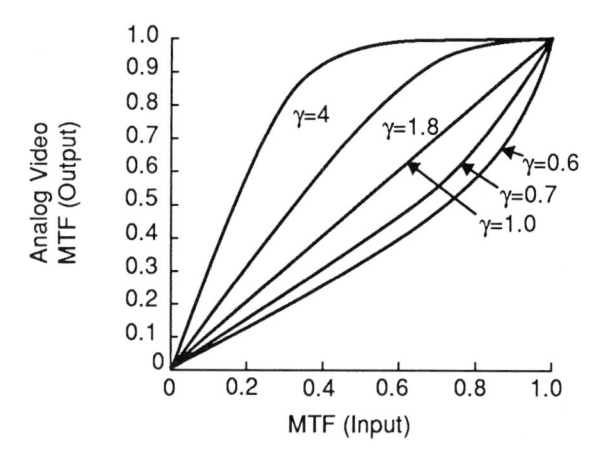

Figure 2-32: Effects of gamma on MTF measured at the analog output video. If $\gamma > 1$, the analog video appears to provide a higher MTF.

2.5. RECONSTRUCTION

The D/A converter provides an output that typically has a stair step appearance. Each step occurs at each digital sample. The reconstruction filter smooths the data and removes the stair step. The output of the reconstruction filter is the analog video. The reconstruction subsystem usually does not introduce any artifacts and is shown in Figure 2-1 (page 22) for completeness.

2.6. MONITORS

The monitor may be an integral part of the infrared system or stand alone. If the monitor is separate from the infrared system, it is important to use a high quality monitor that does not degrade the image. If the system has a gamma correction algorithm, it is important that the monitor selected have the same gamma. Unfortunately, monitor specifications are not always relatable. Monitors may be specified by electronic bandwidth, number of available lines (i.e., RS 170 displays 480 lines) or number of lines resolved (which may or may not include the Kell factor[15]). The resolution[16] also may be determined by a variety of other measures including the shrinking raster method, television resolution considerations or by measuring the MTF. Furthermore, monitors are not consistent[17] and their response should be measured. High quality monitors minimize most of these issues.

Finally, the monitor format must be the same as the system format. It is important when performing subjective evaluation of image quality that the displayed image has the correct aspect ratio[18]. For example, an infrared system may have a 1:1 aspect ratio and has an output equivalent to RS 170. If displayed on a standard monitor (4:3 aspect ratio), the image will be elongated in the horizontal direction such that circles become ellipses. On the other hand, a frame grabber may appropriately digitize a 4:3 aspect ratio image but display the image on a computer monitor that is typically a 1:1 aspect ratio. Here, circles are compressed in the horizontal direction and the circles will again appear as ellipses.

2.7. SUMMARY

The test engineer must be thoroughly knowledgeable in the operation of the infrared system before performing any test. Each subsystem modifies the image uniquely as shown in Table 2-1. These artifacts can affect all input-to-output transformations (Table 2-2). The most dramatic effect is that of phasing. By moving the target $\pm$ ½ IFOV, the output will change from a maximum to a minimum value. Beat frequencies can be avoided by selecting targets whose spatial frequencies follow f_n/k. A maximum output is obtained when these targets are in-phase with the detector array to produce the in-phase contrast transfer function (IPCTF). Although present, beat frequencies are not obvious in oversampled systems.

Table 2-1
TYPICAL INFRARED SYSTEM ARTIFACTS

SUBSYSTEM	ARTIFACT
Optics and scanner	$Cosine^N\theta$ Aberrations Narcissus
Detector and detector electronics	Finite fill factor Temperature variations Different gain/levels AC coupling
Digitization	Phasing effects Quantization
Gain/level normalization	Incomplete removal of FPN
Image formatting	Loss of vertical or horizontal resolution
Gamma correction	Nonlinear system response when measured at the analog video
Reconstruction	Typically none
Monitors	Loss or resolution and distortion

IR IMAGING SYSTEM OPERATION 53

Table 2-2
ARTIFACTS AFFECT ALL TEST RESULTS

MOST PREVALENT ARTIFACT	Resolution	SiTF	ATF	NEDT	CTF	MTF	MDT	MRT	Nonuniformity
Cosine$^N\theta$		x		x			x	x	
AC coupling					x	x			x
Phasing	x		x		x	x	x	x	
Finite fill factor	x		x		x	x	x	x	
Gamma		x		x	x	x			
Detector Temperature variations		x	x	x	x	x	x	x	x

Image processing algorithms enhance certain image characteristics or compensate for existing nonlinear processes. Remapping of gray scales is sometimes called tonal transfer curves and an image processing algorithm can be a tonal transfer function. Gamma correction is one type of tonal transfer curve.

For convenience, targets are labeled as either hot or cold with respect to the immediate background. For MWIR and LWIR systems, the term *thermal* is misleading. Infrared imaging systems with photoconductive or photovoltaic detectors do not sense warmth or cold (they are not thermometers) but sense the radiation emitted by an object. All objects emit radiation in the *thermal* regions. *Hot* refers to targets that appear warmer than its immediate background and *cold* means the target appears cooler than its immediate background. The choice of hot objects appearing white and cold objects appearing black is arbitrary. With electronic polarity reversal available, either "white hot" or "black hot" targets can be created. With "black hot," *hot* objects appear black or dark gray against a neutral background. As the object becomes *hotter*, its representation on a monitor becomes blacker. The reverse is true for "white hot". With "white hot", the target becomes "whiter" as its apparent temperature increases relative to the background. The output also can be mapped into pseudo-colors. Keeping with human feelings, *cold* objects are often represented as blue and *hot* objects as red.

Finally, this chapter discussed the operation of systems that are derivatives of US common module technology and some staring array systems. These systems tend to employ similar optical designs, detectors and electronics and therefore tend to have similar artifacts. As new technologies emerge such as uncooled detectors and quantum well detectors, new artifacts[19] may be created that would require modifications to test procedures and data analysis techniques.

2.8. REFERENCES

1. J. W. Howard and I. R. Abel, "Narcissus: Reflections on Retroreflectors in Thermal Imaging Systems", Applied Optics, Vol. 21 (18), pp. 3393-3397 (1982).
2. D. E. L. Freeman, "Guidelines for Narcissus Reduction and Modeling", in Simulation and Modeling of Optical Systems, R. E. Fischer and D. C. O'Shea, eds.: SPIE Proceedings Vol. 892, pp. 27-37 (1988).
3. L. O. Vroombout and B. J. Yasuda, "Laboratory Characterization of Thermal Imagers", in Thermal Imaging, I. R. Abel, ed., SPIE Proceedings Vol. 636, pp. 36-39 (1986).
4. T. S. Lomheim, L. W. Schumann, R. M. Shima, J. S. Thompson, and W. F. Woodward, "Electro-Optical Hardware Considerations in Measuring the Imaging Capability of Scanned Time-delay-and-integrate Charge-coupled Imagers", Optical Engineering, Vol. 29(8), pp. 911-927 (1990).

5. G. C. Holst, "Effects of Phasing on MRT Target Visibility", in Infrared Imaging Systems: Design, Analysis, Modeling and Testing, II, G. C. Holst, ed.: SPIE Proceedings Vol. 1488, pp. 90-98 (1991).
6. C. M. Webb, "Results of Laboratory Evaluation of Staring Arrays", in Infrared Imaging Systems: Design, Analysis, Modeling and Testing, G. C. Holst, ed.: SPIE Proceedings Vol. 1309, pp. 271-285 (1990).
7. T. H. Cook, C. S. Hall, F. G. Smith and T. J. Rogne, "Simulation of Sampling Effects in FPAs", in Infrared Imaging Systems: Design, Analysis, Modeling and Testing II", G. C. Holst, ed.: SPIE Proceedings Vol. 1488, pp. 214-225 (1991).
8. J. D. Vincent, Fundamentals of Infrared Detector Operation and Testing, pp. 220-227: John Wiley and Sons, New York (1990).
9. N. Bluzer, "Sensitivity Limitations of IRFPAs Imposed by Detector Nonuniformities", in Infrared Detectors and Arrays, E. L. Dereniak, ed.: SPIE Proceedings Vol. 930, pp. 64-75 (1988).
10. J. M. Mooney, F. D. Shepherd, W. S. Ewing. J. Marguia and J. Silverman, "Responsivity Nonuniformity Limited Performance of Infrared Staring Cameras", Optical Engineering, Vol. 28 (11), pp. 1151-1161, (1989).
11. D. A. Scribner, M. R. Kruer, and C. J. Gridley, "Physical Limitations to Nonuniformity in IR Focal Plane Arrays", in Focal Plane Arrays: Technology and Applications, J. Chatard, ed.: SPIE Proceedings Vol. 865, pp. 185-202 (1987).
12. D. A. Scribner, M. R. Kruer, K. Sarkady and J. C. Gridley, "Spatial Noise in Staring IR Focal Plane Arrays", in Infrared Detector and Arrays, E. L. Dereniak, ed.: SPIE Proceedings Vol. 930, pp. 56-63 (1988).
13. D. A. Scribner, K. Sarkady, M. R. Kruer and J. C. Gridley, "Test and Evaluation of Stability in IR Staring Focal Plane Arrays After Nonuniformity Correction", in Test and Evaluation of Infrared Detectors and Arrays, F. M. Hoke, ed.: SPIE Proceedings Vol. 1108, pp. 255-264, (1984).
14. "Electrical Performance Standards - Monochrome TV", EIA Standard RS-170, Electronic Industry Association, NY, NY.
15. S. C. Hsu, "The Kell Factor: Past and Present", Society of Motion Picture and Television Engineers Journal, Vol. 95, pp. 206-214 (1986).
16. L. M. Biberman, "Image Quality", in Perception of Displayed Information, L. M. Biberman, ed., pp. 13-18: Plenum Press, New York (1973).
17. S. J. Briggs, "Photometric Technique for Deriving a "Best Gamma" for Displays", Optical Engineering, Vol. 20(4), pp. 651-657 (1981).
18. N. Sampan, "The RS170 Video Standard and Scientific Imaging: The Problems", Advanced Imaging, pp. 40-43, Feb. 1991.
19. P. A. Bell, C. W. Hoover Jr., and S. J. Pruchnic Jr., "Standard NEDT Test Procedure for FLIR Systems With Video Outputs", in Infrared Imaging Systems: Design, Analysis, Modeling and Testing IV, G. C. Holst, ed.: SPIE Proceedings Vol. 1969, pp 194-205 (1993).

56 TESTING & EVALUATION OF IR IMAGING SYSTEMS

EXERCISES

1. Figure 2-26 (page 45) illustrates 3 detector outputs as a function of input intensity with 2-point correction. Sketch the output for 1-point, 3-point and 5-point correction.
2. Figure 2-28 (page 47) illustrates system noise when 2-point correction is used. Sketch the noise versus input intensity for 3-point and 5-point correction.
3. Assume that a system has a gamma correction algorithm. Plot the video output voltage as a function of input intensity for $\gamma = 1$ and $\gamma = 2$.
4. Contrast is $C = (L_T/L_B)-1$ where L_T is the target luminance and L_B is the background luminance. Assume that the video output, V, is proportional to the input luminance and that the monitor has a gamma such that the monitor luminance is $L = K V^\gamma$. Derive the relation between the input contrast and output contrast.
5. Using Figure 2-3 (page 24) as a guide, write a specification for the SiTF or responsivity.
6. How does the detector temperature (Figure 2-11: page 31) affect the specifications written in Exercise 5?
7. Figure 2-3b (page 24) illustrates how the signal level decreases according to $\cosine^N \theta$ for a staring array. The responsivity can be made equal for all the detectors by increasing the gain of each detector the appropriate amount. What effect would this have on noise measurements?
8. A staring array consists of 512 x 512 detector elements. If the effective focal length is 40 cm and the detectors are 50 μm x 50 μm on 75 μm centers, what is (a) the detector IFOV, (b) the system FOV and (c) the system Nyquist frequency?
9. Sketch the in-phase and out-of-phase responses for a staring array when $f_x = f_n$, $f_x = f_n/2$, and $f_x = f_n/3$.

3
BASIC CONCEPTS IN IR TECHNOLOGY

Radiometry describes the energy or power transfer from a source to a detector. When the source size is much larger than the projected area of the detector, the source is said to be resolved or the system is viewing an extended source. Equivalently, the detector is flood-illuminated. The system's responsivity or signal transfer function (SiTF) applies to a system viewing an extended source. As the source size becomes smaller, the responsivity must be modified by the aperiodic transfer function (ATF). The ratio of the actual response to the ideal response is the target transfer function (TTF). As the source size approaches a point, the target transfer function approaches the point visibility factor or ensquared power value.

There are two fundamental test configurations: one with the infrared image system focussed onto a target (Figure 3-1) and the other with the infrared imaging system viewing a target in a collimator (Figure 3-2). Collimators are used if the system cannot be focussed at typical laboratory distances or if measurements must be performed with the system focussed at infinity. The SiTF should be calculated before performing the SiTF measurement.

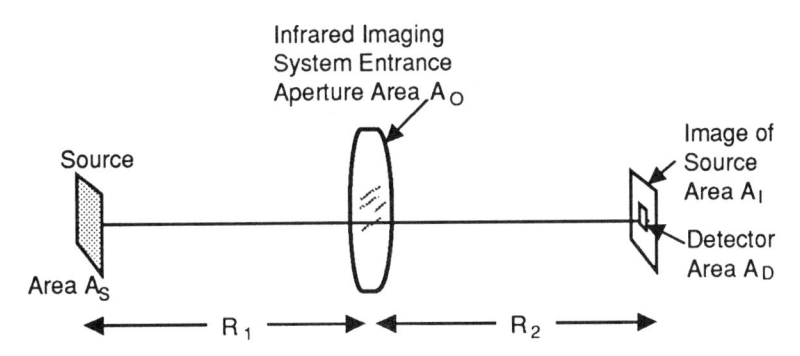

Figure 3-1: Infrared imaging system directly viewing a source.

Although most thermal imaging systems respond to radiant flux differences, it is convenient to specify the radiant flux difference between a target and its immediate background by an equivalent temperature difference, ΔT. Using average values of the various transmittances simplify analyses. However with broad spectral response systems, the average transmittance is a function of the source's spectral radiant exitance, the spectral transmittances of the atmosphere, collimator, and the system as well as the detector spectral response.

58 *TESTING & EVALUATION OF IR IMAGING SYSTEMS*

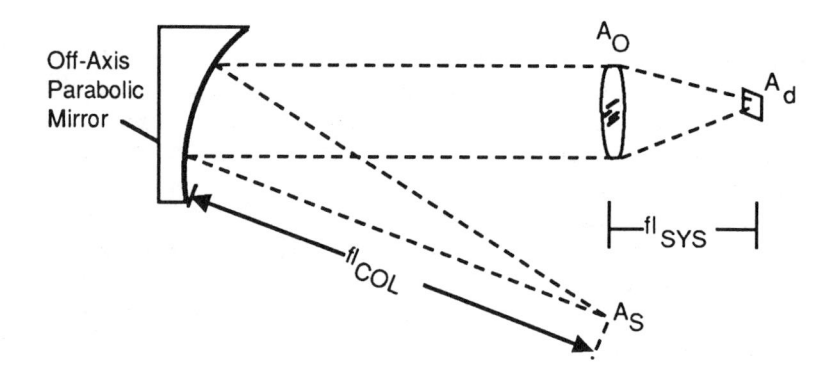

Figure 3-2: Infrared imaging system viewing a source in the focal plane of a collimator. Off-axis collimators provide the highest throughput and maximum flexibility in terms of set up.

The bar patterns used for CTF and MRT measurements are specified by their spatial frequency. Object space spatial frequency (units of cycles/mrad) is used exclusively in this text. Image space, monitor, and observer spatial frequencies are used by optical designers, monitor designers, and visual psychophysicists respectively.

3.1. RADIOMETRY

Radiant sterance, L_e, is the basic quantity from which all other radiometric quantities can be derived (e.g., radiant exitance, radiant flux, or radiant intensity). It contains both the areal and solid angle concept[1] that is necessary to calculate the radiant flux incident onto a system. It is the amount of radiant flux, Φ, radiated into a cone of solid angle Ω from a source whose area is A_s (Figure 3-3):

$$L_e = \frac{\Phi}{A_s \Omega} \qquad \frac{watts}{m^2\text{-}sr} \qquad \text{3-1}$$

BASIC CONCEPTS IN IR TECHNOLOGY 59

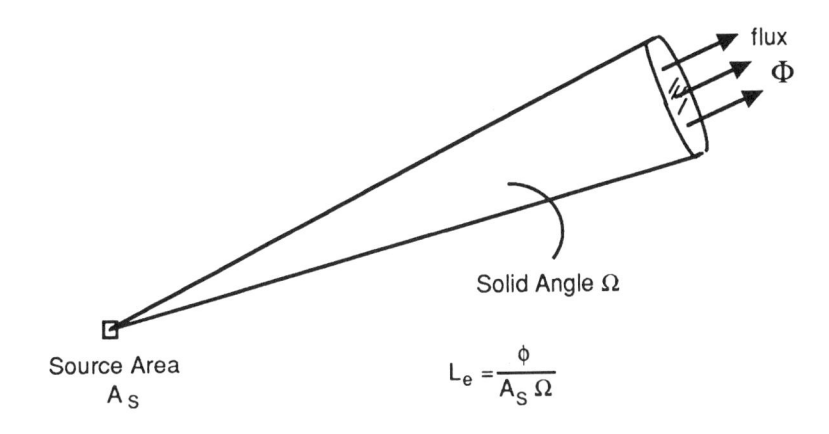

Figure 3-3: Radiant sterance.

The radiant flux emanating from the source is attenuated by the intervening atmosphere, focussed on to the detector by the optical system and then converted into a measurable electrical signal by the detector. Planck's blackbody radiation law describes the source radiant exitance.

3.1.1. PLANCK'S BLACKBODY LAW

The spectral radiant exitance of an ideal blackbody source whose absolute temperature is T, can be described by Planck's blackbody radiation law:

$$M_e(\lambda,T) = \frac{c_1}{\lambda^5} \left(\frac{1}{e^{(c_2/\lambda T)} - 1} \right) \quad \frac{w}{cm^2 - \mu m} \qquad 3\text{-}2$$

where the first radiation constant is $c_1 = 3.7418 \times 10^4$ watt-$\mu m^4/cm^2$ and the second radiation constant is $c_2 = 1.4388 \times 10^4$ μm-K. Figure 3-4 illustrates Planck's blackbody law in logarithmic coordinates and Figure 3-5 provides the curve in linear coordinates.

60 TESTING & EVALUATION OF IR IMAGING SYSTEMS

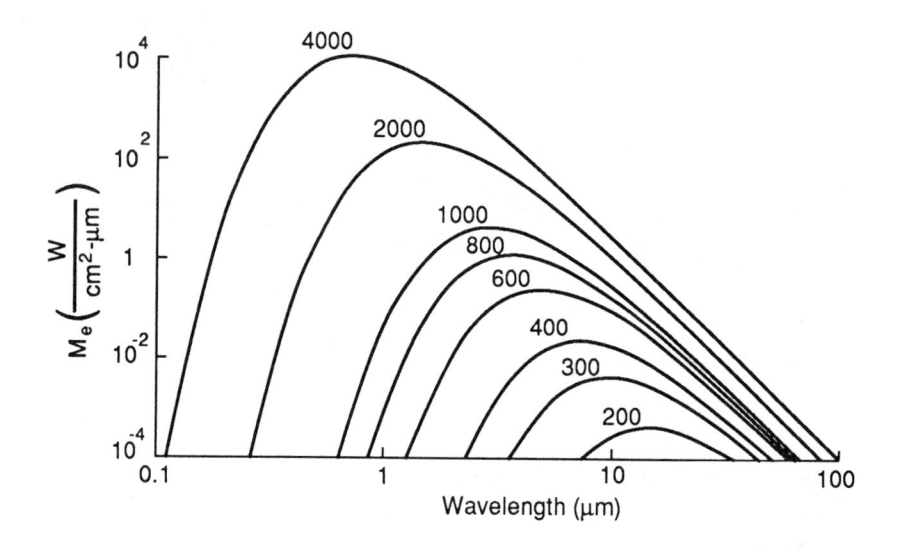

Figure 3-4: Planck's blackbody law plotted in logarithmic coordinates for T = 200, 300, ..., 4000 K.

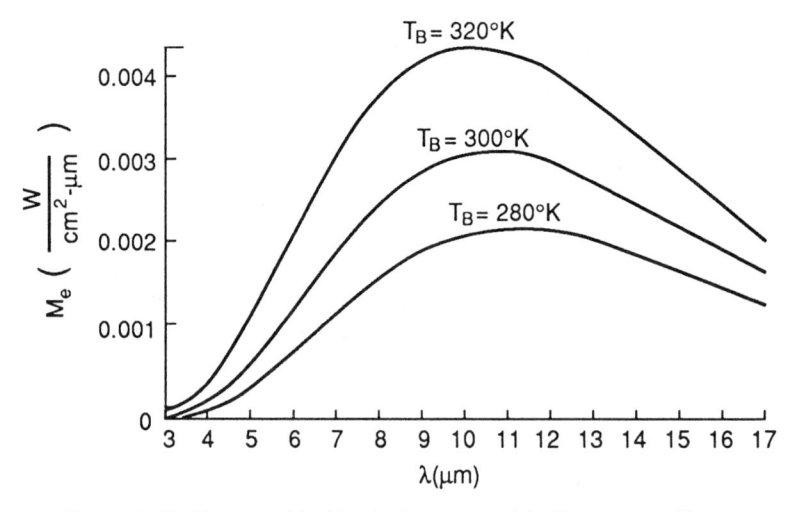

Figure 3-5: Planck's blackbody law plotted in linear coordinates.

BASIC CONCEPTS IN IR TECHNOLOGY

For Lambertian sources, the spectral radiant exitance is related to the spectral radiance sterance by:

$$L_e(\lambda,T) = \frac{M_e(\lambda,T)}{\pi} \quad \frac{w}{cm^2-\mu m-sr} \qquad 3\text{-}3$$

Real blackbodies do not emit all the radiation described by Equation 3-2 but only emit a fraction of it. The ratio of actual radiant sterance to the theoretical maximum is the emittance, ϵ:

$$\epsilon(\lambda) = \frac{M_{actual}(\lambda,T)}{M_{BB}(\lambda,T)} \qquad 3\text{-}4$$

For true blackbodies, $\epsilon(\lambda) = 1$. For gray bodies, $\epsilon(\lambda)$ is a constant.

3.1.2. EXTENDED SOURCE, DIRECT VIEW

If an infrared system is at a distance R_1 from a source (Figure 3-1: page 57), the radiant flux incident onto the optical system of area A_o is:

$$\Phi_{LENS} = L_e \frac{A_o}{R_1^2} A_s T_{ATM} \qquad 3\text{-}5$$

where the small angle approximation was used. T_{ATM} is the intervening atmospheric transmittance between the source and the infrared system and A_S is the source area. The on-axis radiant flux reaching the image plane is:

$$\Phi_{IMAGE} = L_e \frac{A_o}{R_1^2} A_S T_{SYS} T_{ATM} \qquad 3\text{-}6$$

where T_{SYS} is the system's optical transmittance. If A_s is large (extended or resolved source) and the image of A_s, whose area is A_i, is greater then the detector area ($A_i >> A_d$), then the radiant flux incident onto the detector is simply the ratio of the areas:

62 *TESTING & EVALUATION OF IR IMAGING SYSTEMS*

$$\Phi_{DETECTOR} = \Phi_{IMAGE} \frac{A_d}{A_i} \qquad 3\text{-}7$$

By symmetry:

$$\frac{A_s}{R_1^2} = \frac{A_i}{R_2^2} \qquad 3\text{-}8$$

The radiant flux becomes:

$$\Phi_{DETECTOR} = \frac{L_e A_o A_d}{(fl_{SYS})^2 (1+M)^2} T_{SYS} T_{ATM} \qquad 3\text{-}9$$

where the magnification is $M = R_2/R_1$ and R_1 and R_2 are related to the system focal length, fl_{SYS}, by:

$$\frac{1}{R_1} + \frac{1}{R_2} = \frac{1}{fl_{SYS}} \qquad 3\text{-}10$$

Assuming a circular aperture and since $f/\# = fl_{SYS}/D$:

$$\Phi_{DETECTOR} = \frac{\pi}{4} \frac{L_e A_d}{(f/\#)^2 (1+M)^2} T_{SYS} T_{ATM} \qquad 3\text{-}11$$

The voltage produced by a detector is proportional to the detector's responsivity, R:

$$V_d = R\, \Phi_{DETECTOR} \qquad 3\text{-}12$$

The system output, V_{SYS} is simply V_d multiplied by the system gain, G. Since all of the variables are a function of wavelength and the source spectral radiant sterance depends on temperature:

$$V_{SYS} = G\int_{\lambda_1}^{\lambda_2} R(\lambda)\,\frac{\pi}{4}\,\frac{L_e(\lambda,T)A_d}{(f/\#)^2(1+M)^2}\,T_{SYS}(\lambda)\,T_{ATM}(\lambda)\,d\lambda \qquad 3\text{-}13$$

where the system spectral response is band limited from λ_1 to λ_2. Off-axis responses may be reduced by $\cos^N\theta$.

AC coupling suppresses the large uninformative background so that small variations about ambient can be amplified. What is of interest is the signal difference produced by a source (target) at temperature T_T and its immediate background at temperature T_B.

$$\Delta V_{SYS} = G\int_{\lambda_1}^{\lambda_2} R(\lambda)\,\frac{\pi}{4}\,\frac{[L_e(\lambda,T_T) - L_e(\lambda,T_B)]A_d}{(f/\#)^2(1+M)^2}\,T_{SYS}(\lambda)\,T_{ATM}(\lambda)\,d\lambda \qquad 3\text{-}14$$

If the source and background are true blackbodies with unity emittance, then Equation 3-3 can be substituted into Equation 3-14 to yield

$$\Delta V_{SYS} = G\int_{\lambda_1}^{\lambda_2} R(\lambda)\,\frac{[M_e(\lambda,T_T) - M_e(\lambda,T_B)]A_d}{4(f/\#)^2(1+M)^2}\,T_{SYS}(\lambda)\,T_{ATM}(\lambda)\,d\lambda \qquad 3\text{-}15$$

3.1.3. EXTENDED SOURCE IN COLLIMATOR

When the system is viewing an extended source in a collimator of focal length fl_{COL} (Figure 3-2: page 58), the radiant flux incident onto the optical system is:

$$\Phi_{LENS} = L_e\,\frac{A_o}{(fl_{COL})^2}\,A_s\,T_{COL}\,T_{ATM} \qquad 3\text{-}16$$

where the collimator transmittance, T_{COL}, has been added. The radiant flux falling on the image plane is:

64 TESTING & EVALUATION OF IR IMAGING SYSTEMS

$$\Phi_{IMAGE} = L_e \frac{A_o}{(fl_{COL})^2} A_S T_{SYS} T_{COL} T_{ATM} \qquad 3\text{-}17$$

If A_S is large (extended or resolved source) and $A_i >> A_d$ then Equation 3-7 can be used to define the radiant flux incident onto the detector. When viewing a source in a collimator, the system is focussed at infinity so that:

$$\frac{A_s}{(fl_{COL})^2} = \frac{A_i}{(fl_{SYS})^2} \qquad 3\text{-}18$$

Then the radiant flux falling on the detector is

$$\Phi_{DETECTOR} = \frac{L_e A_o A_d}{(fl_{SYS})^2} T_{SYS} T_{COL} T_{ATM} \qquad 3\text{-}19$$

or, for circular apertures,

$$\Phi_{DETECTOR} = \frac{\pi}{4} \frac{L_e A_d}{(f/\#)^2} T_{SYS} T_{TEST} \qquad 3\text{-}20$$

where $T_{TEST} = T_{COL} T_{ATM}$ is due to the test configuration. Equation 3-11 reduces to Equation 3-20 when $R_1 >> R_2$ or equivalently when $M \rightarrow 0$ with the modification of the finite collimator transmittance. The T/# is the f/# divided by the square root of T_{SYS}. The T/# is useful for evaluating infrared systems where the optics may have reduced transmittance.

$$\Phi_{DETECTOR} = \frac{\pi}{4} \frac{L_e A_d}{(T/\#)^2} T_{TEST} \qquad 3\text{-}21$$

All of the above variables are a function of wavelength. By using Equation 3-12, the voltage difference produced by a target and its immediate background is:

BASIC CONCEPTS IN IR TECHNOLOGY   65

$$\Delta V_{SYS} = G \int_{\lambda_2}^{\lambda_1} R(\lambda) \frac{\pi[L_e(\lambda,T_T) - L_e(\lambda,T_B)] A_d}{4 (f/\#)^2} T_{SYS}(\lambda) \, T_{TEST}(\lambda) \, d\lambda \qquad 3\text{-}22$$

If the target and background are true blackbodies with unity emittance, then Equation 3-3 can be substituted into Equation 3-22 so that ΔV_{SYS} can be calculated from Planck's blackbody law:

$$\Delta V_{SYS} = G \int_{\lambda_2}^{\lambda_1} R(\lambda) \frac{[M_e(\lambda,T_T) - M_e(\lambda,T_B)] A_d}{4 (f/\#)^2} T_{SYS}(\lambda) \, T_{TEST}(\lambda) \, d\lambda \qquad 3\text{-}23$$

Off-axis responses may be reduced by $\cos^N\theta$.

☞ ───

Example 3-1
ΔV_{SYS} CALCULATION

What is the expected output voltage difference for a US common module system? The system is viewing an extended 310 K target against a 300 K background in a collimator. Assume the HgCdTe detector has a peak responsivity of 20,000 V/watt. Assume $T_{ATM}(\lambda) = 1$, $T_{COL}(\lambda) = 0.9$, $T_{SYS}(\lambda) = 0.70$ (all independent of wavelength), $f/\# = 3$ and $G = 10,000$. Common module detectors are band limited from 8 to 12 μm. Outside this region the response is zero. The detector element is square with each side equal to 0.002 inches (5.08 x 10^{-3} cm). The detector area is $A_d = 2.58 \times 10^{-5}$ cm². For convenience, the target and its background are considered ideal blackbodies. $fl_{SYS} = 18$ inches.

The responsivity of most photoconductors follows:

$$R(\lambda) = \frac{\lambda}{\lambda_{peak}} R_{peak} \qquad 3\text{-}24$$

Then the output voltage difference for an on-axis detector is:

$$\Delta V_{SYS} = \frac{G \, T_{TEST} \, T_{SYS} \, A_d \, R_{peak}}{4(f/\#)^2} \int_8^{12} \frac{\lambda}{\lambda_{peak}} [M_e(\lambda,310) - M_e(\lambda,300)] \, d\lambda \qquad 3\text{-}25$$

Equation 3-25 can be approximated by:

66 TESTING & EVALUATION OF IR IMAGING SYSTEMS

$$\Delta V_{SYS} = 90.4 \sum_{8}^{12} \frac{\lambda}{\lambda_{peak}} \Delta M_e(\lambda,T) \Delta\lambda \qquad 3\text{-}26$$

where $\Delta M_e(\lambda,T) = M_e(\lambda,310) - M_e(\lambda,310)$. Using Simpson's rule, the integral is evaluated numerically at the center of each increment. For example, when $\Delta\lambda = 1$, the interval [8 μm, 9 μm], is evaluated at 8.5 μm (Table 3-1).

Table 3-1
ΔV_{SYS} CALCULATION

Wavelength (μm)	λ/λ_p	$M_e(\lambda,310)$ x 10^{-3}	$M_e(\lambda,300)$ x 10^{-3}	$\lambda/\lambda_p \cdot \Delta M_e(\lambda,T)$ x 10^{-3}
8.5	0.708	3.60	3.00	0.425
9.5	0.792	3.68	3.12	0.444
10.5	0.875	3.57	3.08	0.429
11.5	0.958	3.35	2.92	0.412
				sum = 1.71x10^{-3}

Then $\Delta V_{SYS} = (90.4)(1.71 \times 10^{-3}) = 154$ mv. The SiTF is $\Delta V_{SYS}/\Delta T = 154/10 = 15.4$ mv/K. The wavelength interval, $\Delta\lambda = 1$ μm, was selected for illustrative purposes only. For numerical integration, at least 20 intervals should be chosen.

3.1.4. POINT SOURCE

As the source area approaches zero, the source becomes an ideal point source. Geometric optics predicts that the image size also will approach zero. However, diffraction and aberrations will limit the minimum image size. The differential system output, ΔV_{SYS}, depends on the relative size of the blur diameter to the detector size. Hereafter, if the source angular subtense is much less than the IFOV, the source will be called a point source.

If the blur diameter is much less than the detector size then:

$$\Phi_{DETECTOR} = \Phi_{IMAGE} \qquad 3\text{-}27$$

If the idealized point source is placed in a collimator, the radiant flux on the detector is

$$\Phi_{DETECTOR} = L_e \frac{A_o}{(fl_{COL})^2} A_S T_{SYS} T_{TEST} \qquad 3\text{-}28$$

Since all objects emit infrared radiation, the detector will sense radiant flux from the point source and the radiant flux from the immediate background within the IFOV. In Figure 3-6, A_{IFOV} is the projected area of an IFOV in object space. The radiant flux is proportional to:

$$\Phi_{DETECTOR} = K[L_e(\lambda,T_T)A_S + L_e(\lambda,T_B)(A_{IFOV} - A_S)] \qquad 3\text{-}29$$

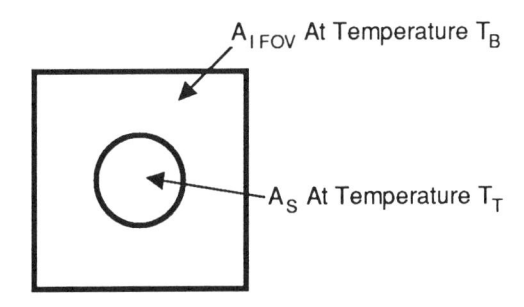

Figure 3-6: Small target inside the projected area of an IFOV. The detector senses radiation from the target and the background.

The radiant flux difference between this detector and an adjoining detector is:

$$\Delta\phi = K\left[[L_e(\lambda,T_T)A_S + L_e(\lambda,T_B)(A_{IFOV} - A_S)] - L_e(\lambda,T_B)A_{IFOV}\right] \qquad 3\text{-}30$$

or

$$\Delta\phi = K[L_e(\lambda,T_T) - L_e(\lambda,T_B)]A_S \qquad 3\text{-}31$$

The voltage difference produced is:

68 TESTING & EVALUATION OF IR IMAGING SYSTEMS

$$\Delta V_{SYS} = G \int_{\lambda_1}^{\lambda_2} R(\lambda) \frac{[L_e(\lambda,T_T) - L_e(\lambda,T_B)] A_S A_o}{(fl_{COL})^2} T_{SYS}(\lambda) T_{TEST}(\lambda) \, d\lambda \qquad 3\text{-}32$$

By symmetry,

$$\frac{A_{IFOV}}{(fl_{COL})^2} = \frac{A_d}{(fl_{SYS})^2} \qquad 3\text{-}33$$

and assuming a circular aperture and true blackbodies with unity emittance,

$$\Delta V_{SYS} = G \int_{\lambda_1}^{\lambda_2} R(\lambda) \frac{\pi \, \Delta L_e(\lambda,T) A_d}{4 \, (f/\#)^2} \frac{A_S}{A_{IFOV}} T_{SYS}(\lambda) T_{TEST}(\lambda) \, d\lambda \qquad 3\text{-}34$$

where $\Delta L_e(\lambda,T) = L_e(\lambda,T_T) - L_e(\lambda,T_B)$.

For extended (resolved) sources, Equation 3-23 is convenient to use since the source area is irrelevant: the amount of radiant flux reaching the detector is limited by the IFOV. For small sources (Equation 3-34), the source area must be known. Figure 3-7 illustrates the ideal (geometric) image area as a function of source area and where Equations 3-23 and 3-34 are applicable. The slope of line is the ratio A_d/A_{IFOV}. When $A_d = A_{IFOV}$ the two equations are identical.

Diffraction and aberrations will limit the smallest image size that can be achieved. In Figure 3-8, three cases are illustrated: (a) the geometric approach where there is no diffraction, (b) when the blur diameter is less than the diameter of a detector, and (c) when the blur is larger than a detector. Figure 3-9 illustrates the relationships depicted in Figure 3-8. Curve A is the ideal case and is identical with Figure 3-7. Curve B illustrates a system where the diffraction blur area is smaller that the detector area. This case is equivalent to Figure 3-8b and the Equations 3-23 and 3-34 are applicable as shown in Figure 3-7. For curve C, diffraction produces a blur diameter that is larger than the detector. Here, Equation 3-23 is only valid when curve C asymptotes to the geometric curve.

BASIC CONCEPTS IN IR TECHNOLOGY 69

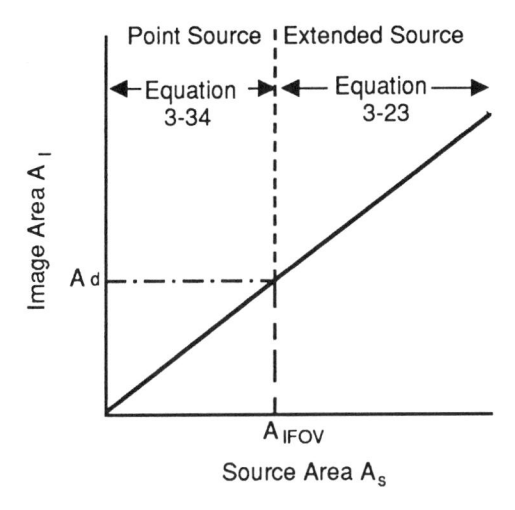

Figure 3-7: Geometric relationship between source area and image area. When $A_s = A_{IFOV}$, Equations 3-23 and 3-34 are identical.

When diffraction is important, Equation 3-34 must be modified by the aperiodic transfer function, ATF:

$$\Delta V_{SYS} = G \int_{\lambda_1}^{\lambda_2} R(\lambda) \frac{\pi \Delta L_e(\lambda,T) A_d}{4 (f/\#)^2} ATF\, T_{SYS}(\lambda)\, T_{TEST}(\lambda)\, d\lambda \qquad 3\text{-}35$$

Equation 3-35 combines Equation 3-23 and 3-34. When the source is resolved, ATF = 1 and Equation 3-23 is obtained. When $A_i < A_d$, Equation 3-34 is obtained. The ideal and system ATFs are shown in Figure 3-10 and the applicable regions for the equations. The ATF is the input-to-output transformation ΔV_{SYS} versus A_S normalized to unity. Calculation of the system ATF is beyond the scope of this book. However, the limiting case, where the image size becomes independent of A_S, is of interest for this is the case when the system is viewing a point source.

70 *TESTING & EVALUATION OF IR IMAGING SYSTEMS*

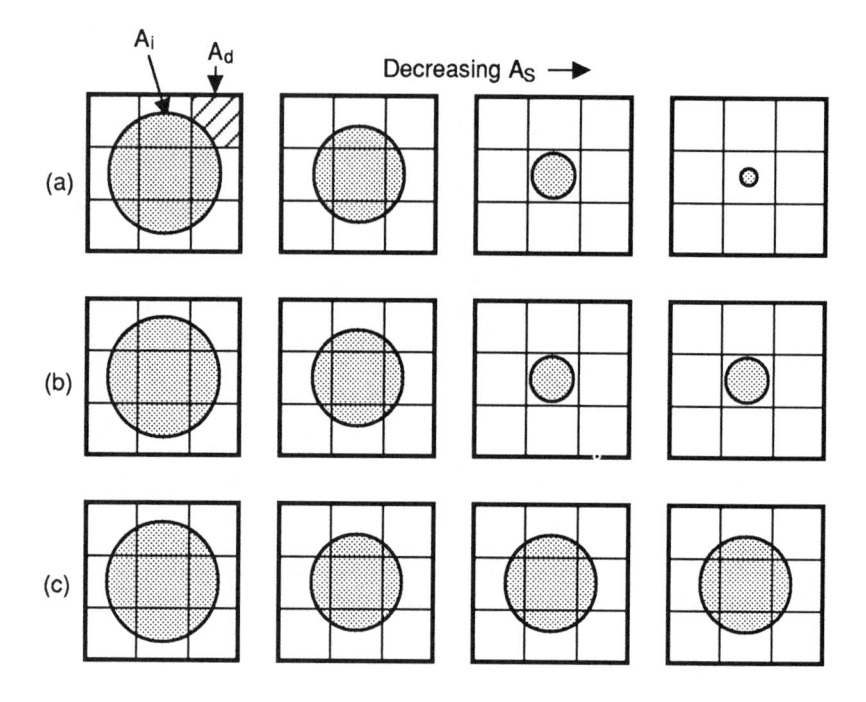

Figure 3-8: Various image sizes. (a) No diffraction, (b) blur diameter less than detector size, and (c) blur diameter greater than detector size.

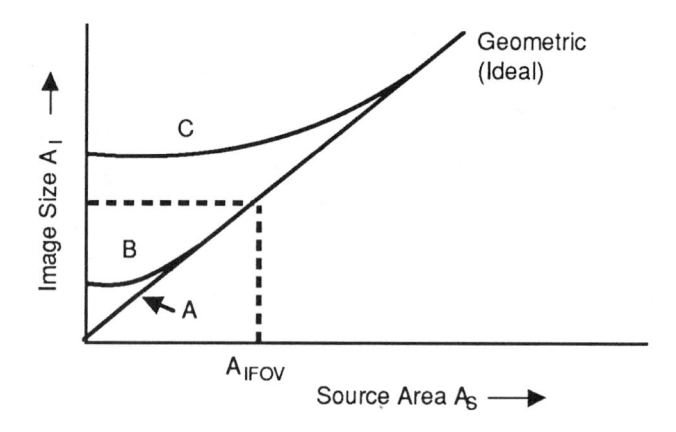

Figure 3-9: Relationship between image size and source size for cases (a), (b) and (c) given in Figure 3-8.

BASIC CONCEPTS IN IR TECHNOLOGY 71

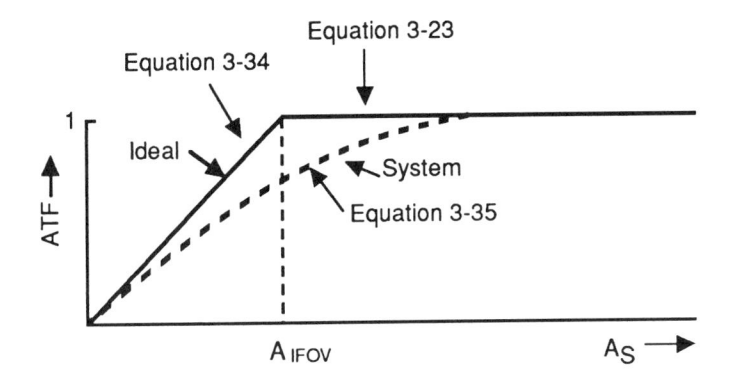

Figure 3-10: Aperiodic transfer function. Equation 3-35 asymptotes to Equation 3-23 for extended sources and to Equation 3-34 for point sources.

The target transfer function (TTF) is:

$$TTF = \frac{ATF_{SYS}}{ATF_{IDEAL}} = ATF_{SYS}\left(\frac{A_{IFOV}}{A_S}\right) \quad \text{when } A_S \leq A_{IFOV} \qquad 3\text{-}36$$

$$TTF = ATF_{SYS} \quad \text{when } A_S \geq A_{IFOV}$$

Substituting 3-36 into 3-35 yields:

$$\Delta V_{SYS} = G\int_{\lambda_1}^{\lambda_2} R(\lambda)\frac{\pi\,\Delta L_e(\lambda,T)\,A_d}{4\,(f/\#)^2}\, TTF\, \frac{A_S}{A_{IFOV}}\, T_{SYS}(\lambda)\, T_{TEST}(\lambda)\, d\lambda \qquad 3\text{-}37$$

As A_S becomes smaller, the TTF approaches a constant that is the point visibility factor, PVF. The PVF is also called the blur efficiency and ensquared power[2]. As A_S approaches zero, it is appropriate to represent $\Delta L_e\, A_S$ by the source radiant intensity $\Delta I_e(\lambda,T) = I_e(\lambda,T_T) - I_e(\lambda,T_B)$:

$$\Delta V_{SYS} = G\int_{\lambda_1}^{\lambda_2} R(\lambda)\,\frac{\pi\,\Delta I_e(\lambda,T)\,A_d}{4\,(f/\#)^2}\, PVF\, \frac{1}{A_{IFOV}}\, T_{SYS}(\lambda)\, T_{TEST}(\lambda)\, d\lambda \qquad 3\text{-}38$$

or equivalently

72 TESTING & EVALUATION OF IR IMAGING SYSTEMS

$$\Delta V_{SYS} = G \int_{\lambda_1}^{\lambda_2} R(\lambda) \frac{A_o \Delta I_e(\lambda,T)}{(fl_{COL})^2} PVF\, T_{SYS}(\lambda) T_{TEST}(\lambda)\, d\lambda \qquad 3\text{-}39$$

For point sources at a distance R_1 from the system (Figure 3-1: page 57):

$$\Delta V_{SYS} = G \int_{\lambda_1}^{\lambda_2} R(\lambda) \frac{A_o \Delta I_e(\lambda,T)}{R_1^2} PVF\, T_{SYS}(\lambda) T_{ATM}(\lambda)\, d\lambda \qquad 3\text{-}40$$

Example 3-2
POINT SOURCE

The system described in Example 3-1 (page 65) is viewing a small target ($A_S = 2 \times 10^{-6}$ cm^2) in a collimator with fl$_{COL}$ = 60 inches. The point visibility factor is 0.65. What is the expected output?

$$\Delta V_{SYS} = \frac{G\, T_{SYS}\, T_{TEST}\, A_d\, A_S\, R_{peak}}{4(f/\#)^2\, A_{IFOV}} PVF \int_{\lambda_1}^{\lambda_2} \frac{\lambda}{\lambda_p} \Delta M_e(\lambda,T)\, d\lambda \qquad 3\text{-}41$$

which is approximated by

$$\Delta V_{SYS} = \left[\frac{G\, T_{SYS}\, T_{TEST}\, A_d\, R_{peak}}{4(f/\#)^2} \sum_{8}^{12} \frac{\lambda}{\lambda_p} \Delta M_e(\lambda,T)\, \Delta\lambda \right] \left[\frac{A_s\, PVF}{A_{IFOV}} \right] \qquad 3\text{-}42$$

From Example 3-1, the first bracket is 154 mv and the second bracket is 0.045. Then $\Delta V_{SYS} = 154 \times (0.045) = 7$ mv. The output is significantly smaller than that obtained for an extended source of the same temperature. When measuring point source response, a high source temperature is usually selected (i.e., 500° C to 1000° C).

3.1.5. ΔT CONCEPT

It is convenient to express the small radiant exitance difference between a target and its background by the temperature difference between the target and its background. It is desirable to express ΔV_{SYS} as a function of ΔT and call the proportionality constant the SiTF:

$$\Delta V_{SYS} = SiTF \, \Delta T \qquad 3\text{-}43$$

This approach assumes that the target and background are ideal blackbodies with unity emittance. The temperature differential is proportional to the partial derivative of Planck's law with respect to temperature ("thermal derivative"):

$$M_e(\lambda, T_T) - M_e(\lambda, T_B) \approx \left[\frac{\partial M_e(\lambda, T_B)}{\partial T}\right] \Delta T \qquad 3\text{-}44$$

where

$$\frac{\partial M_e(\lambda, T)}{\partial T} = M_e(\lambda, T) \frac{c_2 e^{c_2/\lambda T}}{\lambda T^2 \left[e^{c_2/\lambda T} - 1\right]} \qquad 3\text{-}45$$

When substituted into Equation 3-35 yields

$$\Delta V_{SYS} = \left[G \int_{\lambda_1}^{\lambda_2} R(\lambda) \frac{A_d}{4 \, (f/\#)^2} \frac{\partial M_e(\lambda, T_B)}{\partial T} ATF \, T_{SYS}(\lambda) T_{TEST}(\lambda) \, d\lambda \right] \Delta T \qquad 3\text{-}46$$

The thermal derivative is plotted in Figure 3-11. It is a function of wavelength and the background temperature. As a result, all measurements that use ΔT as an input will be affected by the background temperature[3] and system spectral response. The ΔT concept is a matter of convenience. Infrared imaging systems that have photo detectors do not measure temperature but respond to radiance differences. While the ΔT concept may be useful for thermometers, it does not uniquely specify system performance unless both the system spectral response and the background temperature are specified.

74 TESTING & EVALUATION OF IR IMAGING SYSTEMS

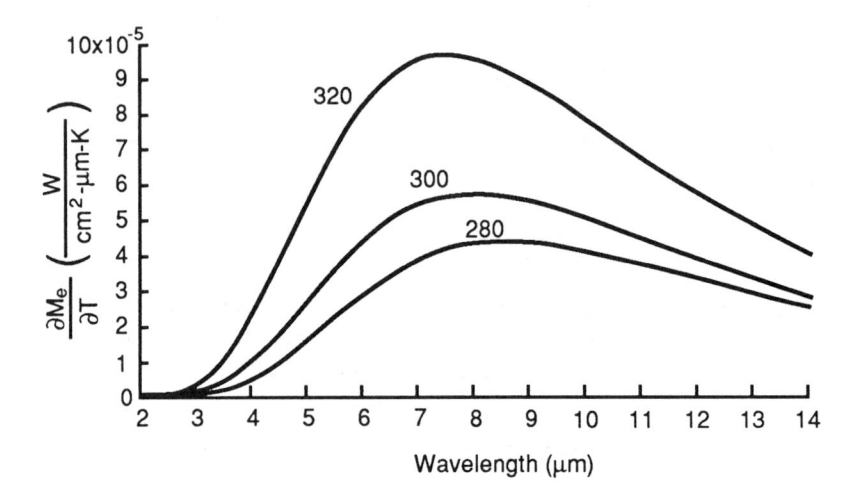

Figure 3-11: Thermal derivative of Planck's blackbody radiation law for three different background temperatures.

The effect of the ambient temperature variation on radiance differences is shown in Figure 3-12 and 3-13. These graphs should be taken as representative: actual values depend upon the spectral response of the particular system. The 3 to 5 μm region is more sensitive to background changes than the 8 to 12 μm region. As indicted in these figures, the difference in radiant flux from 300 to 301 K and from 280 to 281 K are different although the thermometric difference is one degree. For example, in the 8 to 12 μm spectral region, a temperature differential of 1.2 K at 280 K produces the same radiant flux differential as 1 K at 300 K. A drift in ambient of 1 K manifests itself as an ΔT_{eff} change of 0.02 K. Since the MRT and NEDT are inversely proportional to the SiTF, the MRT[4,5,6] (Figure 3-14) and NEDT (Figure 3-15) increase as the ambient temperature decreases. By allowing the ambient temperature to drift from 20 to 25° C the NEDT will have approximately 17% variation in the 3 to 5 μm spectral region and approximately 5% in the 8 to 12 μm spectral region. The linear approximation given by Equation 3-43 is only valid for small excursions about the background temperature. Departures from linearity (Figure 3-16) are apparent when $\Delta T > 10°$ C in the LWIR and $\Delta T > 5°$ C in the MWIR. The nonlinearity can be avoided if all calculations and measurements used watts or photons rather than ΔT. Nevertheless, ΔT has become an industry wide standard and the nonlinearity exists in all input-to-output transformations (e.g., SiTF, NEDT, MRT, and MDT).

BASIC CONCEPTS IN IR TECHNOLOGY 75

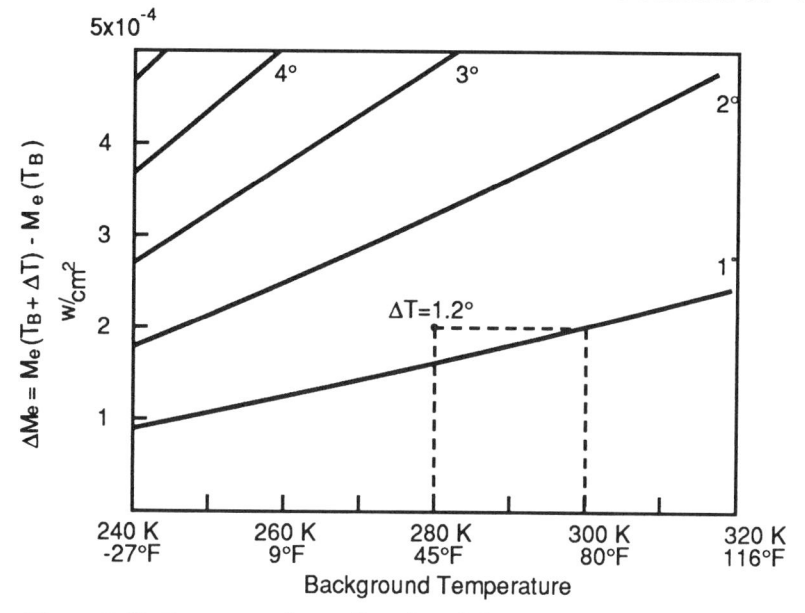

Figure 3-12: Representative radiant flux differences for the 8 to 12 μm region for fixed thermometric differences. Actual flux differences depend upon the specific design.

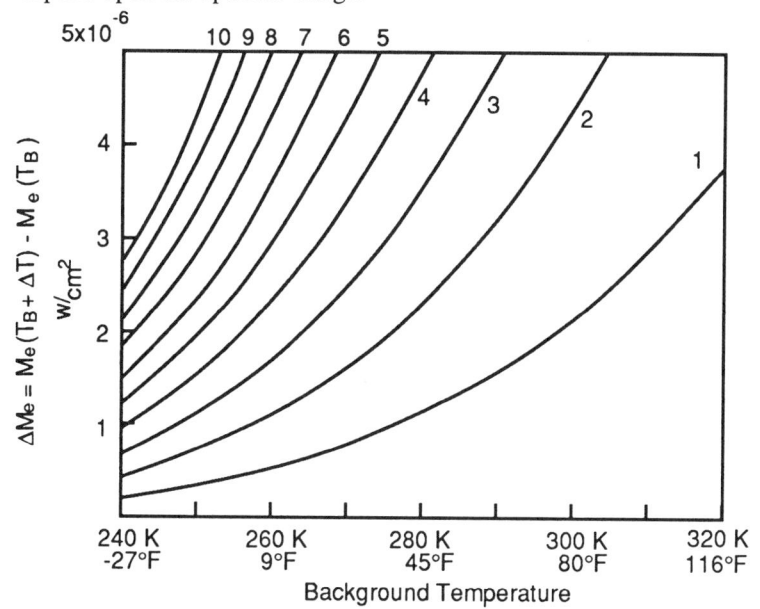

Figure 3-13: Representative radiant flux differences for the 3 to 5 μm region for fixed thermometric differences.

76 *TESTING & EVALUATION OF IR IMAGING SYSTEMS*

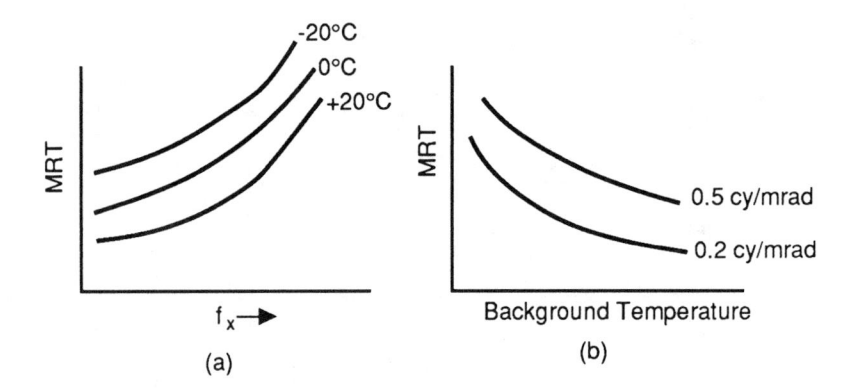

Figure 3-14: Effects of ambient temperature variations on MRT. As the ambient temperature decreases, the apparent MRT increases. (a) As a function of spatial frequency and (b) as a function of background temperature.

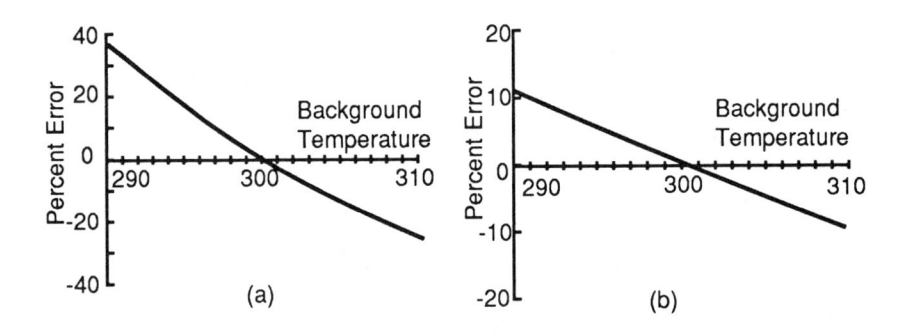

Figure 3-15: Representative effects of ambient temperature variations on NEDT. (a) 3-5 μm region and (b) 8-12 μm region. As the ambient temperature decreases, the apparent NEDT increases.

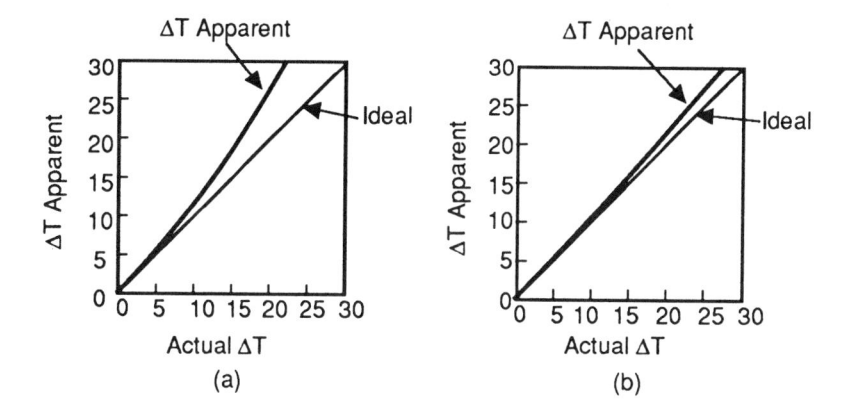

Figure 3-16: Deviations in apparent ΔT as a function of actual ΔT when the background is 300 K. (a) 8-12 μm region and (b) 3-5 μm region.

Since the ΔT concept depends upon the system's spectral response, different LWIR systems (e.g., 7.5 to 11.5 μm and 8 to 12 μm) will produce different responsivity curves when plotted as a function of ΔT. The ΔT concept also assumes that the target and background are ideal blackbodies with unity emittance. When the emittance decreases, reflection increases and then the apparent ΔT depends upon extraneous sources.

Thermometric calibration[7] (ΔT) is appropriate if the atmosphere, source, collimator and system do not have any spectral features over the wavelengths of interest. Radiometric calibration overcomes the difficulties associated with thermometric calibration. However, the difficulty of radiometric calibration can be appreciated when one realizes the many books devoted to the subject[8].

78 TESTING & EVALUATION OF IR IMAGING SYSTEMS

Example 3-3
SiTF

What is SiTF for the system described in Example 3-1 (page 65)?

When using the small radiance difference approximation, the SiTF is independent of the ΔT and numerical integration yields:

$$\text{SiTF} = \frac{\Delta V_{SYS}}{\Delta T} = 90.4 \left[\sum_{8}^{12} \frac{\lambda}{\lambda_p} \frac{\partial M_e(\lambda,300)}{\partial T} \Delta \lambda \right] \qquad 3\text{-}47$$

For illustrative purposes, Equation 3-47 is evaluated with $\Delta \lambda = 1$ μm (Table 3-2)

Table 3-2
SiTF NUMERICAL INTEGRATION

Wavelength (μm)	λ/λ_p	$\dfrac{\partial M_e(\lambda,310)}{\partial T}$	$\dfrac{\lambda}{\lambda_p} \dfrac{\partial M_e(\lambda,300)}{\partial T}$
8.5	0.708	5.64 x 10^{-5}	3.99 x 10^{-5}
9.5	0.792	5.24	4.15
10.5	0.875	4.68	4.09
11.5	0.958	4.05	3.88
			sum = 1.61x10^{-4}

Then $\Delta V_{SYS} = (90.4)(1.61 \times 10^{-4}) = 14.5$ mv/K. The exact calculation (Example 3-1) provided 154 mv for a 10 K delta temperature source whereas the approximation provided only 145 mv for the same source. This is the variation illustrated in Figure 3-16 where the apparent ΔT (used in the calculations) is higher than the actual ΔT. When measuring the SiTF (Chapter 6), ΔV_{SYS} is plotted as function of ΔT. The nonlinearity is obvious in the graph and the SiTF is the linear portion of the responsivity curve.

3.2. NORMALIZATION

"Normalization .... is the process of reducing measurement results as nearly as possible to a common scale"[9]. Normalization is essential to insure that appropriate comparisons are made. Figure 3-17 illustrates the relationship between the spectral response of a system to two different sources. The output of a system depends upon the spectral features of the input and the spectral response of the infrared imaging system. System output does not infer anything about the source other than that of an equivalent blackbody of a certain temperature would provide the same output.

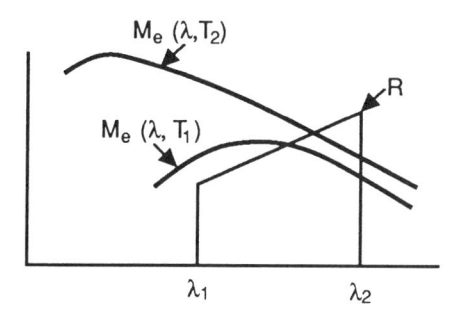

Figure 3-17: Sources with different spectral outputs can produce different system outputs. T_2 provides more radiant flux than T_1. The system output will be higher when viewing T_2.

Variations in output can also occur if "identical" systems have different spectral responses (Figure 3-18). Spectral mismatch is a major contributor to fixed pattern noise in focal plane arrays when each detector has a different spectral response[10]. Equation 3-35 is integrated over the interval $[\lambda_1,\lambda_2]$ for System 1 and over the interval $[\lambda_3,\lambda_4]$ for System 2. Due to the spectral mismatch, $\Delta V_{SYS-1} \neq \Delta V_{SYS-2}$. For example, an infrared imaging system whose spectral response is 8 to 12 μm may have a different responsivity than a system that operates 7.5 to 11.5 μm although both systems are labeled as LWIR systems. Systems can be made to appear as equivalent or one can be made to provide better performance by simply selecting an appropriate source. Spectral mismatch among sources, collimators and atmospheric conditions may account for the different results obtained at various laboratories.

80 TESTING & EVALUATION OF IR IMAGING SYSTEMS

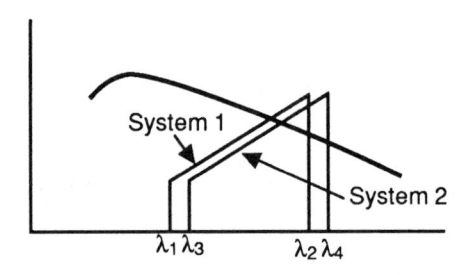

Figure 3-18: Different spectral response systems can produce different outputs when viewing the same source.

It is sometimes useful to discuss average responsivity or average transmittance. The term *average* represents the mean value of a function. If f(x) is weighted by another function g(x), the average of f(x) over an interval [a,b] is:

$$f_{ave} = \frac{\int_a^b f(x)g(x)dx}{\int_a^b g(x)dx} \qquad 3\text{-}48$$

Using this methodology, the average responsivity becomes:

$$R_{AVE} = \frac{\int_{\lambda_1}^{\lambda_2} R(\lambda)\,\Delta M_e(\lambda,T)\,T_{SYS}(\lambda)\,T_{TEST}(\lambda)\,d\lambda}{\int_{\lambda_1}^{\lambda_2} \Delta M_e(\lambda,T)\,T_{SYS}(\lambda)\,T_{TEST}(\lambda)\,d\lambda} \qquad 3\text{-}49$$

Similarly, the average collimator transmittance, system transmittance or atmospheric transmittance also can be calculated. Equation 3-49 shows that the *average* response depends on the source's spectral characteristics (e.g., the source temperature). Examples 3-1 through 3-3 assumed average transmittances for the system, collimator and atmosphere.

Example 3-4
AVERAGE OPTICAL TRANSMITTANCE

What is the average optical transmittance of a lens system for the LWIR system described in Example 3-1 (page 65)? The optical spectral transmittance, $T_{SYS}(\lambda)$, is given in Table 3-3.

The average optical transmittance, T_{AVE}, is

$$T_{AVE} = \frac{\int_{\lambda_1}^{\lambda_2} T_{SYS}(\lambda)\, R(\lambda)\, \Delta M_e(\lambda, T)\, d\lambda}{\int_{\lambda_1}^{\lambda_2} R(\lambda)\, \Delta M_e(\lambda, T)\, d\lambda} \qquad 3\text{-}50$$

which is approximated by

$$T_{AVE} = \frac{\sum_{\lambda_1}^{\lambda_2} T_{SYS}(\lambda)\, R(\lambda)\, \Delta M_e(\lambda, T)\, \Delta\lambda}{\sum_{\lambda_1}^{\lambda_2} R(\lambda)\, \Delta M_e(\lambda, T)\, \Delta\lambda} \qquad 3\text{-}51$$

In Table 3-3, Simpson's rule is used to numerically evaluate the integral at the center of each interval ($\Delta\lambda = 1$). $\Delta M_e(\lambda, T)$ is obtained from the values given in Table 3-1.

82 *TESTING & EVALUATION OF IR IMAGING SYSTEMS*

Table 3-3
T_{AVE} OBTAINED BY NUMERICAL INTEGRATION

λ	R(λ)	ΔM_e ×10⁻⁴	T_SYS(λ)	R(λ)ΔM_e(λ,T)T_SYS(λ) ×10⁻³	R(λ)M_e(λ,T) ×10⁻³
8.5	0.708	6.0	0.8	3.398	4.248
9.5	0.792	5.6	0.8	3.548	4.435
10.5	0.875	4.9	0.7	3.001	4.287
11.5	0.958	4.3	0.6	2.471	4.119
				sum = 12.42	sum = 17.09

The average transmittance is 12.42/17.09 = 0.727. The average transmittance could be substantially different if the blackbody source had a different absolute temperature. As with the previous examples, Δλ = 1 μm was chosen for illustrative purposes. Δλ should be much smaller when performing the calculation.

3.3. SPATIAL FREQUENCY

There are four different spatial frequency concepts. They are object space, image space, monitor and observer spatial frequencies. Simple equations relate all the spatial frequencies.

Object space spatial frequency (Figure 3-19) is used to characterize system response. It is used by electro-optical analysts and system test personnel. This text exclusively uses object space spatial frequency. Using the small angle approximation, the angle subtended by one cycle (one bar and one space) is d/R_1 where d is the spatial extent of one cycle and R_1 is the distance from the infrared system entrance aperture to the target. When using a collimator, the collimator focal length replaces R_1 so that targets placed in the collimator's focal plane can be described in object space. The horizontal object space spatial frequency, f_x, is the inverse of the horizontal target angular subtense and is usually expressed in cycles/mrad:

$$f_x = \frac{1}{1000}\left(\frac{R_1}{d}\right) \quad \frac{cycles}{mrad} \qquad 3\text{-}52$$

or

$$f_x = \frac{1}{1000}\left(\frac{fl_{COL}}{d}\right) \quad \frac{cycles}{mrad}$$

Optical designers typically use image space spatial frequency to specify the resolving capability of lens systems. It is the object space spatial frequency divided by the system focal length:

$$f_I = \frac{f_x}{fl_{SYS}} \quad \frac{line\text{-}pairs}{mm} \quad or \quad \frac{cycles}{mm} \qquad 3\text{-}53$$

Although used interchangeably, line-pairs suggest square wave targets and cycles suggest sinusoidal targets. To maintain dimensionality, if f_x is measured in cy/mrad then the focal length must be measured in meters to obtain cy/mm. f_I is the inverse of one cycle in the focal plane of the lens system.

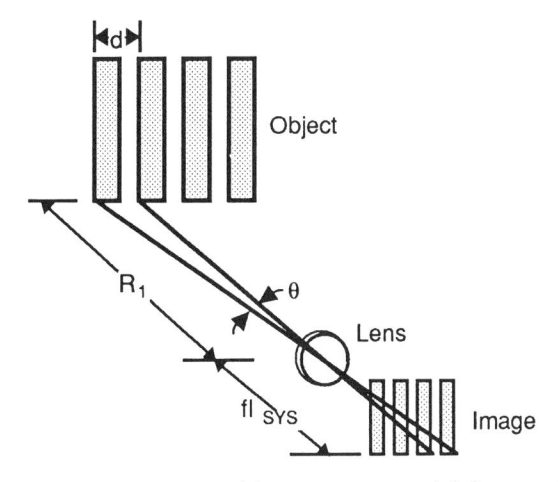

Figure 3-19: Object and image space spatial frequencies.

An equivalent equation exists for the vertical object spatial frequency f_y. Although the units suggest sinusoids, cycles/mrad is used for both sinusoids and square waves.

The resolving capability of a monitor can be specified by the number of lines seen (Figure 3-20). By convention, although the bar width, Δx, is measured in the horizontal direction, the number of lines is related to the picture height, PH. The number of lines may be specified as TV lines per picture height, lines per picture height or simply lines:

$$N_m = \frac{PH}{\Delta x}$$

When specifying the monitor resolution, N_m is modified by the Kell factor[11] which accounts for possible phasing effects. The Kell factor is often assumed to be 0.7. For typical commercial Tvs that display 480 lines, the resolution is reported as 336 lines. There are two TV lines per cycle. Therefore, the monitor spatial frequency is

$$f_m = \frac{N_m}{2} = \frac{PH}{2\Delta x} \quad \frac{cycles}{picture\ height} \qquad 3\text{-}54$$

The size of the image can be measured with a ruler. Using the monitor aspect ratio, α, f_m can be converted to a spatial frequency normalized to the monitor width:

$$f_{mx} = \alpha f_m \quad \frac{cycles}{picture\ width} \qquad 3\text{-}55$$

For most computer monitors, $\alpha = 1$. For commercial Tvs, $\alpha = 4/3$ and for high definition Tvs, $\alpha = 16/9$.

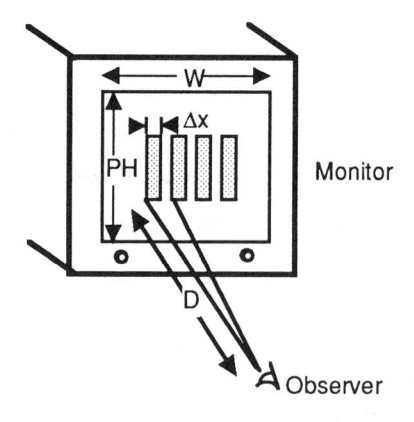

Figure 3-20: Monitor and observer spatial frequencies.

The spatial frequency presented to the observer depends upon the observer's distance to the monitor and the image size on the monitor. In the visual psychophysical literature, the usual units are cycles/deg:

$$f_e = \frac{1}{2\tan^{-1}\left(\frac{\Delta x}{D}\right)} \quad \frac{cycles}{deg} \qquad 3\text{-}56$$

The small angle approximation may not be valid for the observer spatial frequency since the angle subtended can become quite large as the observer moves toward the monitor. The total number of cycles, N_c, which can be sensed is f_x HFOV where HFOV is the horizontal field-of-view in mrad. If the monitor width is W then:

$$\Delta x = \frac{W}{2N_c} = \frac{W}{2f_x \text{HFOV}} \qquad 3\text{-}57$$

3.4. SUMMARY

Four different spatial frequency concepts are used. They are object space, image space, monitor and observer spatial frequencies (Table 3-4). The test engineer must be conversant in all 4 spatial frequencies.

Table 3-4
SPATIAL FREQUENCY DEFINITIONS

SPATIAL FREQUENCY	USER	DEFINITION (typical units)
Object space	E-O analysts, system test engineers	$f_x = R_1/(1000d)$ (cy/mr)
Image space	Optical designers	$f_i = f_x/fl$ (cy/mm)
Monitor space	Monitor designers[*]	$f_m = PH/(2\Delta x)$ (cy/picture height)
Observer space	Visual psychophysicist	$f_e = 1/(2\cdot\tan^{-1}(\Delta x/D))$ (cy/deg)

[*] Monitor designers tend to use lines per picture height that often includes the Kell factor.

The spectral radiant exitance, M_e, of an ideal blackbody source whose absolute temperature is T, can be described by Planck's blackbody radiation law:

$$M_e(\lambda,T) = \frac{c_1}{\lambda^5}\left(\frac{1}{e^{(c_2/\lambda T)} - 1}\right) \quad \frac{w}{cm^2-\mu m} \qquad 3\text{-}58$$

where the first radiation constant is $c_1 = 3.7418 \times 10^4$ watt-$\mu m^4/cm^2$ and the second radiation constant is $c_2 = 1.4388 \times 10^4$ μm-K. For Lambertian blackbody extended sources, the radiant sterance is:

$$L_e(\lambda,T) = \frac{M_e(\lambda,T)}{\pi} \quad \frac{watts}{cm^2-\mu m-sr} \qquad 3\text{-}59$$

For extended sources in a collimator, the differential output of the system is:

$$\Delta V_{SYS} = G \int_{\lambda_2}^{\lambda_1} R(\lambda) \frac{\Delta M_e(\lambda,T) A_d}{4\,(f/\#)^2} T_{SYS}(\lambda)\, T_{TEST}(\lambda)\, d\lambda \qquad 3\text{-}60$$

where $T_{TEST}(\lambda) = T_{COL}(\lambda) T_{ATM}(\lambda)$. As the source size decreases, the target transfer function must be added:

$$\Delta V_{SYS} = G \int_{\lambda_1}^{\lambda_2} R(\lambda) \frac{\pi \Delta L_e(\lambda,T) A_d}{4\,(f/\#)^2} TTF \frac{A_S}{A_{IFOV}} T_{SYS}(\lambda)\, T_{TEST}(\lambda)\, d\lambda \qquad 3\text{-}61$$

For small sources where the target transfer function asymptotes to the PVF:

$$\Delta V_{SYS} = G \int_{\lambda_1}^{\lambda_2} R(\lambda) \frac{A_o \Delta L_e(\lambda,T) A_S}{(fl_{COL})^2} PVF\, T_{SYS}(\lambda)\, T_{TEST}(\lambda)\, d\lambda \qquad 3\text{-}62$$

In a well-designed test configuration, the limiting radiometric aperture is the system aperture. If the collimator aperture is smaller than the system aperture, the equations must be modified by the ratio of the aperture areas.

It is convenient to express the output as a function of target-background temperature differential. However, this depends upon the spectral response of the system and the background temperature.

BASIC CONCEPTS IN IR TECHNOLOGY 87

$$\Delta V_{SYS} = \left[G \int_{\lambda_2}^{\lambda_1} R(\lambda) \; \frac{\partial M_e(\lambda, T_B)}{\partial T} \; \frac{A_d}{4(f/\#)^2} \; T_{SYS}(\lambda) \; T_{TEST}(\lambda) \; d\lambda \right] \Delta T \qquad 3\text{-}63$$

The input-to-output transformation, $\Delta V_{SYS}/\Delta T$, is the signal transfer function (SiTF). The SiTF should be calculated before performing any measurements. If the measured SiTF does not equal the calculated value, both the measurement technique and the calculation should be carefully examined. The SiTF is not strictly linear when plotted as a function of ΔT because most infrared imaging systems respond to photons rather than temperature differences. When measuring the SiTF (Chapter 6), ΔV_{SYS} is plotted as function of ΔT. The nonlinearity is obvious in the graph and the SiTF is the linear portion of the responsivity curve. Since the output is a function of wavelength, it is important to verify the spectral transmittance of the collimator and intervening atmosphere. Since the atmospheric spectral transmittance depends on path length, the distance from the source to the system aperture is important.

System output does not infer anything about the source other than that of an equivalent blackbody of a certain temperature would provide the same output. This is true no matter what output units are used (volts, amps or any other arbitrary unit). These units, by themselves, are not very meaningful for system-to-system comparison. Manufacturers' specifications are offered as a guide to the system's performance but the test conditions are not always fully described. As such, it is dangerous simply to compare system response based upon only a few numbers.

Because of spectral dependency, the average responsivity is:

$$R_{AVE} = \frac{\int_{\lambda_1}^{\lambda_2} R(\lambda) \; \Delta M_e(\lambda, T) \; T_{SYS}(\lambda) \; T_{TEST}(\lambda) \; d\lambda}{\int_{\lambda_1}^{\lambda_2} \Delta M_e(\lambda, T) \; T_{SYS}(\lambda) \; T_{TEST}(\lambda) \; d\lambda} \qquad 3\text{-}64$$

Similar equations exist for average collimator transmittance, average atmospheric transmittance and average optical transmittance. It is convenient to express the optical transmittance and collimator transmittance by average values since the target and background temperatures do not change very much. However, the atmospheric transmittance can be strongly dependent upon the path length and the average value should be used with extreme caution.

3.5. REFERENCES

1. C. L. Wyatt, Radiometric System Design, Chapter 3: Macmillan Publishing Co. New York, NY (1987).
2. L.M. Beyer, S. H. Cobb and L. C. Clune, "Ensquared Power for Circular Pupils with Off-center Imaging", Applied Optics, Vol. 30(25), pp. 3569-3574 (1991).
3. R. G. Driggers, G. L. Boylston and G. T. Edwards, "Equivalent Temperature Difference with Respect to Ambient Temperature Difference as a Function of Background Temperature", Optical Engineering, Vol. 31(6), pp. 1357-1361 (1992).
4. American Society of Heating, Refrigeration and Air-conditioning Engineers, "Applications of Infrared Sensing Devices to the Assessment of Building Heat Loss Characteristics", ANSI/ASHRAE Standard 101-1981, ASHRAE, Atlanta, GA (1983).
5. Y. M. Chang and R. A. Grot, "Performances Measurements of Infrared Imaging Systems Used to Assess Thermal Anomalies", in Thermal Imaging, I. R. Abel, ed.: SPIE Proceedings Vol. 636, pp. 17-30, (1986).
6. G. B. McIntosh and A. F. Filippone, "Minimum Resolvable Temperature Difference (MRTD) Testing: Equipment Specifications for Building Performance Diagnostics", in Thermosense IV, R. A. Grot and J. Wood, eds.: SPIE Proceedings Vol. 313, pp. 102-111 (1981).
7. P. Richardson, "Radiometric vs Thermometric Calibration of IR Test Systems: Which is Best?", in Infrared Imaging: Design, Analysis, Modeling and Testing II, G. C. Holst, ed.: SPIE Proceedings Vol. 1488, pp. 80-88 (1991).
8. See for example: C. L. Wyatt, Radiometric Calibration: Academic Press, Orlando, Fl (1978) and C. L. Wyatt Radiometric System Design: Macmillan Publishing, NY (1987).
9. F. E. Nicodemus, "Normalization in Radiometry", Applied Optics, Vol. 12(12), pp. 2960-2973 (1973).
10. N. Bluzer, "Sensitivity Limitations of IRFPAs Imposed by Detector Nonuniformities", in Infrared Detectors and Arrays, E. L. Dereniak, ed.: SPIE Proceedings Vol. 930, pp. 64-75 (1988).
11. S. C. Hsu, "The Kell Factor: Past and Present", Society of Motion Picture and Television Engineers Journal, Vol. 95, pp. 206-214 (1986).

EXERCISES

1. A thermal imaging system has an IFOV of 10 mrad, FOV of 30 mrad x 30 mrad and a focal length of 10 inches. The observer stands 24 inches from a 14 inch monitor (measured diagonally with a 1:1 aspect ratio). Fill in the following table:

Object spatial frequency (cy/mrad)	Image spatial frequency (lp/mm)	Monitor spatial frequency (cy/PH)	Monitor spatial frequency (cy/picture-width)	Observer spatial frequency (cy/deg)
1				
2				
5				
10				

BASIC CONCEPTS IN IR TECHNOLOGY 89

2. Using the approach in Example 3-1, calculate ΔV_{SYS} for a source temperature of 50° C above an ambient background of 27° C. Plot the results as a function of ΔT. Discuss the difference in this calculated SiTF compared to the values given in Examples 3-1 (page 65) and 3-3 (page 78).
3. For the system described in Example 3-1 (page 65), calculated the SiTF for a target temperature of 280 K on a 270 K background.
4. Using the approach in Example 3-1 (page 65), calculate ΔV_{SYS} for a MWIR system whose spectral response is from 3 to 5 μm for a 320 K target against a 300 K background. Assume R_{peak} = 30,000 v/w, T_{ATM} = 0.9, T_{COL} = 0.9, T_{SYS} = 0.8, f/# = 4, G = 20,000, A_d = 4 x 10^{-5} cm² and fl_{SYS} = 10 inches.
5. Using the approach given in Example 3-3 (page 78), calculate the SiTF for the MWIR sensor described in Exercise 4. Compare the answer with that obtained in Exercise 3.
6. Calculate T_{AVE} using the data given in Example 3-4 (page 81) but assume the target temperature is 800 K. Compare the answer with that obtained in Example 3-4.

4

GENERAL MEASURING TECHNIQUES

The source, target and collimator combination projects standardized targets of known size and intensity onto the infrared imaging system. Blackbody sources provide the flux that the infrared imaging system detects. With the target plate, its function is to provide a known geometric shaped target with known characteristics against a known background. The source is usually viewed through cutouts in the target plate. By convention, the cutout is called the target and the radiation that appears to emanate from the target plate, the background. The background radiant exitance may be due to the natural blackbody emission of the target plate or from another source reflected by the target plate. For most tests, it is the difference between the target and background radiant exitances that is of interest. For convenience, this difference is measured by the effective temperature difference, ΔT. Both the target and the background are constructed to simulate blackbodies.

Collimators are lens assemblies that optically place targets at infinity. Collimators may contain either refractive or reflective elements. The combination of the source, target, and collimator is sometimes called a target simulator or target projector. Collimator transmittance and atmospheric transmittance reduce the signal levels reaching the infrared imaging system. Atmospheric turbulence distorts detail. Therefore, turbulence must be minimized to resolve detail.

All physical and subjective tests are input-to-output transformations. The collimator and blackbody provide the input. Many methods are available to measure the output. Thess include measuring the analog video output with a transient digitizer, frame grabber or an oscilloscope. The monitor luminance can be measured with a solid state camera or a scanning microphotometer. Appropriate statistical analyses increase data validity. The process of recording, analyzing, interpreting, presenting and archiving data must be done rigorously and faithfully.

4.1. BLACKBODIES

The classical blackbody is a hollow sphere made of an opaque material and it has a small aperture to view the radiation. The radiation passing through the small aperture is nearly ideal blackbody radiation. Commercial sources behave as ideal blackbodies and therefore may be called blackbody simulators. A real blackbody is an approximation to a theoretical blackbody because it does not emit the radiant exitance predicted by Planck's blackbody law. The ratio of actual to theoretical radiant exitance is the emittance. The emittance is always less than unity but, for well-designed sources, can range from 95% to 99%. For opaque material, energy conservation requires

$$\alpha(\lambda) + \rho(\lambda) = 1$$

where α is the absorption coefficient and ρ is the reflectivity. According to Kirchoff's law, when opaque materials are at thermal equilibrium, the amount of energy absorbed must equal the amount of energy emitted at each wave length:

$$\alpha(\lambda) = \epsilon(\lambda)$$

where ϵ is the emittance. As a result it is appropriate to say "good absorbers are good emitters." For good blackbodies, the emittance approaches unity for all wavelengths. Therefore, blackbody sources visually appear black. Equivalently, good absorbers are poor reflectors. Therefore, a true blackbody has a very diffuse surface. If the surface appears shiny, the emittance is less than unity and would not be a good source.

There are many types of blackbody sources[1]. The most common are point sources, differential sources, high temperature extended area sources and ultra-high temperature blackbodies. Each source lends itself to different types of tests.

Point source blackbodies are typically constructed in a cavity shape with an aperture stop (Figure 4-1). They are equivalent to the classical blackbody.

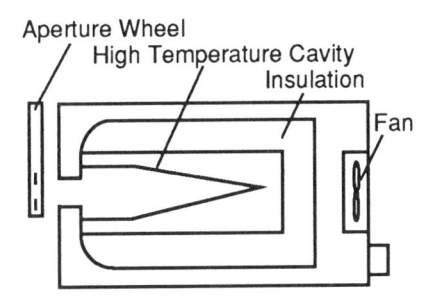

Figure 4-1: Construction of a point source blackbody with an aperture wheel.

92 TESTING & EVALUATION OF IR IMAGING SYSTEMS

They are slow to heat up and cool very slowly due to its thermal. Temperature ranges vary but are typically available from 50°C to 1000°C. The aperture size is typically less then 2 inches with 1 inch being a *standard* size. Larger sizes are available but are rare due to manufacturing constraints. The cavity design provides high emissivity but temperature uniformity is an issue with these sources. Aperture wheels, choppers, filter wheels and shutters are typically used with the source for IRST system testing and detector testing.

The differential blackbody source is primarily for infrared system testing to provide a low thermal contrast between the target plate and blackbody source (figure 4-2). These systems are typically thermoelectrically heated or cooled to provide both positive and negative ΔTs. They typically have a large emitting surface and therefore are extended sources. Temperature resolution of 0.001°C is commercially available. The target is mounted in front of the source and its temperature is monitored. As the ambient temperature changes, the blackbody source temperature changes such that the target-background ΔT is held constant. The target plate is assumed to be at ambient temperature. By convention the target plate is called the background and the target holes the source even though the radiation emanates from the blackbody behind the cutouts.

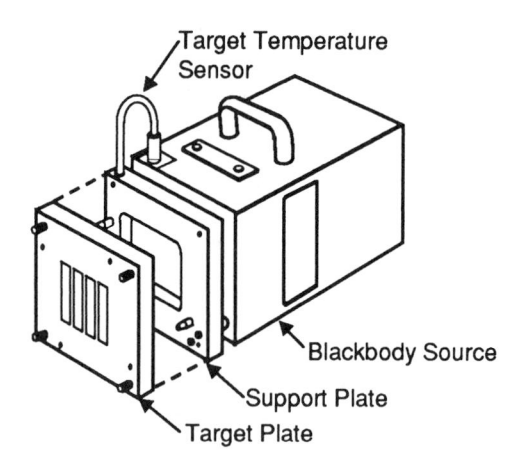

Figure 4-2: Construction of a differential blackbody source. The electronic controller (not shown) is a separate piece of equipment that can be rack mounted. The target plate is usually called the target.

The typical limits for the blackbody source are 0°C to 100°C and this provides a ΔT of approximately -25 to +75°C about ambient. The ambient temperature around the thermoelectric heat sink limits the lowest achievable temperature. The low temperature limit can be extended by cooling the heat sink with a liquid chiller or putting the blackbody in a low temperature chamber. However, as the temperature becomes cooler, the blackbody temperature may be below the ambient dew point and then moisture will form on the emitting surface: an undesirable situation since the emissivity of moisture is not well known. Surface emissivity is a major concern with these sources.

These systems often have a joy stick for temperature slew control. This useful feature decreases MRT and MDT test time. The source also can be computer controlled for automated testing.

High temperature extended area blackbodies are used in applications requiring a high flux level with a large field size such as gain/level corrections on focal plane arrays. These sources typically operate up to 600°C. Finding high emissivity paints that can survive the temperature is a challenge. Temperature uniformity can be a big problem with these sources. Because of the large thermal mass, temperature slewing is slow and heat up time is slow.

Ultra-high temperature blackbodies are used for testing short wave IR systems. Temperatures range between 1500° and 3000°C. These sources are typically calibrated from the UV to near infrared region of the spectrum. Even higher temperature sources are used for the visible but these are typically filaments within a quartz envelope.

4.2. TARGETS

Targets provide the geometric shapes used for the various tests. Target size can be expressed in either angular or linear dimensions. The relationship is

$$\theta = 2 \tan^{-1}\left(\frac{d_{target}}{2 fl_{COL}}\right)$$

where d_{TARGET} is the physical target size, fl_{COL} is the collimator focal length and θ is the angular subtense of the target. Since fl_{COL} is usually much larger than d_{TARGET}, the small angle approximation will be used from now on:

$$\theta = \frac{d_{TARGET}}{fl_{COL}}$$

The target angular subtense is often compared to the infrared imaging system's instantaneous-field-of-view (IFOV):

$$IFOV = \frac{d}{fl_{SYS}}$$

where d is the detector size and fl_{SYS} is the infrared imaging system's focal length. For example, the target size should be greater than 10 times the IFOV when measuring the SiTF.

The infrared community has been using emissive targets for years and, therefore, these targets have become standard. These targets can be manufactured by machining holes in an opaque material or by etching thin material. As new technologies emerge, new targets have evolved. Targets can now reproduce complex scenes. Because differential sources rely on the temperature of the target, varying ambient temperature can create radiance differences even when the ΔT is fixed (Section 3.1.5: page 73). To minimize ambient drift problems, reflective targets can be used. All targets can be mounted on target wheels to minimize handling and to speed up the test time. Although not used for quantitative testing, passive targets offer a method simply to verify that the system is functioning properly.

4.2.1. STANDARD EMISSIVE TARGETS

Figure 4-3 shows an enlarged view of an emissive target plate. Since thermal conductivity is proportional to the thickness of the material, it is important to keep the target plate as thick as possible. In competition with this requirement is the manufacturing problem of that detailed targets can only be made from very thin materials. The back surface (the side facing the source) should be highly reflective to prevent the source from heating the plate. The plate rake angle must be cut at an angle greater than the collimator acceptance angle so the edges cannot be seen by the infrared imaging system.

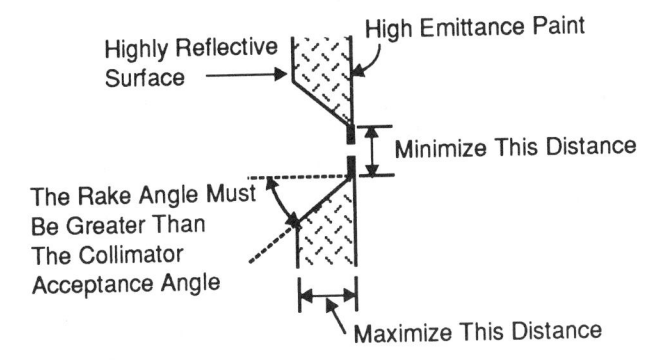

Figure 4-3: Target plate construction. Emissive target plates are coated with diffuse, flat black paint. The blackbody is placed to the left and the target plate is viewed from the right.

Figure 4-4 illustrates an individual target plate that must be inserted into the target holder. Some target plates have mounting screws located on the front surface. By constantly handling the plates, the emissive paint may be slowly worn away. Skin oils and dirty fingers also can change the emittance. It is important to keep the targets as clean as possible. If handling is required, hands must be clean. Gloves offer a safer handling method.

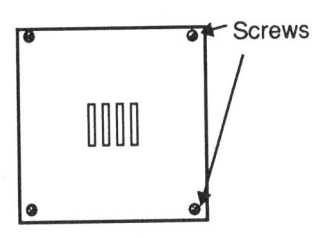

Figure 4-4: Typical single emissive target plate held in place by four screws. Constant handling can damage the emissive coating.

The system output is related to the temperature difference between the target and its *immediate* background. Hardware and manufacturing considerations prevent placing the temperature probe right next to the cutouts and therefore the temperature is measured at a finite distance away. (Note location of temperature probe in Figure 4-2: page 92). With very thick target plates, there is some assurance that the measured temperature is equivalent to the temperature of the immediate background. If a fixed finite temperature exists

96 TESTING & EVALUATION OF IR IMAGING SYSTEMS

between the probe location and the cutout location, then the data analysis techniques remove this fixed bias. If the bias varies due to variations in ambient conditions, then the test results will fluctuate.

An example of a test configuration using an emissive target is shown in Figure 4-5. The emissive target plate temperature, T_B, is monitored. The blackbody source temperature, T_T, varies as T_B varies to maintain a constant ΔT. The temperature probe that measures T_B is usually placed in the target plate holder (Figure 4-2: page 92). Although the emissive target plate may have been in the lab, handling it can change its temperature a minute amount. This small change can be seen with a very sensitive infrared imaging system (gradients as small as $0.05°$ C can be discerned with modern equipment). The target plate must reach thermal equilibrium with its immediate surroundings before making measurements and this may take several minutes.

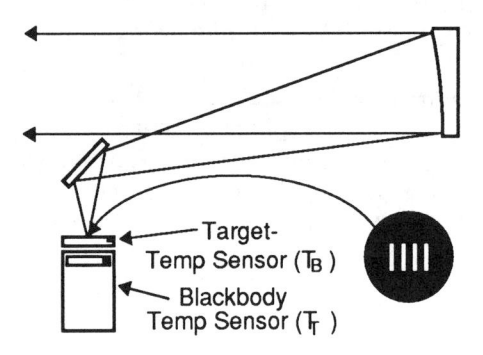

Figure 4-5: Emissive target plate in an off-axis parabolic collimator. $\Delta T = T_T - T_B$. As the ambient temperature fluctuates, T_B fluctuates. To maintain a constant ΔT, T_T changes as T_B changes.

Target plates with less than unity emittance will produce apparent target-background intensity differences that can depend upon the environment. As the emittance decreases[2], the reflectance increases. An emittance of 98% represents a reflectance of 2%. If hot external sources (e.g., people in the lab, electronics etc.) are present, then these can be reflected and can greatly affect the measured target-background intensity. Baffling (Figure 4-6) can minimize the effects of reflectivity if the baffles are approximately at the same temperature as the immediate background around the target.

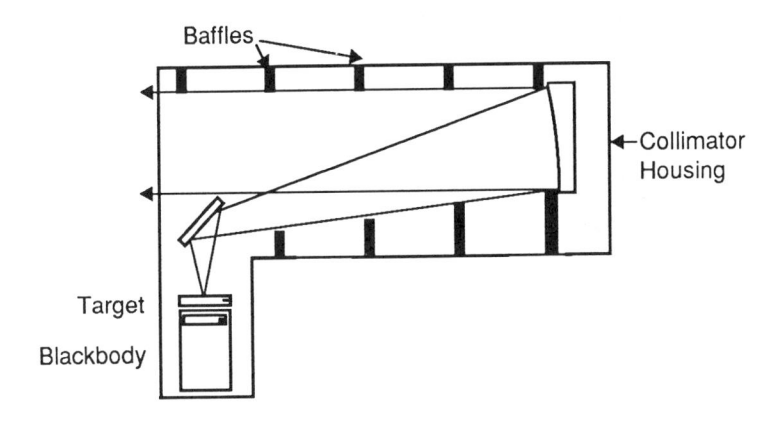

Figure 4-6: Using baffles and an enclosure to reduce stray radiation.

The term *ambient temperature* implies the room temperature, but it is the target plate temperature that is of interest. In an isothermal environment, the plate temperature will be equal to the room temperature. But the plate temperature can be affected by the location of air conditioning ducts, air circulation fans, air conditioning cycling, and any electronics or motors near the plate. These extraneous environmental effects can be minimized by placing the test configuration in a curtained area, using baffles and carefully placing heat sources away from the test configuration. Emittance variations and variations in background temperature affect all input-to-output transformations that rely on the measured ΔT. As a result, different setups and different measuring conditions can lead to different SiTF, NEDT, MRT and MDT values. If the ambient temperature is not specified nor recorded then results cannot be compared.

Emissive targets typically are made by machining holes in copper or aluminum plates. These low cost targets provide excellent thermal conductivity and this provides good thermal uniformity. Targets are typically painted with a flat black paint to produce a high emissivity surface. Painting small target features can be very difficult. The limiting factor is that complex targets or small features cannot be machined cost-effectively.

Complex and small patterns can be manufactured by photochemically etching a thin metal sheet (typically BeCu). This sheet is then bonded to a support ring (Figure 4-7). The disadvantage is that the thermal transfer from the emitting surface to the target ring tends to be poor through the bond line. Since thin targets have very low thermal mass, these targets are more susceptible to temperature variations due to thermal loads from the blackbody or ambient conditions. Etched targets also can be produced by growing thin masks though an electro-deposition process on a mandrel. These targets are identical with the etched targets but the process provides tighter tolerances and smaller feature sizes.

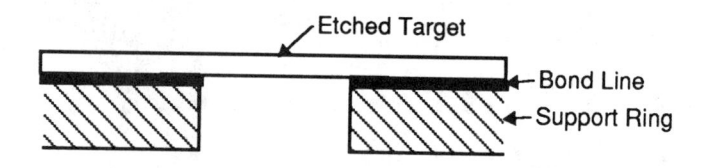

Figure 4-7: An etched target mounted onto a support ring.

4.2.2. NOVEL EMISSIVE TARGETS

Complex targets that simulate real world scenes can be created by a halftone process or by photolithography. The halftone process is identical with that used for magazines and newspapers to create pictures. The newspaper method provides dots of ink that are below the resolution of the eye and the eye blends the dots to create a gray scale. Similarly, the infrared imaging system must blend the target edges to create a gray scale. That is to say, the target edges must be smaller than the infrared imaging systems resolution.

The halftone targets are created by one of two methods. An IR opaque material (typically tungsten or chromium) is deposited on an IR transmissive substrate. The differential temperature is established when the deposited material blocks energy. These targets are typically expensive and some have problems associated with low emissivity of the deposited material and self heating from the blackbody. Halftones also can be created by etching small holes in a thin metal substrate. However, if too many holes are used to create a high transmittance portion, the target in that area is weakened. The dynamic range in terms of gray scale is limited and provides less than 15 gray levels.

A gray scale with 290 steps has been achieved[3] with photolithography when the individual pixel sizes are 64 x 64 μm. As with the halftone process, the infrared imaging system's IFOV must be larger than the target pixel angular subtense. For both manufacturing methods, the scenes can be generated photographically or digitally.

These complex scenes are not scene generators. Scene generators generally describe devices that create complex scenes that vary in real time. These targets contain only one scene. If a different scene is required, a new target must be inserted into the collimator. By using moving mirrors or moving beam splitters, these complex targets appear to move across the infrared imaging system's field-of-view. Since they are not scene generators, the target is simply superimposed onto another complex background. For example, when the target transmittance is 40%, 60% of the background flux is reflected and combined with the target signature. Target visibility is dependent upon the background temperature. While these scenes appear natural, it is difficult to calibrate them and therefore difficult to accurately specify the ΔT. Nevertheless, these scenes provide a method of testing image processing algorithms on the actual target of interest.

As targets become more complex, it is useful to try out new concepts before purchasing an expensive target. Many thin plastics are fairly transparent in the infrared with overhead transparency material being one. Any design can be placed on the transparency using only a simple copying machine. The copy machine toner is black and therefore has a different emissivity then the clear plastic. Overhead transparencies also can be made with a laser printer. Images can be read into the computer via an optical scanner, modified by software and then deposited onto an overhand transparency by a laser printer. Image resolution is limited by the copy machine or laser printer resolution. While the image may appear to be a good gray scale rendition visually, a sensitive infrared imaging system may detect nonuniformities in the test target. While it is impossible to accurately measure the ΔT, the purpose is simply to view a complex target for evaluation. Furthermore it may be difficult to hold the flexible material taught and therefore parts may appear out of focus. This concept works well if the blackbody temperature is not too high since plastic melts at a relatively low temperature. If the plastic sheets provide the desired image, then a more durable target can be made by etching a metal substrate or depositing the image on a transparent substrate. Computer generated images can easily be transferred to the photolithography machine or used for the deposition process.

4.2.3. REFLECTIVE TARGETS

Reflective target plates[4] minimize temperature gradient effects. The only difference in the construction of a reflective target plate from the emissive plate (Figure 4-3: page 95) is that the front surface is highly reflective. Gold coating produces a good reflective surface. In Figure 4-8, the background source is a homogeneous thick block enabling accurate measurement of its temperature. The effects of any residual thermal nonuniformities are further reduced by placing the source outside the collimator focal plane. This results in an extremely uniform background. Target plates can easily be changed during testing since it

100 *TESTING & EVALUATION OF IR IMAGING SYSTEMS*

is no longer necessary to wait for the target plate to reach thermal equilibrium: all target plates reflect flux emanating from the same background.

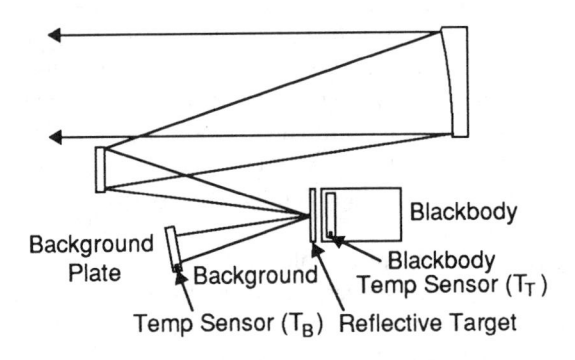

Figure 4-8: Reflective targets. The background is a thermally massive plate (By courtesy of Santa Barbara Infrared).

The flux difference when specified by ΔT depends on T_B. Since T_B is not controlled with the emissive plate (Figure 4-5: page 96), or by the emissive background (Figure 4-8), variations in T_B result in variations in the system output even though the ΔT is fixed. These variations are avoided by using a second blackbody source to control T_B (Figure 4-9). With the 2 sources any background temperature can be selected and carefully controlled.

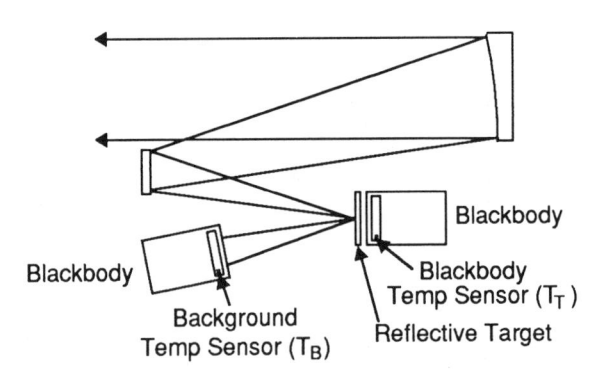

Figure 4-9: Reflective targets. The second blackbody source controls the background temperature. (By courtesy of Santa Barbara Infrared).

4.2.4. TARGET WHEELS

Target wheels are useful devices to position multiple targets quickly and easily in the collimator focal plane (Figure 4-10). This decreases test time and protects the targets from man-handling (i.e., prevents greasy, dirty fingers from touching the targets). Target wheels are typically computer controlled. Designs incorporate shielding, baffling and high thermal mass to achieve high thermal stability and uniformity. With target wheels, location of the temperature probe can become critical.

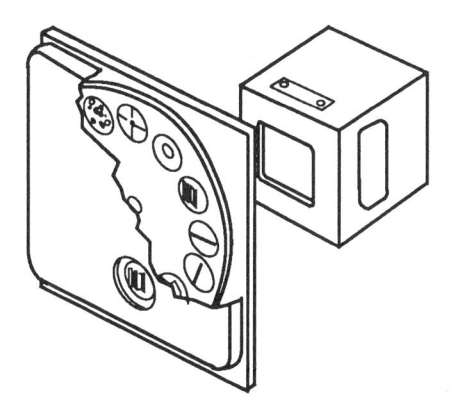

Figure 4-10: Typical target wheel. The background temperature probe may not be located near the target currently being used (By courtesy of Santa Barbara Infrared).

4.2.5. PASSIVE TARGETS

Passive targets are used for qualitative evaluation of system performance. They are not meant for quantitative testing. Passive targets rely on the fact that materials with different emissivities can appear to have different apparent temperatures. By Kirchoff's law, the amount of energy emitted depends upon the emittance and the amount of energy reflected depends upon the surroundings. In an isothermal environment, all materials appear to be the same temperature because the reflected energy is at the same temperature as the emitted energy.

Passive targets can only be detected when there is a finite temperature difference present. The differences may be created by of air conditioning cycling

102 TESTING & EVALUATION OF IR IMAGING SYSTEMS

or other air ducts. To insure target visibility, auxiliary heaters are placed on the back of the target. One method to fabricate these targets, is to use anodized aluminum and mill out the pattern. Figure 4-11 illustrates a large target with milled out lines that represent the field-of-view requirements. This approach is viable for those systems that can be focussed at normal laboratory distances.

Another method is to photoetch a metal plate. With this process, the target looks like an old Daguerreotype photographic plate. Small targets, which also require heaters, can be place in a collimator. With these targets, as with any passive target, the target temperature is usually unknown and therefore cannot be used for calibrated characterization.

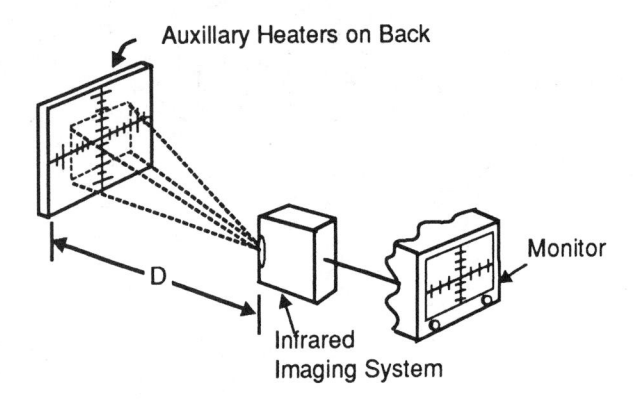

Figure 4-11. Using a large passive target to measure the field-of-view. The passive target must have a ΔT with respect to the surroundings. This can be achieved by placing heaters on the back.

4.2.6. SOURCES AS TARGETS

Since all objects emit infrared radiation, an appropriately designed source also can be used as a target. Nichrome wire or transformer wire can be placed in a variety of shapes and an electrical current can be passed through the wire to produce heat. Since metal expands when heated, springs are necessary to hold the wire taut (Figure 4-12). The temperature of these source targets is usually unknown or, at the minimum, difficult to control. They are useful for those tests that simply require a target of known shape. It the wire thickness is much less that the infrared imaging system's IFOV, it approximates an ideal line source and can be used to measure the MTF.

GENERAL MEASURING TECHNIQUES 103

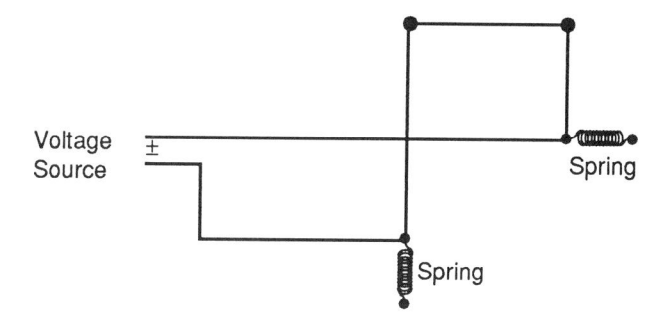

Figure 4-12. Heated wires can be used as a target. The springs compensate for thermal expansion.

4.2.7. SPECIAL CONSIDERATIONS

Target plates can be manufactured with several target cutouts on one plate. For MRT and MDT testing, it recommended that only a single target be placed on each target plate. Multiple targets on a single target plate may confuse and disturb the observer. This confusion can increase test time and possibly affect test results. These effects become more noticeable when viewing small targets near large easily resolved targets. AC coupling and insufficient angular separation of the targets produce the largest effects.

Due to phasing effects (Section 2.3: page 36), the target orientation and location with respect to the detector array is critical. Slits should be aligned to the detector axis and its location with respect to a detector should be known. A target plate can be mounted on an x, y, θ translation/rotation stage (Figure 4-13). Since phasing effects are seen over $\pm$ ½ IFOV, the equivalent x and y incremental movements must be much less than this. As an alternative, the infrared imaging system can be mounted on a rotary table (See Section 4.4). In certain situations, multi-phase patterns may be appropriate. In Figure 4-14 each succeeding bar pattern is offset by a subpixel amount. Here, an IFOV is considered a pixel. With this arrangement, there will be one pattern that is essentially in-phase while the remainder may be out-of-phase.

104 *TESTING & EVALUATION OF IR IMAGING SYSTEMS*

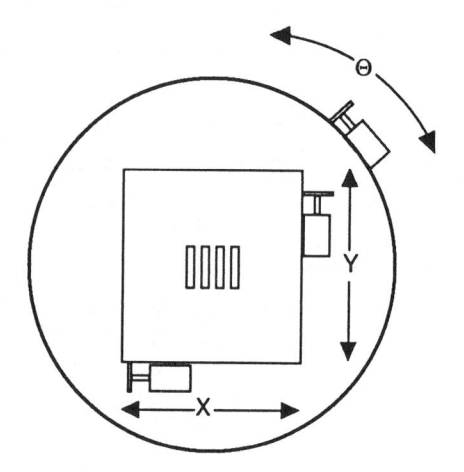

Figure 4-13: Target plate holder with three degrees of freedom. Critical alignment requires micropositioning capability.

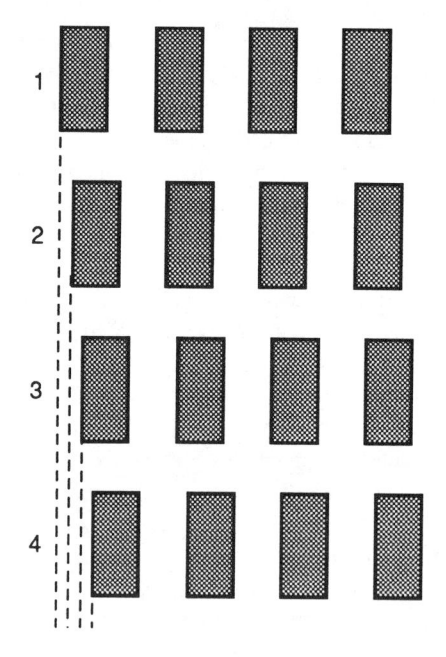

Figure 4-14: Multi-phase target pattern. Each pattern is offset by ¼ pixel. At least one pattern should be in phase.

If the infrared imaging system has an automatic gain, it is desirable to disable the gain mechanism. If that is not possible, the AGC may be controlled by placing a large target in the field of view. Here, the large target controls the AGC and not the small test pattern. Figure 4-15a illustrates a design that places the AGC controller source in the lower portion of the FOV. Figure 4-15b depicts the image seen. The AGC pattern should not be placed in a location where it can interact with the system's response to the test target. For example, if a system has AC coupling, the AGC pattern should not be on the same line as the target. The AGC target shape may be tailored to the particular system being tested. Figure 4-15a schematically illustrates 2 blackbodies. Because of their size, a careful optical layout is required.

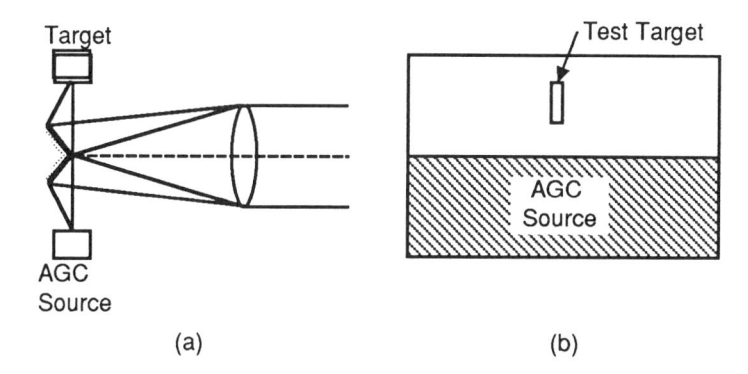

Figure 4-15: AGC controller. (a) Location of the test target and AGC controlling source, and (b) image seen on the monitor.

Always verify that the correct target has been selected. When purchasing a target, measure the target carefully to insure that it is the correct size. Mark it with the physical dimensions, the effective spatial frequency and the collimator focal length for which the spatial frequency applies. This labeling avoids confusion in laboratories that may have several different focal length collimators.

4.3. COLLIMATORS

Collimators are lens assemblies that optically place targets at infinity. Collimators may contain either refractive or reflective elements. The combination of the source, target, and collimator is sometimes called a target simulator or target projector. The collimator clear aperture should be much greater than the system clear aperture. Otherwise the system output will be reduced by the ratio of the aperture areas and the measured MTF will be modified by the collimator MTF. With small aperture collimators, the unobstructed infrared imaging system aperture can sense radiation from elsewhere. This radiation is superimposed onto the source radiation. In principle, correction factors[5] can be applied but this approach is not recommended because the correction factors are system dependent.

Collimator aberrations should be significantly less than the aberrations of the infrared imaging system. Since aberrations are inversely proportional[6] to the f/# or to powers of the f/#, the collimator f/# should be larger than the infrared imaging system f/#. Equivalently, the focal length of the collimator should be much longer than the focal length of the infrared imaging system. As a general guideline, the collimator focal length should be at least 5 times that of the infrared imaging system. Reflective mirrors do not have chromatic aberrations. Therefore collimators with reflective elements than refractive elements tend to be the collimator of choice. Parabolic mirrors are preferred since they do not have spherical aberrations. The transmittance of collimators with refractive elements depends upon the lens material used and the anti-reflection coatings placed on each surface. Furthermore, refractive collimators can produce a narcissus signal.

Figure 4-16 illustrates the location of a target, source and infrared imaging system when using an off-axis parabolic mirror. Maximum collimator throughput is obtained with an off-axis collimator (maximum flux reaching the infrared imaging system). The offset must be large enough to allow room for target plate holders, other fixtures and cabling so that they do not block the main beam. The closest the infrared imaging system can be placed to the mirror is approximately at a distance equal to the focal length. If placed any closer, the infrared system may partially block the flux emanating from the source. To avoid problems incurred with large sources whose housing may block the primary beam, a fold mirror may be used (Figure 4-17). Reflective mirrors may be made from aluminum, gold, silver, gold, copper, rhodium or platinum. The most common is aluminum with a protective overcoat of SiO. Reflectivity curves supplied by the manufacturer provide an estimate of the collimator transmittance. These curves are usually for normal incidence. At 45° (typical fold mirror angle), SiO protected aluminum mirrors exhibit reduced reflectance in the 8 to 9 μm region that depends upon the coating thickness[7,8,9]. Reflectivity is polarization dependent. The measured reflectance for unpolarized light (the usual case) is shown in Figure 4-18. An infrared imaging system, which is responsive in the interval 8 to 10 μm, may be affected by the fold mirror spectral transmittance.

GENERAL MEASURING TECHNIQUES 107

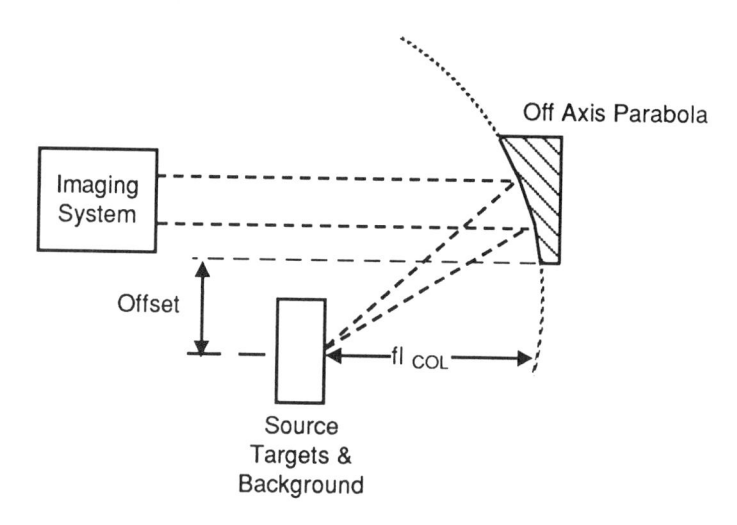

Figure 4-16: Off-axis parabolic collimator illustrating the mirror offset and location of the target plate.

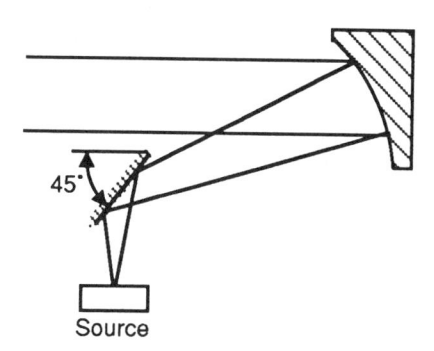

Figure 4-17: Off-axis parabolic collimator with a fold mirror. The Newtonian mount places the fold mirror at 45° with respect to the parallel rays. If overcoated with SiO, the fold mirror may exhibit reduced reflectivity in the 8-9 μm region.

108 *TESTING & EVALUATION OF IR IMAGING SYSTEMS*

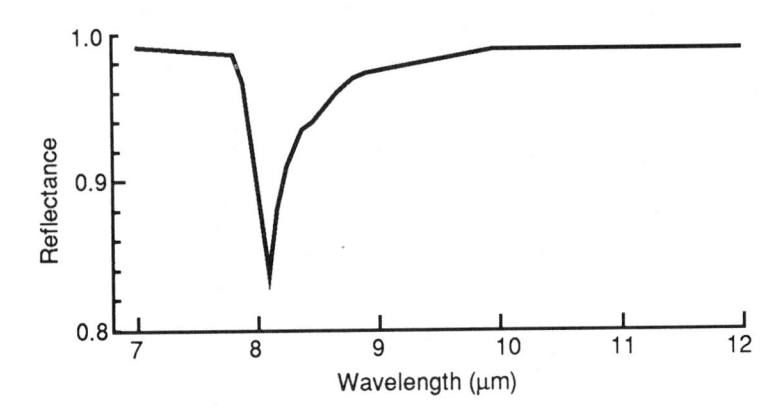

Figure 4-18: Reflectance of an aluminum mirror overcoated with SiO. The angle of incidence is 45°. The reflectivity dip depends upon the SiO thickness (By courtesy of Space Optics Research Labs).

To reduce size, a collimator may have several reflective mirrors. The transmittance of the collimator is the product of the reflectance of each mirror. If the reflectivity, ρ, is 0.96, then a collimator with 4 mirrors has a transmittance of $\rho^4 = 0.849$. The collimator transmittance can be measured using the methodology illustrated in Figure 4-19. In Figure 4-19a, the infrared imaging system's responsivity is obtained by measuring the output for a variety of source intensities. The collimator is then removed and the responsivity is again measured for a variety of intensities (Figure 4-19b) with the source at the same distance. A least-squares line is fit to the two data sets (Figure 4-19c). The collimator transmittance is the ratio of the responsivities or ratio of the slopes. This procedure works well if the infrared imaging system can focus onto the source. By using the system under test to estimate the transmittance, the spectrally weighted average transmittance is obtained (Equation 3-50: page 81). Because of differences in detector responsivities, $\text{cosine}^N \theta$ shading and vignetting, it is important to insure that the source subtends the same detector elements in both tests. If the collimator transmittance is measured with a separate calibrated sensor, it is extremely important that the calibrated sensor have the same spectral response as the system under test.

GENERAL MEASURING TECHNIQUES 109

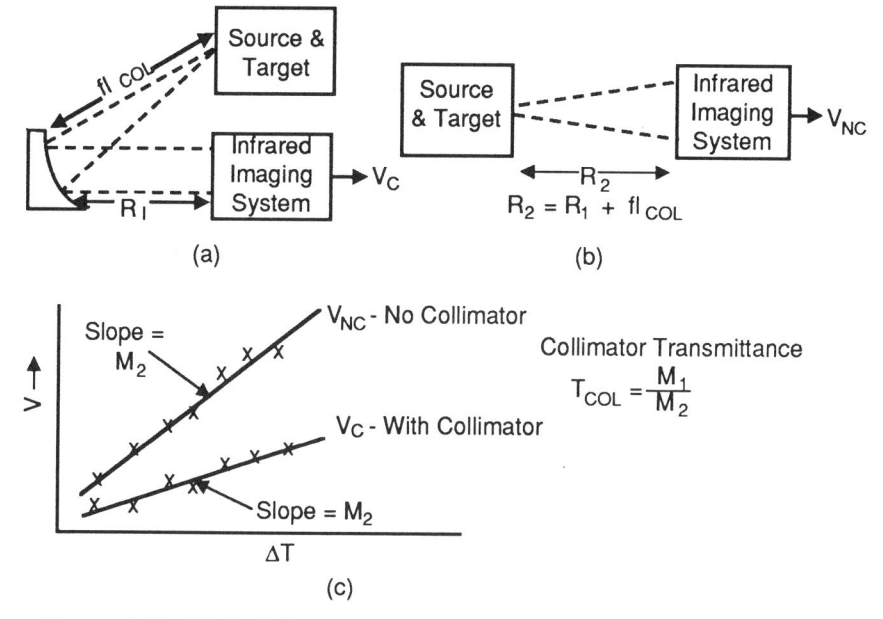

Figure 4-19: Methodology to measure the collimator transmittance. (a) the SiTF is measured with the collimator; (b) the SiTF is measured without the collimator; and (c) the transmittance is the ratio of SiTFs. The distance from the blackbody source and infrared imaging system should be the same.

Although a point source produces parallel light, an extended source can only be seen in a well-defined region (Figure 4-20). The maximum distance at which the imaging system can be placed from the collimator is:

$$R_{MAX} = \frac{fl_{COL}}{d_T}(D_{COL} - D) \qquad 4\text{-}1$$

and

$$L = \frac{fl_{COL}}{d_T} D_{COL} \qquad 4\text{-}2$$

where the collimator field-of-view is d_T/fl_{COL}. D_{COL} is the collimator clear aperture, D is the infrared imaging system clear aperture, and d_T is the largest dimension of the target. For square and rectangular targets, the largest linear

110 *TESTING & EVALUATION OF IR IMAGING SYSTEMS*

dimension is the diagonal. If the infrared imaging system is placed at range R_{MAX} or less, the entrance aperture of the system will collect all the rays projected by the collimator. The target edges will appear at reduced intensity if the system is placed between R_{MAX} and L. If the system placed at a distance greater than L, the target outer edges will be totally clipped and only the central portion of the target will be seen. The largest anticipated size (target and background) should be used for the calculation. This target size represents the lowest spatial frequency that can be produced by the collimator. The collimator aperture should be sufficiently large so that no extraordinary care is required to center the infrared imaging system onto the collimator. Although Figure 4-20 illustrates a refractive collimator, the equations apply to a reflective collimator where R_{MAX} and L are distances from the primary (output) mirror. For reflectance collimators, the minimum distance is approximately equal to the focal length of the collimator. Figure 4-6 (page 97) illustrates a typical collimator lay out where $R_{MIN} \approx fl_{COL}$.

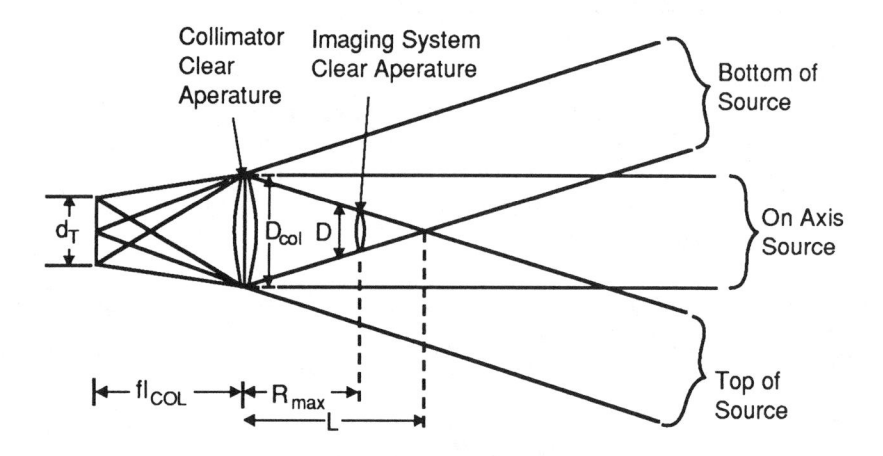

Figure 4-20: Collimator working distance. The infrared imaging system must be placed at a distance less than R_{MAX}. Although a refractive collimator is shown, the distance R_{MAX} and L equally apply to reflective collimators.

So far, it has been assumed that the collimator FOV is larger than the system FOV. However, if the infrared imaging system has a large FOV, it can sense radiation from elsewhere. This radiation will not be superimposed upon the target but will be in the periphery. This stray radiation may confound the experiment and baffles are recommended to minimize the extraneous radiation (Figure 4-21). The baffles should be at the same temperature as the background.

GENERAL MEASURING TECHNIQUES 111

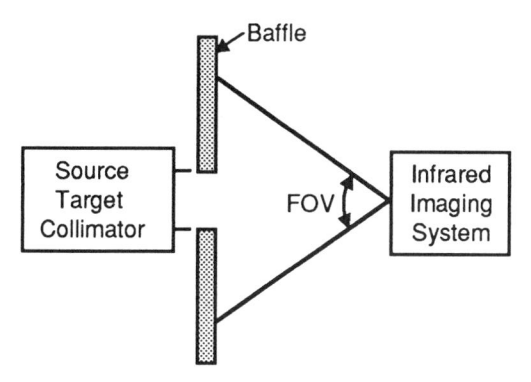

Figure 4-21: Baffles are required to reduce stray radiation when the infrared imaging system FOV exceeds the collimator FOV.

Many collimator mirrors are diamond turned aluminum. This produces a relatively low scatter finish. They are then coated with aluminum to improve reflectivity and to cover any imperfection left by the diamond turning. A SiO overcoat may be added to protect the mirror surface. The primary mirror often contains a reference flat for alignment purposes (Figure 4-22). Another alignment feature is a three-point mount (bulk head reference plane) that assists alignment to the infrared imaging system. An athermal design insures that boresight accuracy will be maintained over widely varying temperatures. A cylindrical housing takes advantage of the inherent stiffness of a rigid tube.

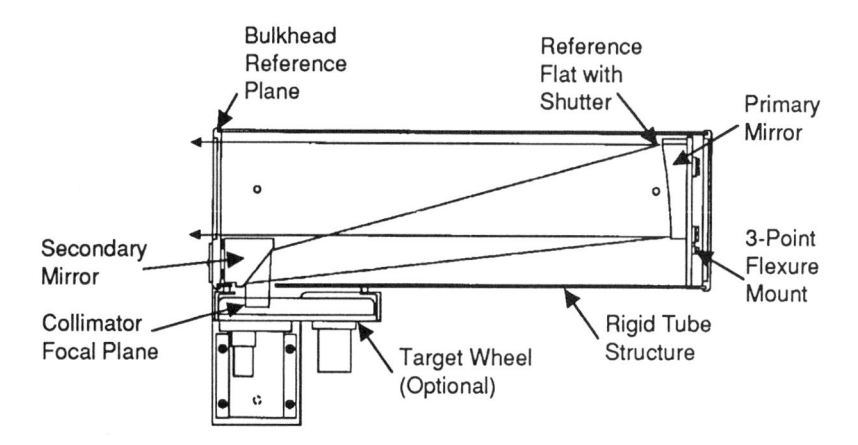

Figure 4-22: Typical commercially available collimator (By courtesy of Santa Barbara Infrared)

112 *TESTING & EVALUATION OF IR IMAGING SYSTEMS*

Example 4-1
COLLIMATOR WORKING DISTANCE

What is the maximum distance that an infrared imaging system can be placed from a collimator whose focal length is 40 inches and clear aperture is 6 inches? The infrared imaging system's entrance aperture is 5 inches. The largest linear angular dimension (target plus background) to be viewed subtends 20 mrad and the FOV is square.

The target diagonal is 28.2 mrad. The target size is $(fl_{COL})(\theta) = (40)(28.2 \times 10^{-3}) = 1.13$ inches and the maximum distance is

$$R = \frac{fl_{COL}}{d_T}(D_{COL} - D) = \frac{40}{1.13}(6-5) = 35.4 \ inches$$

At this distance, the infrared imaging system must be precisely on-axis with the collimator. In practice, the infrared imaging system should be placed at a shorter distance to alleviate alignment difficulties.

If using an enclosed reflective collimator (Figure 4-6: page 97), then the minimum distance to the primary mirror is approximately equal to the focal length or 40". With this physical constraint, the largest target that can be fully seen is:

$$d_T = \frac{fl_{COL}}{R}(D_{COL} - D_{SYS}) = \frac{40}{40}(6-5) = 1 \ inch$$

or an angular subtense of $1/40 = 25$ mrad.

4.4. ATMOSPHERIC TRANSMITTANCE and TURBULENCE

Atmospheric transmittance and the collimator transmittance reduce the flux reaching the infrared imaging system (Equation 3-60: page 86). Figure 4-23 illustrates the atmospheric transmittance as a function of wavelength for a typical laboratory distance of 10 feet. The transmittance was calculated by LOWTRAN7 assuming a mid-latitude summer atmospheric model and an urban (10 km visibility) aerosol model. Actual transmittances may deviate significantly depending upon the aerosols present, temperature, atmospheric pressure, and the absolute humidity. The distance shown is a reasonable value for the laboratory since the path length is the distance from the source to the infrared imaging system. Referring to Figure 4-6 (page 97), the minimum distance from the infrared imaging system entrance aperture to the source is approximately twice the focal length. Therefore, a 5-foot focal length off-axis collimator places the infrared imaging system at least 10 feet from the source.

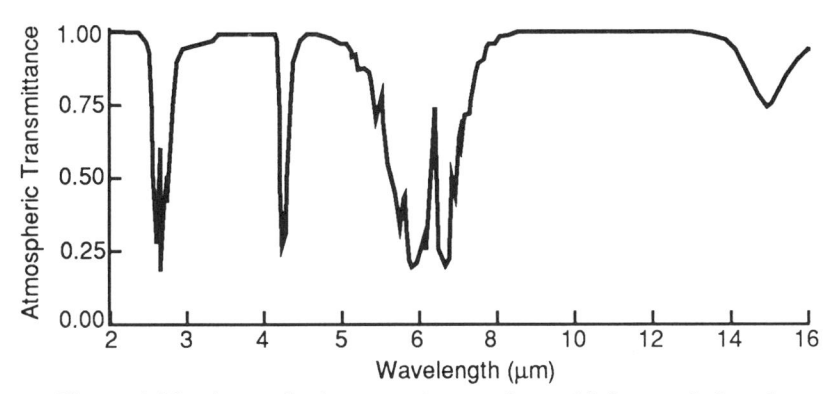

Figure 4-23: Atmospheric transmittance for a 10-foot path length calculated from LOWTRAN7. The nonlinear scale accentuates the absorption bands.

It is convenient to represent the system output as

$$\Delta V_{SYS} \approx SiTF\,(T_{COL}\,T_{ATM}\,\Delta T) \qquad 4\text{-}3$$

or

$$\Delta V_{SYS} = SiTF\,\Delta T_{ENTRANCE\ APERTURE} \qquad 4\text{-}4$$

where T_{COL} is the spectrally averaged collimator transmittance and SiTF is the system's signal transfer function.

The temperature reaching the infrared imaging system's entrance aperture is

$$\Delta T_{ENTRANCE\ APERTURE} = T_{COL}\,T_{ATM}\,\Delta T \qquad 4\text{-}5$$

where ΔT is the true temperature difference between the target and its background. The correction factor, $T_{COL}T_{ATM}$ is used for SiTF, MRT and MDT measurements. For infrared systems that operate in the regions where the atmospheric transmittance is constant (e.g., 3 - 4.1 μm, 4.3 - 5 μm and 8 - 13 μm),

114 TESTING & EVALUATION OF IR IMAGING SYSTEMS

$T_{ATM} \approx 1$ over typical laboratory path lengths. If the infrared imaging system's spectral response extends past these regions, an atmospheric transmittance correction may be required. T_{ATM} expressed in Equation 4-3 is a spectrally averaged transmittance that includes the spectral response of the system and the target and background characteristics:

$$T_{ATM} = \frac{\int_{\lambda_1}^{\lambda_2} T_{ATM}(\lambda)\, R(\lambda)\, [M_e(\lambda,T_T) - M_e(\lambda,T_B)]\, d\lambda}{\int_{\lambda_1}^{\lambda_2} R(\lambda)\, [M_e(\lambda,T_T) - M_e(\lambda,T_B)]\, d\lambda} \qquad 4\text{-}6$$

A representative T_{ATM} is illustrated in Figure 4-24 for an infrared imaging system with a HgCdTe detector that is sensitive from 2 to 12 μm. Water vapor is the primary absorber in the 5 to 8 μm band and therefore the relative humidity affects the transmittance. The relative humidity changes as the temperature changes for a fixed water vapor concentration. Figure 4-24 is unique to the assumed system spectral response and path length. A different system or a different path length will have a different T_{ATM} versus relative humidity function. The atmospheric attenuation can be minimized by placing the source, target, collimator and infrared imaging system in an inclosed chamber that has been purged with dry nitrogen or other inert gas. This will minimize the attenuation due to CO_2 at 4.2 μm and the various water vapor bands[10].

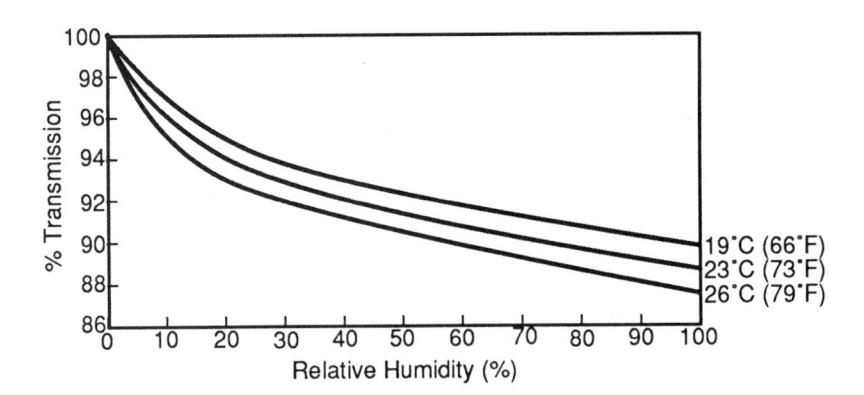

Figure 4-24: Spectrally weighted atmospheric transmittance, T_{ATM}, for a 18-foot path length. The detector can sense radiation from 2 to 12 μm. The graph is unique to the system spectral response and background blackbody temperature selected.

Turbulence[11] results from random fluctuations in the atmospheric refractive index stemming principally from random changes in air pressure and temperature. These changes, ever so slight, causes random refraction of light and cause the incoming light to arrive at the receiver at changing angles, thus blurring the image. To reduce turbulence, it is necessary to reduce all air currents by enclosing the infrared imaging system and collimator in a chamber. Air conditioning ducts, hot electronic equipment and the source can cause air currents. Turbulence may affect those input-to-output transformations where it is important to discern detail such as resolution measures, MTF, CTF, MRT and MDT.

4.5. MOUNTING FIXTURE

For many tests, it is important to align the infrared imaging system axis parallel to the target axis. This can be achieved by either moving the target in the x, y and θ directions (Figure 4-13: page 104) or equivalently by rotating the infrared imaging system about its yaw, pitch and roll axis respectively. The pivot point must be on the first nodal plane of the optical system. The incremental movement and resolution of the rotary table must be much smaller $\pm \frac{1}{2}$ IFOV.

Vibration can significantly affect all input-to-output transformations where detail is important (resolution, MTF, CTF, MRT and MDT). To minimize vibrational effects, the source, target, collimator and the infrared imaging system should be placed on a vibration-isolated optical table.

4.6. DATA ACQUISITION

Data may be acquired by visual observation, measuring the monitor brightness, measuring the analog video signal or any other suitable measuring point (Figure 4-25). Although rarely stated, the test engineer is relied upon to discern any malfunction, test result abnormality and all transient effects. Only the test engineer, who has seen hundreds of similar units, can rapidly identify subtle changes. Every time the test engineer sees an image, he subconsciously evaluates the imagery according to his internal rating scale. A modified Cooper-Harper scale (Section 1.2.2: page 9) can be created to standardize the objective rating scale. The test engineer should be considered the primary *data acquisition system*.

116 TESTING & EVALUATION OF IR IMAGING SYSTEMS

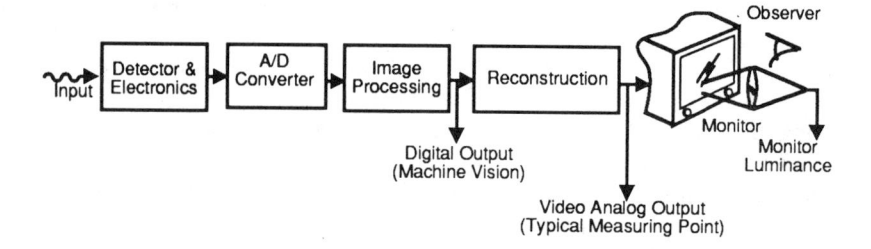

Figure 4-25: Typical measuring points. The system output can be ADUs, analog video voltage, monitor brightness or an observer's impression of image quality.

It is important when performing subjective evaluation of image quality such as MRT and MDT that the displayed image has the correct aspect ratio[12]. For example, an infrared imaging system may have a 1:1 aspect ratio and has an output equivalent to RS 170. If displayed on a standard monitor (4:3 aspect ratio), the image will be elongated in the horizontal direction such that circles become ellipses. On the other hand, a frame grabber may appropriately digitize a 4:3 aspect ratio image but display the image on a computer monitor that typically has a 1:1 aspect ratio. Here, circles are compressed in the horizontal direction and the circles will appear as ellipses.

Figures 4-26 and 4-27 illustrate two methods to measure the analog video. The measurement equipment bandwidth must be greater than the infrared imaging system bandwidth to insure that all the data is collected and properly analyzed. It is prudent to verify the system bandwidth before selecting measuring equipment. The output video may be in a standard format in terms of timing (e.g., RS 170), but may have a much wider bandwidth than the standard video format (e.g., *modified* RS 170). The sampling rate of the image capture board or frame grabber is usually fixed and is sometimes too slow to match the pixel rate of the infrared imaging system. As a result, the frame grabber shown in Figure 4-27 is appropriate only for those measurements that deal with low frequency signals such as responsivity. The frame grabber has its own internal clock to capture, say, 640 pixels. If the frame grabber's clock is not precisely matched to the active line-time, information may be lost (line-time too long) or the line may be represented by, say, 636 pixels (line-time too short). In the latter case, the remaining 4 pixels will remain black and it will not be visually obvious that there is a potential problem. The same problem occurs in the vertical direction. RS 170 displays 485 lines but most frame grabbers only collect 480 lines with 5 being dropped. Although an imaging system may

GENERAL MEASURING TECHNIQUES 117

produce 480 lines (Section 2.4.2: page 48), it may not be the same 480 lines collected by the frame grabber. For example, the imaging system may black-out the first 2½ lines and the last 2½ lines with 480 active lines in between. On the other hand, the frame grabber may neglect the first ½ line and collect the next 480 lines. Here, the frame grabber will collect 2 black lines and 478 active video lines. The remaining 2 active lines are lost.

The frame grabber is calibrated as follows. The system views a target whose angular size is equal to the FOV. The image size, as created by the frame grabber is analyzed. The number of frame grabber pixels created are simply counted. For systems with a digital output, it is possible to match the digital clock with the frame grabber clock such that every pixel is collected (i.e., the frame grabber is synchronized to the system output).

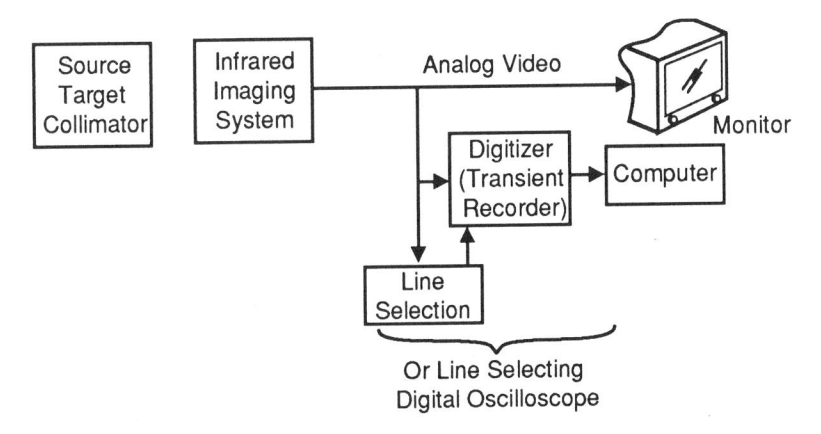

Figure 4-26: Measuring a single line of analog video with a digitizer or transient recorder. The computer calculates the mean, variance and graphs the data.

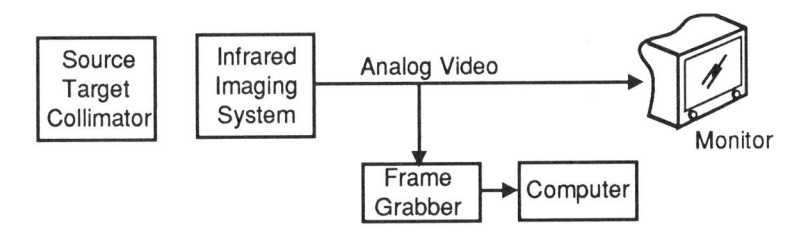

Figure 4-27: Measuring a single line of analog video with a frame grabber or image capture board. The computer calculates the mean, variance and graphs the data.

118 *TESTING & EVALUATION OF IR IMAGING SYSTEMS*

Figure 4-28 illustrates a less precise measurement method. A selected line of the analog video is displayed on an oscilloscope and an observer *estimates* the signal and noise values. The oscilloscope can be used to verify that signal levels are within the transient recorder or frame grabber dynamic range. The oscilloscope provides a "quick look" data analysis capability. Signals can be viewed to verify that they look *correct* (i.e., no glitches, unusual noise spikes etc.).

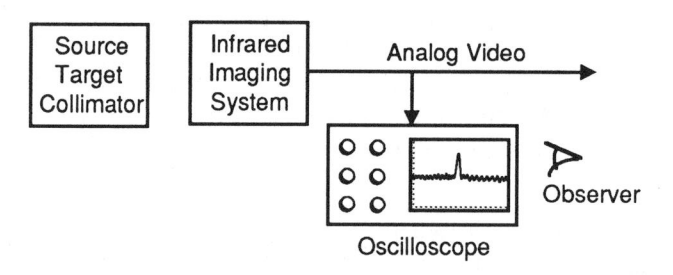

Figure 4-28: Estimating the signal and noise displayed on an oscilloscope.

The disadvantage of measuring the analog video is that it does not include any of the effects introduced by the monitor. The monitor bandwidth may reduce the noise bandwidth and thereby reduce the visually perceived noise. As a result, it may be appropriate to simulate the monitor bandwidth with an external filter (Figure 4-29).

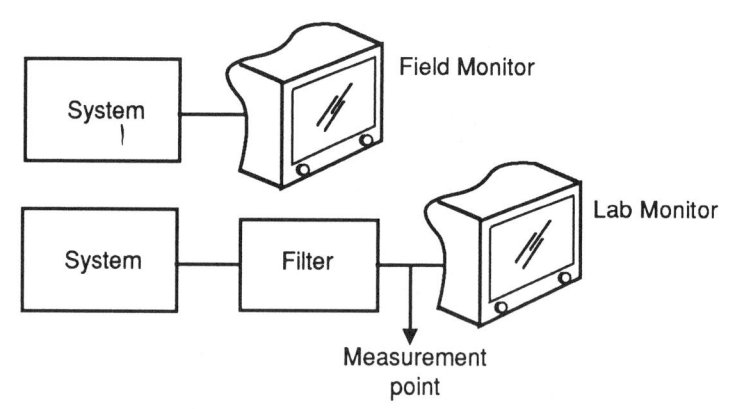

Figure 4-29: Using an external filter to approximate the field monitor bandwidth. The lab monitor bandwidth must be equal to or greater than the field monitor bandwidth.

GENERAL MEASURING TECHNIQUES 119

The monitor luminance can be measured with a scanning microphotometer, stationary solid state camera or movable fiber optics (Figures 4-30 to 4-32). With these methods, the noise and signal as presented to the observer is measured and it includes the performance parameters of the monitor. This method is appropriate for those systems that have an integral monitor (direct view devices) or have specialized unique monitors. When measuring low frequency response such as the responsivity, the photometer update rate and the time to traverse the monitor should be slow so that measurements are averaged over many TV frames. By averaging, framing effects are minimized because the addition or loss of one TV frame does not significantly affect the accuracy of the responsivity measurement. Averaging lines increases the signal-to-noise ratio. On the other hand, for noise measurements, the photometer time constant must be consistent with the monitor line rates. All data must be corrected for the lens transmittance (if used) and the photometer response. The solid state camera should be of high quality with minimal cross talk. It must be verified that the solid state camera is operating in a linear range and that the dynamic range is fully used to avoid quantization errors. The photometer or solid state camera can be tested by all the methods described in this text.

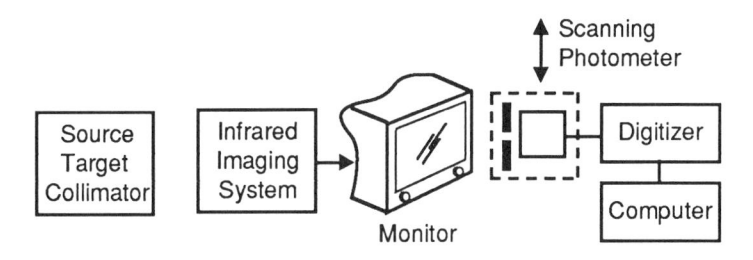

Figure 4-30: Measuring the monitor with a scanning photometer. The computer calculates the mean, variance and graphs the data.

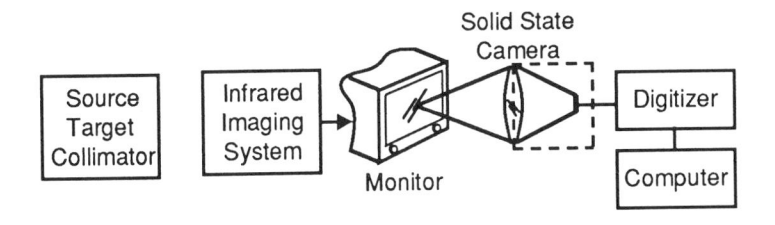

Figure 4-31: Measuring the monitor with a solid state camera. The computer calculates the mean, variance and graphs the data.

120 *TESTING & EVALUATION OF IR IMAGING SYSTEMS*

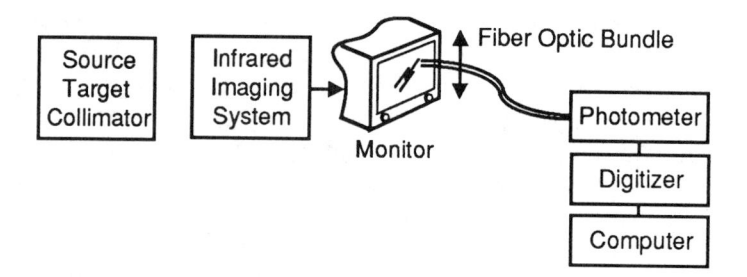

Figure 4-32: Measuring the monitor with a scanning fiber optic bundle. The computer calculates the mean, variance and graphs the data.

The advantage of measuring the monitor luminance is its greater adaptability. With the large number of video standards, the use of a solid state camera does not change the measuring techniques or equipment used whereas the analog video test procedure may require modification due to the different video line rates. The disadvantage of this method is that monitor performance is not consistent[13] and each monitor should be calibrated before starting any test.

It is important that the test equipment have sufficient resolution without introducing quantization errors. This may pose a problem if the test equipment gain is fixed. If the test equipment has sufficient latitude to measure the entire system dynamic range, it may not have enough gain to measure the noise[14]. For example, if the electronic system has a 10-bit A/D converter then the measurement equipment should have more to ensure that the noise can be measured accurately. Alternatively, the measuring equipment could have a variable gain adjustment so that the infrared imaging system output signal fills the dynamic range of the measuring equipment. Although the infrared imaging system output may be stated to be a standard format, the actual voltage levels may be somewhat different. It is wrong to assume that there is a perfect match between that system output and the test equipment input. As shown in Figure 4-33, the black level and white level voltage levels may not completely fill the dynamic range of the test equipment. The pure black level may not appear as zero counts and pure white may not appear as 255 counts on the frame grabber output. This mismatch reduces test accuracy. It is essential to calibrate the test equipment against known system outputs.

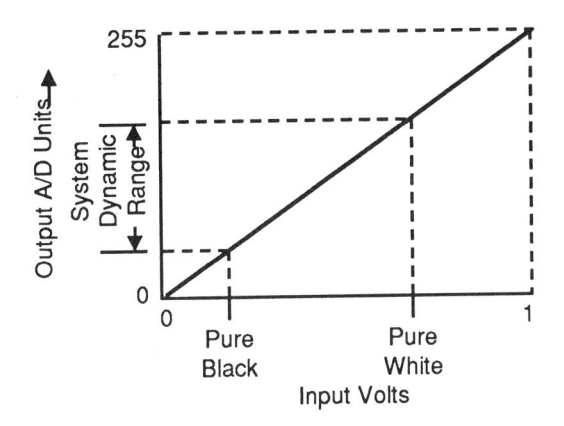

Figure 4-33: Calibration of a frame grabber with an 8-bit A/D converter. The video voltage range representing the black to white range may not fill the dynamic range of the A/D converter.

There are two different sampling systems that are present. There is the digitization internal to the infrared imaging system and the sampling system within the test equipment. While the test engineer has no control over the internal digitizer, he does control the test equipment used. It is important to reproduce the amplitude and pulse width of the signal for appropriate data analysis. With this consideration, the bandwidth of the measuring equipment must be much wider than the infrared imaging system. Phasing problems are bothersome when digitizing detail such as a slit target response for MTF measurements, automatic MRT measurements or determining the CTF.

According to sampling theory, at least 2 samples per period are required to reconstruct the input frequency. However, with low sampling rates, the input amplitude and pulse width may be altered. Figure 4-34a illustrates the signal appearance for an input frequency (input to the test equipment) after being sampled at 2, 3, ..., 10 samples per period. Even up to 9 samples per period, the digitized signal has a slight variation in pulse width. In Figure 4-34b, the relative phase between the test equipment sampler and system's output signal was varied while operating at 5 samples per period. The phase was varied in increments of 0.1 cycles. As shown, phasing dramatically affects the appearance of signals. The test equipment (digital oscilloscope, transient recorder or frame grabber) should operate at a sampling frequency that is much higher than the input signal frequency to avoid these phasing effects.

122 *TESTING & EVALUATION OF IR IMAGING SYSTEMS*

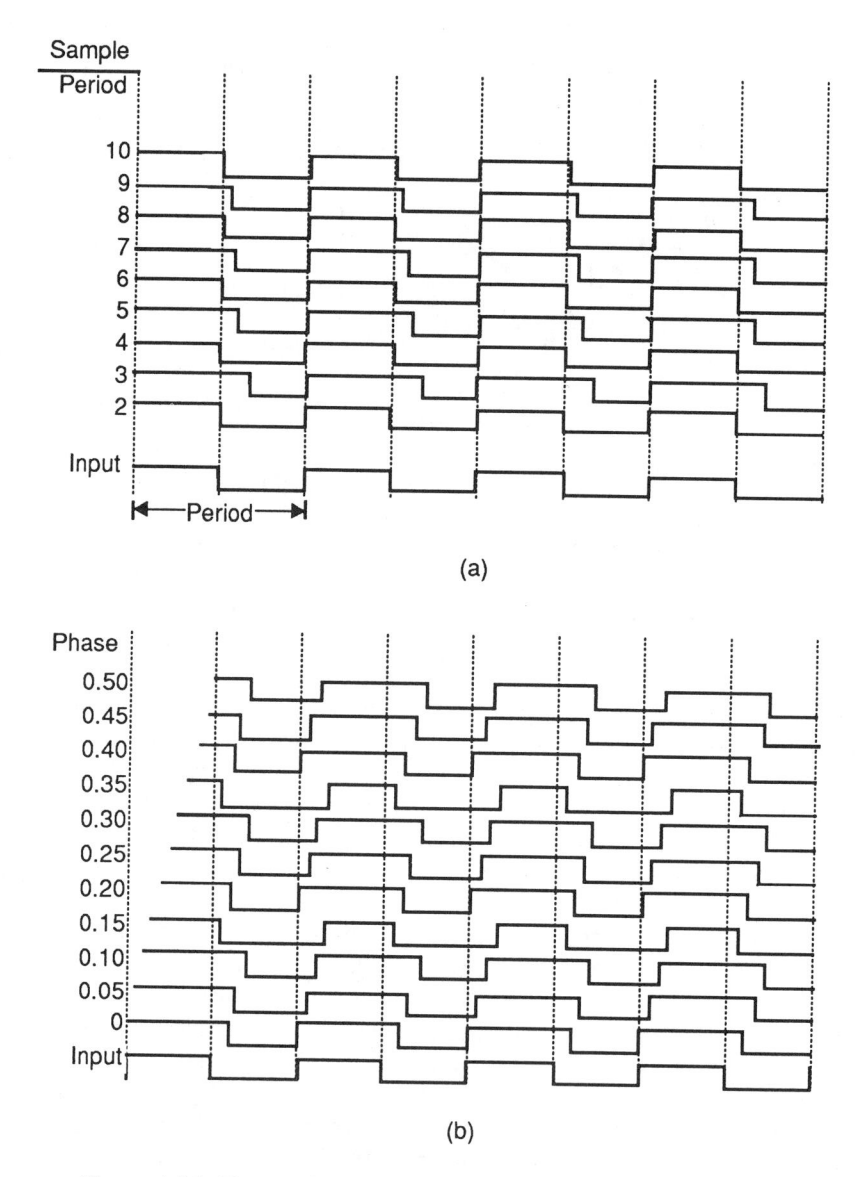

Figure 4-34: Test equipment digitization and phasing effects. (a) For a fixed input frequency, the output varies depending upon the number of samples per frequency. The output nearly replicates the input when there are at least 8 samples per frequency; and (b) at 5 samples per frequency, phasing dramatically affects the output.

The line traces in Figure 4-34a, suggest that about 8 samples per highest frequency are required to reproduce faithfully the signal intensity. This high requirement affects the applicability of frame grabbers for many tests. For example, many frame grabbers digitize the analog video into 512 samples horizontally. Suppose the infrared imaging system's FOV was 30 mrad and it had a 512 x 512 detector array. The highest spatial frequency that can be adequately reproduced is approximately 512/(8·30) = 2.13 cy/mrad whereas the system's Nyquist frequency is 512/(2·30) = 8.5 cy/mrad. Directly accessing the digital signal avoids these potential problems (Figure 4-25: page 116).

A whole wealth of information is available by just looking at the data. This information becomes apparent when the data is graphed rather than placed in a tabular form. For example, a pass/fail criterion says nothing about the *quality* of the data nor can it suggest if the system is operating properly. Pass/fail is not unique to a specific data set. A large variety of data sets after the appropriate mathematical manipulation (e.g., calculating the mean or the variance) can produce the same result. It is the raw (unprocessed) data that provides the information. For example, an infrared imaging system with a digital data line grounded may produce adequate imagery. By plotting the histogram of the digital data (Figure 4-35), the digital line problem becomes immediately obvious. The grounded line did not significantly affect the calculated mean or variance. Figure 4-36 illustrates an infrared imaging system that has a very nonlinear responsivity curve. This nonlinearity would have become obvious after the entire responsivity curve was obtained. However, the same information can be inferred from only one line trace by noting the variation in noise levels. The output amplitude is less in the nonlinear region. As the test engineer becomes more familiar with the test equipment limitations and operation of the infrared imaging system, he will quickly notice subtle data anomalies.

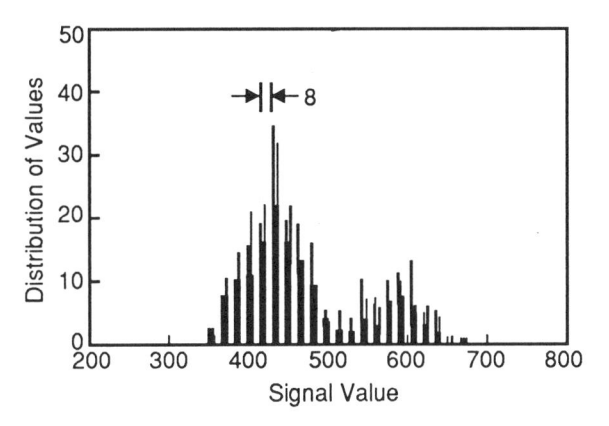

Figure 4-35: Histogram of 10-bit digital data with the 3rd LSB grounded. The imagery visually appeared adequate.

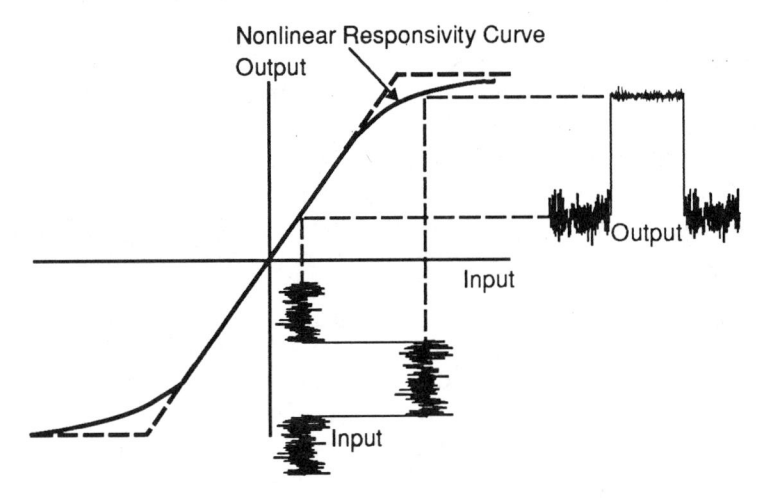

Figure 4-36: Effect of nonlinear responsivity on noise values. The noise is reduced when the signal enters the nonlinear region.

4.7. STATISTICAL ANALYSIS

The output of a system is not a number, but a range of numbers. This range usually follows a well-defined statistical distribution. It is not possible to perform exhaustive testing; an inference is made about the population mean and variance based upon a finite data set. The purpose of statistical analysis is to obtain the best estimate of a particular parameter. Accuracy is the statistical variation of a measured parameter with respect to its true value. It includes bias, precision, and repeatability. A variety of statistical analyses are available for infrared imaging systems[15]. General data analyses can be found texts such as in Dixon and Massey[16] and Bendat and Piersol[17]. Formal error analysis techniques can also be found in a variety of texts[18]. For the tests presented in this text, only the mean, variance, and least-squares fit are used.

Figure 4-37a illustrates a histogram of experimentally obtained data that is unbiased and precise. It is unbiased because the mean of the data set is identical with the true value. It is precise because the standard deviation, σ_p, of the data set is small. That is, each datum value is very close to the true value. Figure 4-37b illustrates an unbiased data set that is imprecise because there is a wide range of data values about the mean. An imprecise, biased data set is shown in Figure 4-37c. A single test produces a single data set. Data set means and variances may vary from test-to-test (Figure 4-37d). The range of the mean values, σ_R, is the repeatability of the measurement. The mean of the means is the best estimate of the population mean[19,20].

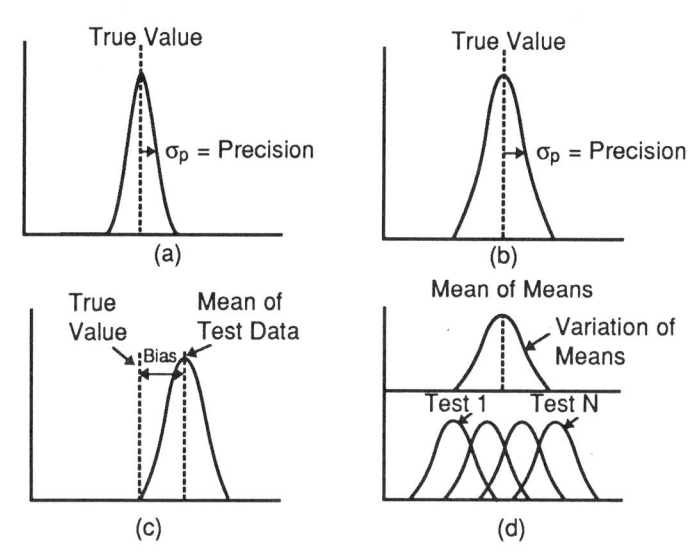

Figure 4-37: Precision, bias and repeatability. (a) An unbiased, precise data set; (b) an unbiased, imprecise data set; (c) a biased, imprecise data set; and (d) repeatability is the standard deviation of means obtained from multiple tests.

Accuracy is:

$$ACCURACY = Bias \pm t\sigma \qquad 4\text{-}7$$

where

$$\sigma = \sqrt{\sigma_P^2 + \sigma_R^2} \qquad 4\text{-}8$$

and t is the percentile of the student t-distribution*. The bias is the sum of all systematic errors whereas random errors, σ, are added in quadrature (root-sum-of-the-squares).

Unfortunately, the full advantage of statistical analysis cannot be exploited when there is insufficient data. When the mean is calculated, it is done almost subconsciously without fully understanding why the data is averaged. Furthermore, the range of data (error bars) or the precision of the data set is often

*The student t-distribution can be found in most statistics text books.

126 TESTING & EVALUATION OF IR IMAGING SYSTEMS

overlooked. The underlying distribution must be known to apply the appropriate statistical analyses. Because of the widespread usage of Gaussian statistics, it is usually assumed that the data set is Gaussian distributed (although this may not be true). The Gaussian distribution is also called the normal distribution or normal probability distribution. The Gaussian probability density distribution is:

$$p(x) = \frac{1}{\sqrt{2\pi}\,\sigma_o} e^{-\frac{1}{2}\left(\frac{x-\mu}{\sigma_o}\right)^2} \qquad 4\text{-}9$$

The sample mean, m, is an estimate of the population mean, μ. For a sample data set, $\{x_1, ..., x_N\}$, the mean is

$$m = \frac{1}{N}\sum_{i=1}^{N} x_i \qquad 4\text{-}10$$

σ_o^2, the population variance, is estimated by the sample variance s^2:

$$s^2 = \frac{1}{N-1}\sum_{i=1}^{N}(x_i - m)^2 \qquad 4\text{-}11$$

or equivalently:

$$s^2 = \frac{N\sum_{i=1}^{N} x_i^2 - \left(\sum_{i=1}^{N} x_i\right)^2}{N(N-1)} \qquad 4\text{-}12$$

As $N \rightarrow \infty$, $m \rightarrow \mu$ and $s^2 \rightarrow \sigma_o^2$. When $N > 30$, the distribution of measured variances approaches a Gaussian distribution. If the variance is measured k times, the mean of this distribution is the best estimate of σ_o^2:

$$s_{ave}^2 = \frac{s_1^2 + s_2^2 + ... + s_k^2}{k} \qquad 4\text{-}13$$

where it is assumed that N data points were used to calculate each s_i^2.

The best estimate of the standard deviation is:

$$\sigma = \sqrt{s_{ave}^2} = \sqrt{\frac{s_1^2 + \ldots + s_k^2}{k}} \qquad 4\text{-}14$$

For noise measurements, σ is the rms noise value although the units 'rms' are seldom used. It is important to note that the average rms noise is the square root of the average variance. It is incorrect to average measured rms noise values.

To reduce further the measured variability, a least-squares line can be placed through the data points. If there are N data sets $[V_i, I_i]$ where V_i is the system output for input, I_i, the least-squares slope is:

$$m = \frac{N\sum_{i=1}^{N} V_i I_i - \sum_{i=1}^{N} V_i \sum_{i=1}^{N} I_i}{N\sum_{i=1}^{N} (I_i)^2 - \left(\sum_{i=1}^{N} I_i\right)^2} \qquad 4\text{-}15$$

The mean values of I_i and V_i are

$$\bar{I} = \frac{\sum_{i=1}^{N} I_i}{N} \qquad 4\text{-}16$$

and

$$\bar{V} = \frac{\sum_{i=1}^{N} V_i}{N} \qquad 4\text{-}17$$

This provides the least-squares polynomial:

$$V_{SYS} = mI + V_{OFFSET} \qquad 4\text{-}18$$

where

$$V_{OFFSET} = \bar{V} - m\bar{I} \qquad 4\text{-}19$$

128 TESTING & EVALUATION OF IR IMAGING SYSTEMS

It is usually assumed that time and ensemble statistics are identical[21]. When this is true, the mean and variance will be the same whether only one detector output is monitored for a long time (time statistics) or whether the outputs of many identical detectors are measured simultaneously (ensemble statistics) (Figure 4-38). For a stationary process, the mean is independent of how the data is collected:

$$m = \frac{1}{N}\sum_{i=1}^{N} V(t_i) = \frac{1}{M}\sum_{i=1}^{M} V_i(t_c) \qquad \text{4-20}$$

Time and ensemble averages will be different for any system composed of nonidentical detectors or if the system changes its operating conditions in time. Infrared imaging systems with gain and level corrections that are updated periodically are not stationary systems.

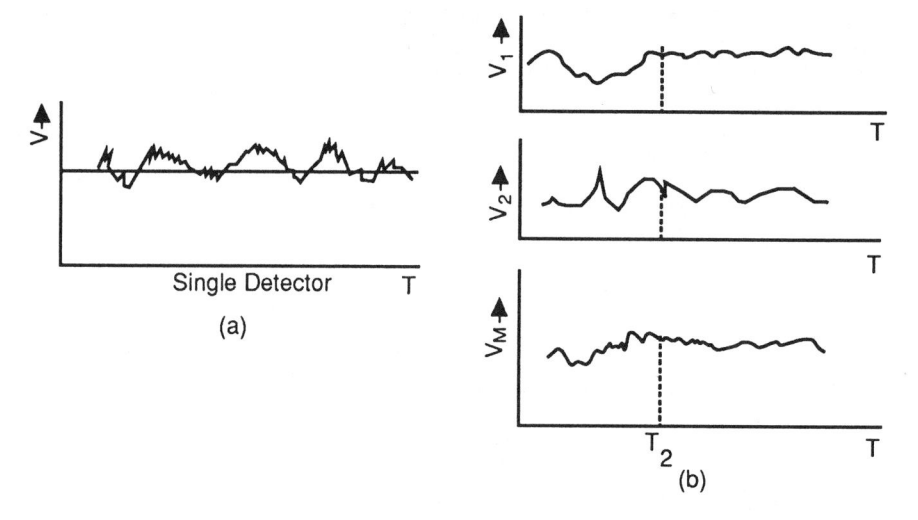

Figure 4-38: Single output and an ensemble of outputs. A system is stationary when the statistics associated with a single output (a) are identical with the statistics of an ensemble of detectors (b).

4.8. SUMMARY

As infrared imaging system performance increases, finding appropriate test equipment continues to challenge the test engineer. Equipment that was satisfactory for yesterday's infrared imaging systems may not be adequate for today's systems. The test engineer must be fully aware of the anticipated infrared imaging system performance and should not blindly use existing test equipment or test procedures because "*We always did it that way.*" He must thoroughly understand the operation of the infrared imaging system before performing any test. The test plan must contain the test configuration, test conditions and data analysis methodology.

The infrared imaging system operation must be fully understood in terms of spectral response, field-of-view, instantaneous-field-of-view, data rates and analog video format so that the test engineer can select appropriate test equipment. The test point (analog video, monitor or visual observation) dictates the equipment type in generic terms. Test specifics and system design guide test equipment requirements (Table 4-1). Because of the rapidly evolving infrared imaging system technology and test equipment capabilities, it is impossible to describe specific test equipment. Rather, the test equipment has been described in generic terms with generic requirements. Table 4-2 lists the governing or limiting parameters for the various tests. Although the measuring techniques discussed are for system level testing, the methodology can be used to characterize subsystem performance.

Table 4-1
TEST EQUIPMENT

TEST POINT	TEST EQUIPMENT	APPLICATION
Analog video	Transient recorder	High resolution: MTF
	Frame grabber	Low resolution: SiTF, CTF, noise
	Oscilloscope	Estimate signal and noise
Monitor	Scanning photometer Solid state camera Scanning fiber optics	All tests
Visual observation	An observer	MRT, MDT, modified Cooper-Harper scale

Table 4-2
SPECIAL CONSIDERATIONS

TEST PARAMETER or REQUIREMENT	GOVERNING/LIMITING PARAMETER
ΔT at the entrance aperture	Atmospheric spectral transmittance Collimator spectral transmittance Ambient temperature changes Target plate emissivity
Reproducible pulse width	Phasing effects in the system Phasing effects in the test equipment
Visibility of detail	Mechanical vibrations Turbulence
No stray radiation	Collimator aperture size Collimator FOV Baffle location

Automating the test configuration (Figure 4-39) speeds up the measuring procedure, reduces the chance for human error and provides a convenient method of recording and plotting results. The results should include at a minimum, the laboratory name, time, date, test number or other identifying nomenclature and system description including the serial number. Ambient conditions, applicable documents referenced, and any other pertinent measured data also should be recorded. Data should be in a format useful to multiple users and multiple analysts. With the widespread use of personal computers, it is prudent to store all data in a format that is consistent with common spreadsheets. If the target selection is computer controlled, the observer need not handle the targets and thereby avoid possible damage. Most laboratories are used by a variety of groups, each of which require different test configurations. It is prudent to check the entire configuration before running a test. For example, are the blackbody source and target wheels in the correct location or have they been moved? Are the optical elements clean with no dust or fingerprints? Are all the baffles in place? Has the configuration been disassembled, moved or modified by another program? Verify with the previous users that all the equipment functioned properly.

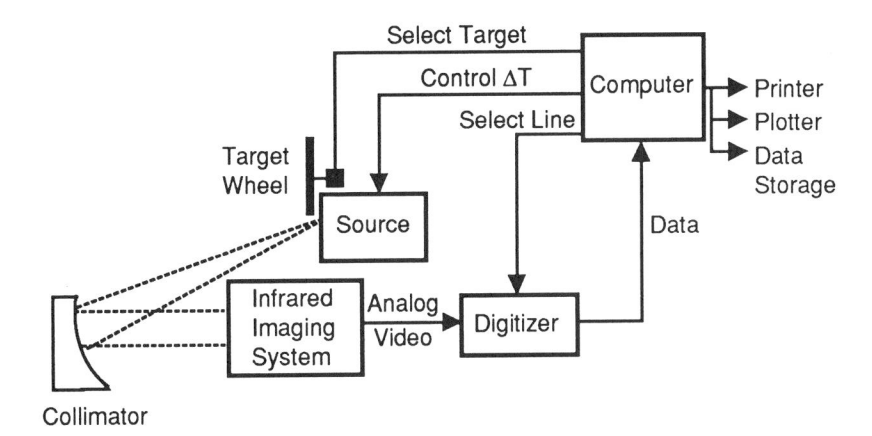

Figure 4-39: Generic automated test configuration. To minimize vibrational effects, the collimator and infrared imaging system should be mounted on a vibration-stabilized optical table. To avoid turbulence effects, the entire configuration should be placed in a curtained area.

The test engineer is relied upon to discern all subtle effects that are often missed during the traditional data collection activity. Only the test engineer, who has seen hundreds of similar units, can rapidly identify subtle changes:

THE TEST ENGINEER SHOULD BE CONSIDERED
THE PRIMARY *DATA ACQUISITION SYSTEM.*

4.9. REFERENCES

1. A. J. LaRocca, "Artificial Sources", in Sources of Radiation, G. J. Zissis, ed., Volume 1 of The Infrared & Electro-Optical Systems Handbook: Environmental Research Institute of Michigan, Ann Arbor, Mich, (1993).
2. W. L. Wolfe, "Errors in Minimum Resolvable Temperature Difference Charts", Infrared Physics, Vol. 17(5), pp. 375-379, (1977).
3. D. Cabib, J. Eliason, B. Hermes, E. Ben-David, S. Ghilai and R. Bracha, "Accurate Infrared Scene Simulation by Means of Microlithography Deposited Substrate" in Infrared Technology XVIII, B.F. Andresen and F. D. Shepherd, eds.: SPIE Proceedings Vol. 1762, pp. 376-384 (1992).
4. S. W. McHugh, "High Performance FLIR Testing Using Reflective Target Technologies", Photonics Spectra, Vol. 25(7), pp. 112-114 (1991).
5. T. L. Williams, "A Portable MRTD Collimator System for Fast In-situ Testing of FLIRS and Other Thermal Imaging Systems", in Infrared Imaging Systems: Design, Analysis, Modeling and Testing, G. C. Holst, ed.: SPIE Proceedings Vol. 1309, pp. 296-304 (1990).

6. W.J. Smith, "Optical Elements - Lenses and Mirrors", in The Infrared Handbook, Revised edition, W. L. Wolfe and G. J. Zissis, eds.: Environmental Research Institute of Michigan, Ann Arbor, Michigan, Chapter 9 (1985).
7. J. T. Cox and W. R. Hunter, "Infrared Reflectance of Silicon Oxide and Magnesium Fluoride Protected Aluminum Mirrors at Various Angles of Incidence from 8 μm to 12 μm", Applied Optics, Vol. 14(6), pp. 1247-1250 (1975).
8. J. T. Cox and G. Hass, "Protected Al Mirrors with High Reflectance in the 8-12 μm Region from Normal to High Angles of Incidence", Applied Optics, Vol. 17(14), pp. 2125-2126 (1978).
9. S. F. Pellicori, "Infrared Reflectance of a Variety of Mirrors at 45° Incidence", Applied Optics, Vol. 17(21), pp. 3335-3336 (1978).
10. M. E. Thomas and D. D. Duncan, "Atmospheric Transmission", in Atmospheric Propagation of Radiation, F. G. Smith, ed., Volume 2 of The Infrared & Electro-Optical Systems Handbook: Environmental Research Institute of Michigan, Ann Arbor, Mich., (1993).
11. N.S. Kopeika, "Imaging Through the Atmosphere for Airborne Reconnaissance", Optical Engineering, Vol. 26(11), pp. 1146-1154 (1987).
12. N. Sampan, "The RS 170 Video Standard and Scientific Imaging: The Problems", Advanced Imaging, pp. 40-43, Feb. 1991.
13. S. J. Briggs, "Photometric Technique for Deriving a "Best Gamma" for Displays", Optical Engineering, Vol. 20(4), pp. 651-657 (1981).
14. J. D. Vincent, Fundamentals of Infrared Detector Operation and Testing, pp. 230-234: John Wiley and Sons, New York (1990).
15. G. R. McNeill and H. B. Macurda, "Infrared (IR) Calibration Measurement Requirements: Development from System Requirements", in Infrared Systems, R. Sanmann, ed.: SPIE Proceedings Vol. 256, pp. 84-90 (1980).
16. W. J. Dixon and F.J. Massey, Introduction to Statistical Analysis, 2nd edition: McGraw-Hill, New York, NY (1957).
17. J. S. Bendat and A. G. Piersol, Random Data: Analysis and Measurement Procedures 2nd edition: John Wiley and Sons, New York, NY (1986).
18. For example: J. D. Vincent, Fundamentals of Infrared Detector Operation and Testing: John Wiley and Sons, New York (1990).
19. D. S. Fraedrich, "Methods in Calibration and Error Analysis for Infrared Imaging Radiometers", Optical Engineering, Vol. 30(11), pp. 1764-1770 (1991).
20. R. B. Johnson and J. D. Jones, "Uncertainties in Calibration", in Long Wavelength Infrared, E. Wolfe, ed.: SPIE Proceedings Vol. 67, pg. 111-114 (1975).
21. M. Schwartz, Information Transmission, Modulation, and Noise, 2nd edition, pp. 390-395: McGraw-Hill, New York (1970).

EXERCISES:

1. An off axis collimator has an entrance aperture of 6 inches and a focal length of 60 inches. What is the minimum rake angle for target design?
2. To conserve space, a reflective collimator is designed with 3 fold mirrors at 45°. What is the collimator transmittance at 5, 8.1 and 10 μm? Assume that the fold mirrors and primary mirror are aluminum with an SiO protective coating. Assume that the reflectivity is 99% outside the anomalous reflection region.

GENERAL MEASURING TECHNIQUES

3. When the test engineer examines a target, he notices that although the targets are black, they appear shiny. What should he do?
4. Design a special target that can be used to verify the system FOV. The FOV is 20 mrad. The target will be placed 15 feet from the infrared system. The least expensive method is to use a large metal plate with heating strips. Sketch the design.
5. Discuss the advantages and disadvantages of using a small (2 inch aperture) collimator to test a system that has an entrance aperture of 6 inches? Include diffraction effects in your discussion.
6. The MRT test target consists of 4 bars (4 bars with 3 spaces). What is the lowest spatial frequency that can be obtained when using the collimator described in Example 4-1? The collimator design accepts targets up to 2 inches in diameter.

5

FOCUS and SYSTEM RESOLUTION

Resolution and focussing techniques are presented as a unit because, typically, the same targets and test techniques are used for both. The only difference is that the targets are "calibrated" for resolution testing. On the surface, focussing a sensor may appear obvious and straight forward. But what does *in focus* really mean and is it directly related to *good* image quality? *In focus* can be defined as the ability to see sharp edges, obtain the maximum system output when viewing a source, obtain the highest MTF, obtain the highest output of an edge detection image processing algorithm or the ability to resolve specific spatial frequencies. For well-behaved linear systems, these quantities tend to be equivalent. On the other hand, infrared imaging systems are nonlinear: the detectors are discrete, the image is digitized, and the image may be noisy. When these processes are present, the definition of *in focus* becomes ill-defined.

Good image quality can be determined by simple visual inspection but best focus cannot. Focus may be considered a dynamic test in that the focus lens-assembly must be varied over a range before the best focus can be found. Focus data are not normally recorded nor presented in any particular format. Rather, the system is simply labeled *in focus*.

Best focus may be a function of field angle. Best focus for on-axis images may be different from off-axis images. This can be ascertained by, for example, measuring the MTF on-axis and off-axis and noting the variation at specific spatial frequencies. Large area detector arrays may be difficult to focus due to field lens curvature. After a system has been assembled, good image quality may be difficult to achieve due to decentration or tilting of lenses. That is, with a non-optimized system, best focus may still provide poor imagery and *best* focus may still be unacceptable.

Because of a myriad of definitions, resolution can be confused with other image quality metrics. The system operation and application determine which definition is appropriate. A large variety of resolution measures exist and the various definitions may not be readily relatable because they are generated by different disciplines. As infrared imaging systems are incorporated into new disciplines, it must be specified in terms used by those industries. For example, photo interpreters use the ground resolved distance (GRD) when evaluating photo reconnaissance imagery captured on film. But as infrared imaging systems replace the wet film process, the GRD, by default, becomes an infrared imaging system resolution measure. Once defined, resolution provides valuable information regarding the finest detail that can be discerned. However, a single measure of spatial resolution cannot be satisfactorily used to compare all sensor systems. Resolution does not provide information about total imaging capability nor how contrast is affected. The contrast transfer function and modulation transfer function provide contrast information. Resolution is not affected by noise and is not related to sensitivity.

There are four different aspects of resolution: (1) temporal resolution, which is the ability to separate events in time; (2) gray scale resolution, which is determined by the A/D converter design, noise floor, or the monitor capability; (3) spectral resolution; and (4) spatial resolution. An imaging system operating at 30 Hz frame rate has a temporal resolution of 1/30 sec. Gray scale resolution is a measure of the dynamic range discussed in Chapter 6. The spectral resolution is simply the spectral band pass (e.g., SWIR, MWIR or LWIR) of the system. This chapter discusses spatial resolution.

5.1. TEST METHODOLOGY

Resolution and focussing may be more difficult to determine than expected. The degree of accuracy depends significantly upon the test configuration and the target selected. Vibration affects high spatial frequency targets (fine detail) before affecting low spatial frequency components. Therefore, any resolution measure will depend upon the vibrational level present. For critical focussing and resolution testing, the source, target, collimator and infrared imaging system should be placed on a vibration-isolated optical table.

Best focus and resolution can be easily ascertained by many techniques if no noise were present. Noise introduces ambiguities in measured values and the variations may exceed the expected change in the parameter being measured. If possible, it is appropriate to reduce the system gain to reduce the noise voltage and to increase the signal by increasing the source intensity. For systems with dynamic AGC and image processing capability, image sharpness may be a function of the image size or the image intensity. If possible, all nonlinear image processing subsystems should be turned off. Saturation also mimics good focus where the edges of the saturated target appear sharp. Simply stated, focus and resolution should be tested only when the system is operating in a linear region.

Any method in which detail is being examined can be affected by phasing effects. An important aspect of focus and resolution is the relationship between the image size compared to the size of each detector. When the image size is much smaller than the detector, small changes in focus do not affect the detector output nor does it change the resultant displayed image size.

Measurement results depend upon the particular system under test and the test criterion. Thus the operation of the specific system as well as the limitation of the measurement technique must be understood. Any appropriate test configuration described in Chapter 4 can be used. The test procedure is generic for focussing and determining system resolution (Table 5-1). Since most tests are subjective in nature, the test engineer should have corrected 20/30 vision or better. However, concentration and visual acuity tend to deteriorate with fatigue. In Chapter 10 a variety of human state-of-awareness issues are discussed that affect an individual's ability to perform a test.

Many different types of test targets can be used for visual determination of best focus. Most of them consist of varying sized black and white bars, slits, pinholes, sweep frequency arrays, stars or portions of stars (wedges). Many targets used for focussing also can provide resolution information when appropriately calibrated.

Table 5-1
FOCUS/RESOLUTION TEST PROCEDURE

1. Fully understand the infrared imaging system operation (Chapter 2).
2. Establish a test philosophy, criteria for success and write a thorough test plan (Section 1.3).
3. Verify that the test equipment is in good condition and that the test configuration is appropriate. Ask previous users if any problems were noticed.
4. Verify the spectral response of the system and its relationship to the source characteristics, collimator spectral transmittance and atmospheric spectral transmittance (Chapter 3).
5. Insure that the system is operating in a linear region.
6. Adjust system gain and source intensity to maximize the signal-to-noise ratio.
7. Insure that the infrared imaging system has reached operating equilibrium before proceeding.
8. Adjust the target location to minimize phasing effects (Section 2.3).
9. Adjust focus or measure resolution.
10. Fully document any test abnormality and all results.

5.2. FOCUS TESTS

The most common method of verifying focus is by visual inspection of black and white bars, slits, pinholes, sweep frequency targets (linear or logarithmic), stars or portions of stars (wedges). Perceived image quality depends significantly on the monitor characteristics. As the image moves away from the center of the monitor, image quality degrades. Best focus may appear nonuniform over the monitor and vary from monitor-to-monitor. Observer variability also increases the test time. To avoid these problems, the methods of optimizing the focus via image amplitude, MTF methodology, or edge detection algorithms offer an alternative because the video signal is evaluated before it reaches the monitor.

5.2.1. VISUAL METHOD

When viewing low spatial frequency targets, the eye subconsciously focuses on an edge. But the eye is not sensitive to slight deviations of edge sharpness[1]. The system is focussed by adjusting the focus lens-assembly in one direction and then in the other direction and selecting a point approximately midway between the two defocus positions. The distance the focus lens-assembly has moved could be considered the lens-assembly depth of focus. This distance is dependent upon the target size and the distance the observer is from the monitor. Large targets (low spatial frequency) provide a large depth and small targets (high spatial frequency) provide a small depth. This distance is not related to the classical text book definition of depth of focus. Instead it is a measure of the eye's sensitivity to edge sharpness.

Stars, wedges and sweep frequency arrays should be designed such that the highest spatial frequency in the target is near the system spatial frequency cutoff. When viewing targets above system cutoff, the targets can no longer be resolved and appear as a neutral gray. High spatial frequency targets are needed because slight defocus affects the visibility of these spatial frequencies the most. The focus lens-assembly position that provides the highest visible spatial frequency is the best focus position. With stars, the highest spatial frequency is at the center (Figure 5-1). Assuming no astigmatic aberrations, best focus is that focus lens-assembly position that minimizes the blurred diameter. Wedges, which are just portions of stars, provide the same effects as stars. In principle, the star patterns can merge to a point but manufacturing constraints often produce a solid core that limits the highest spatial frequency available for testing.

138 *TESTING & EVALUATION OF IR IMAGING SYSTEMS*

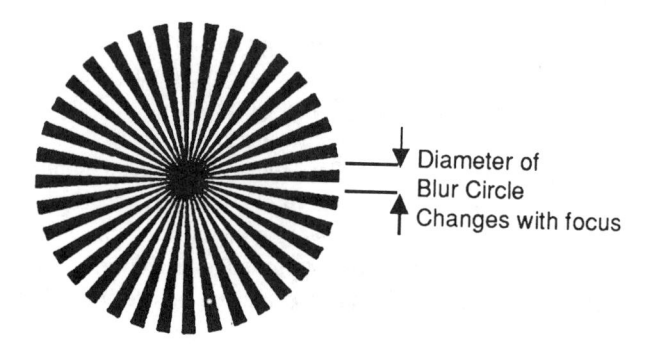

Figure 5-1: Star focus pattern. The focus lens-assembly is adjusted to minimize the central blur diameter.

Stars and wedges may present Moire patterns (aliasing) resulting from the undersampled nature of infrared imaging systems. A raster pattern also creates Moire patterns (Figure 5-2). Moire patterns may be modified with line-to-line interpolation and image processing algorithms. While aliasing is well known to the system designer, it may be too confusing to the laboratory tester[2]. Aliasing confounds the interpretation of best focus.

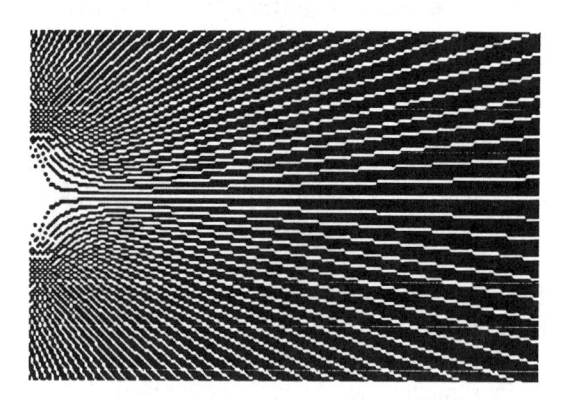

Figure 5-2: Moire pattern produced by a raster scan system.

With a linear sweep frequency target (Sayce target) the raster scan problem does not occur when the target is appropriately aligned. Figure 5-3 illustrates how focus affects the high spatial frequencies but does not significantly alter the low spatial frequencies. The focus lens-assembly is adjusted so that the highest possible spatial frequency is visible. A large amount of defocussing is necessary to soften visually the edges of the low spatial frequency bars. The focus lens-assembly movement required to affect high spatial frequency bars is slight compared to the movement required to affect low spatial frequency bars. This again illustrates that visibility of detail is very sensitive to focus adjustment.

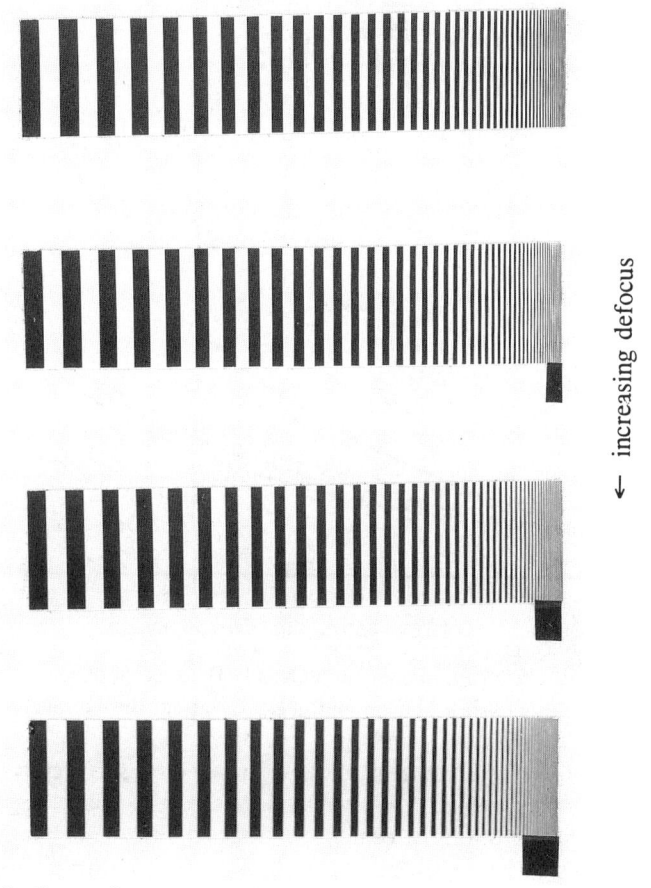

Figure 5-3: Sweep frequency or Sayce target. As defocus increases, the high spatial frequency bars disappear (indicated by the bar). A large amount of defocussing is required to affect the low spatial frequency bars. Phasing effects, if present, will affect target visibility.

The USAF 1951 (MIL-STD-150A) tri-bar targets can be used for both focussing and resolution measurements. Best focus is achieved when the smallest possible target is resolved (Figure 5-4). Target dimensions are given in Section 5.3.2.

140 *TESTING & EVALUATION OF IR IMAGING SYSTEMS*

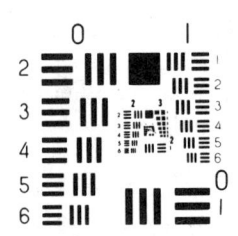

Figure 5-4: USAF tri-bar resolution chart.

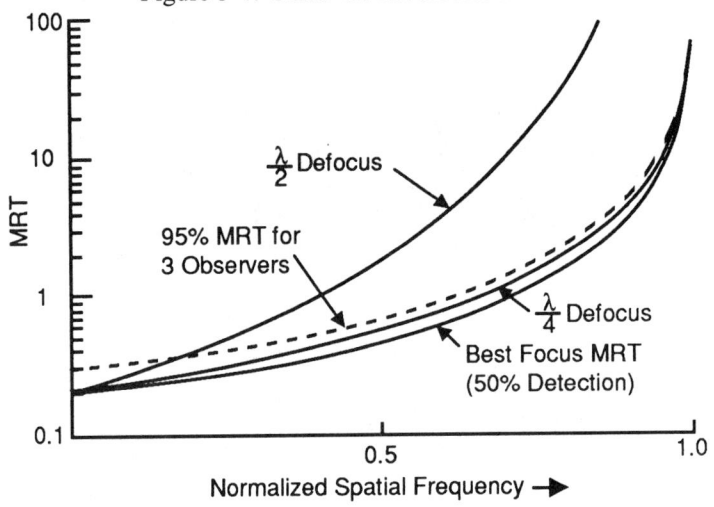

Figure 5-5: Effect of defocussing on MRT. Because of observer variability, slight defocussing will not significantly affect the measured MRT.

It is sometimes assumed that MRT test results are sufficiently sensitive to detect focus errors. While MRT values increase with increasing amounts of defocus, slight focus errors cannot be detected due to the large observer variability (discussed in Chapter 10). In Figure 5-5, the theoretical MRT increases with $\lambda/4$ defocussing[*]. The dashed line shows the range of detection thresholds for 95% of the observer population. The observer variation exceeds the expected difference and therefore, the increased MRT due to slight defocus probably will not be obvious. A large amount of defocus ($\lambda/2$ wave front error or greater) does become apparent in the MRT test results. However, the system will definitely appear out of focus before performing the MRT tests and the test probably will not be performed until the system is refocused.

[*]*It is assumed that a wave front deviation of $\lambda/4$ can be just perceived.*

5.2.2. ANALOG VIDEO AMPLITUDE METHOD

The video signal amplitude can be measured by selecting a video line passing through a target and displaying it on an oscilloscope. The focus lens-assembly is adjusted until a peak output is obtained. This is a dynamic test in that the system must be tested at a variety of focus positions until best focus (maximum amplitude) is achieved. Targets that can be used include slits and high spatial frequency periodic targets. By measuring the amplitude of a slit image, the amplitude of the line spread function is maximized. Simultaneously, the MTF is maximized. For staring arrays and undersampled scanning systems, it is necessary to adjust the target phase so that the maximum amplitude is obtained. It is sometimes difficult to ascertain that the target is centered on a detector. Centering can be verified by viewing the output of adjoining detectors (Figure 5-6). Once centered, best focus occurs when the center peak is maximized and the adjoining detector outputs are minimized. By maximizing the center pixel output, the ensquared power is maximized. If the periodic target is a square wave, the contrast transfer function is maximized at that spatial frequency. For scanning systems, the spatial frequency of the periodic square wave target selected should be well above 1/3 of the system spatial frequency cutoff (See Section 8.1.). For larger targets (lower spatial frequency), the target amplitude approaches a maximum. Slight defocussing does not affect the amplitude but only softens the edges (See, for example, the low spatial frequency bars in Figure 5-3: page 139). For staring arrays, the square wave spatial frequency should equal the array Nyquist frequency.

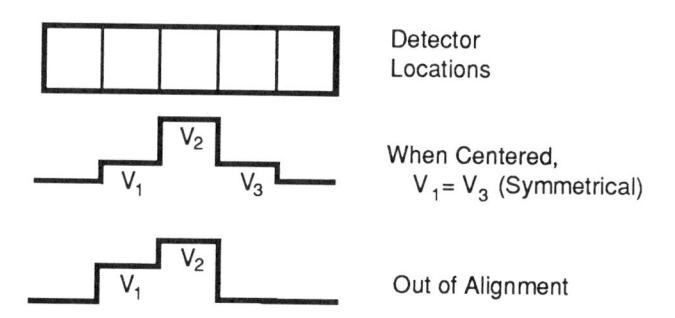

Detector Locations

When Centered, $V_1 = V_3$ (Symmetrical)

Out of Alignment

Figure 5-6: Proper alignment of images. When a small target is centered on a detector, the adjoining detectors will have equal outputs.

142 *TESTING & EVALUATION OF IR IMAGING SYSTEMS*

The amplitude can be verified visually or plotted. When plotted as a function of focus lens-assembly position, a curve fit will determine the best focus (Figure 5-7). Excessive noise may make the analog method difficult to perform. Signal averaging can reduce the noise to an acceptable level.

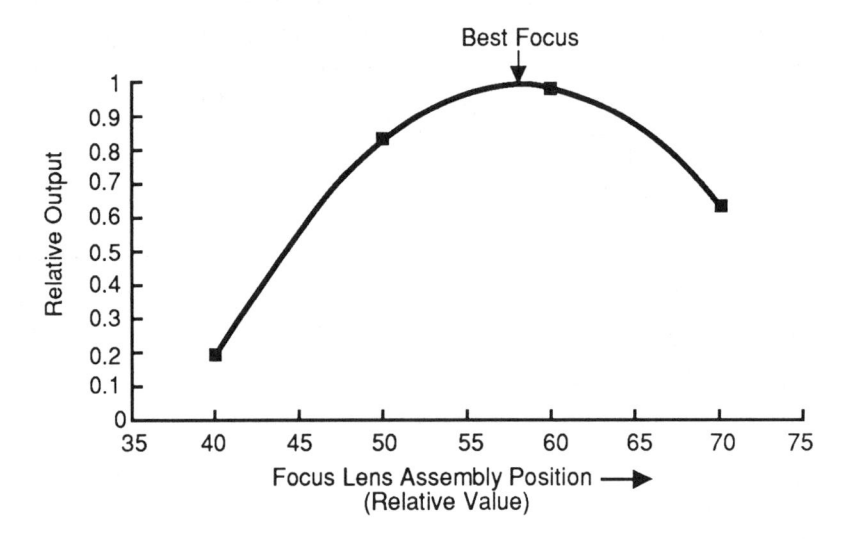

Figure 5-7: System output as a function of focal lens-assembly position. Data was collected at positions 40, 50, 60 and 70. A curve fit to the data provides best focus at position 58.

5.2.3. MTF METHOD

The relationship between good image quality and best focus depends upon the MTF of the system[3], the noise power spectral density, the spatial frequency presented to the eye and the scene content. For a well-behaved system, best focus is synonymous with highest MTF.

The MTF can be plotted as a function of focus lens-assembly position. That position that provides the highest MTF is considered best focus. Rather than plot the entire MTF, it is convenient to track a specific spatial frequency. It is desirable to select a spatial frequency at which the *absolute* change in MTF is very sensitive to slight changes in focus. This typically occurs at spatial frequencies between 0.3 and 0.5 of the system cutoff or equivalently where the MTF is between 0.4 and 0.6. In Figure 5-8a, the system MTF, for illustrative purposes, consists only of an optical MTF and a detector MTF. The defocussed curve represents λ/4 wave front deviation which is typically regarded as that value where defocus can just be perceived. Figure 5-8b illustrates the difference between the ideal and defocussed systems. For this example, the maximum absolute change in MTF occurred at a spatial frequency of 0.35 times the detector spatial frequency cutoff. This figure should only be used as a guide. The expected MTF degradation should be plotted for the specific system and then the appropriate spatial frequency is selected. Chapter 8 provides methods to measure the MTF.

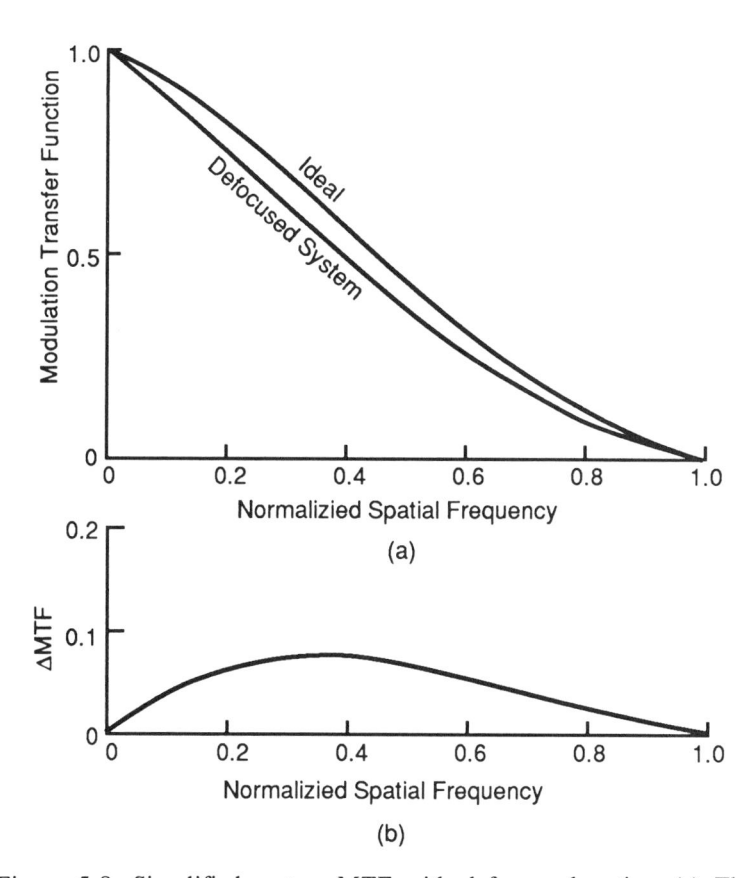

Figure 5-8: Simplified system MTF with defocussed optics. (a) The theoretical maximum and defocussed MTFs for λ/4 wave front error and (b) the difference between the theoretical maximum and the defocussed MTF.

144 *TESTING & EVALUATION OF IR IMAGING SYSTEMS*

5.2.4. EDGE DETECTION ALGORITHMS

Image processing algorithms provide an alternate method of adjusting focus. Algorithms such as the Sobel operator provide an output that is proportional to edge sharpness. Lourens et. al.[4] demonstrated that the Sobel operator output magnitude is approximately proportional to subjective evaluation of best focus for TV images.

The edge detection algorithm does not require any particular target but the target should contain many edges or high frequency detail. This target is often called a complex target. The Sobel operator output is very sensitive to noise and does not provide the desired output when significant noise is present.

5.3. SYSTEM RESOLUTION

System resolution depends on diffraction, optical aberrations, detector angular subtense, digitization, electronic bandwidth, and resolution of the monitor. The vertical and horizontal system resolution can be significantly different. Calibrated test targets are used to determine system resolution. These include Sayce, stars and wedges with fiducial marks and bar targets such as the USAF 1951 tri-bar resolution chart. Phasing effects (Section 2.3: page 36) can dramatically alter the appearance of periodic targets. The target phase should be adjusted for maximum visibility to provide the highest resolution value.

5.3.1. DEFINITIONS

Resolution provides valuable information regarding the finest detail that can be discerned. Each discipline extracts its own type of information from data and each discipline has its own requirements for image quality. Therefore a variety of definitions exist (Table 5-2). Each definition has its own merits. Because of the complexity of imaging systems, the various measures of resolution are not easily relatable.

Resolution may be defined from optical considerations. Diffraction produces the smallest possible spot size. Diffraction measures include the Rayleigh criterion, Sparrow criterion and the Airy disk diameter. The Rayleigh and Sparrow criteria are a measure of the ability to distinguish 2 closely spaced objects, CSO, where the objects are point sources. Optical aberrations and focus limitations will increase the diffraction diameter to the blur diameter. Optical designers using ray trace programs usually calculate the blur diameter. If the blur diameter is larger than a single detector, the detector output is less than if the blur diameter were smaller than a detector. The ratio of the center detector

FOCUS and SYSTEM RESOLUTION 145

output to the sum of all the detector outputs is the ensquared power[5]. Ensquared power is important to systems that are used for point source detection such as IRST systems. It can be measured or may be obtained from the aperiodic transfer function (See Section 6.2). The ensquared power value also is called the point visibility factor or blur efficiency.

Example 5-1
ENSQUARED POWER

A pinhole whose angular subtense is 1/10 of the IFOV simulates a point source. The system's average level is set low so the output of every detector is 5 ADUs (Figure 5-9b) when covered with a black opaque cloth. When the pinhole is centered on a detector, the output is shown in Figure 5-9a. What is the ensquared power?

Subtracting the ambient values (Figure 5-9b) from the image (Figure 5-9a) provides the signal produced only by the source (Figure 5-9c). The ensquared power is

$$Enquared\ power = \frac{center\ pixel\ output}{sum\ of\ all\ pixels} = \frac{220}{384} = 57.3\% \qquad 5\text{-}1$$

The peak ensquared power is 57.3%. If the pinhole was moved, phasing effects would reduce the ensquared power. The smallest ensquared power will be obtained when the pinhole is centered on a corner where 4 detectors are joined.

5	5	6	5	5
5	20	30	20	5
6	30	225	30	6
5	20	30	20	5
5	5	6	5	5

(a)

5	5	5	5	5
5	5	5	5	5
5	5	5	5	5
5	5	5	5	5
5	5	5	5	5

(b)

0	0	1	0	0
0	15	25	15	0
1	25	220	25	1
0	15	25	15	0
0	0	1	0	0

(c) = (a) - (b)

Figure 5-9: Ensquared power. (a) System output when the point source image is centered on a detector, (b) system output with no target, and (c) difference between (b) and (a).

Table 5-2
MEASURES OF RESOLUTION

RESOLUTION	DESCRIPTION	TEST (usual units)
Rayleigh Criterion	Ability to distinguish 2 point sources	$\theta = 1.22\ \lambda/D$ (mrad) (Calculated)
Sparrow Criterion	Ability to distinguish 2 point sources	$\theta = \lambda/D$ (mrad) (Calculated)
Airy Disk	Diffraction limited diameter produced by a point source	$\theta = 2.44\ \lambda/D$ (mrad) (Calculated)
Blur Diameter	Actual minimum diameter produced by point source	Calculated from ray tracing (mrad)
Ensquared power Point visibility factor	Single detector output produced by a point source	Calculated or measured (%)
IFOV, DAS or geometric IFOV	Angle subtended by one detector element	$\alpha = d/fl_{SYS}$ (mrad) (measured or calculated)
Limiting Resolution	Spatial frequency at which MTF = 0.02 to 0.05	Measured or calculated (cy/mrad)
EIFOV	One-half of the reciprocal of the spatial frequency at which MTF = 0.5	Measured or calculated (mrad)
TV limiting resolution	Ability to resolve square waves	Measured (TV lines per picture height)
Imaging resolution	Angular subtense at which SRF = 0.5	Measured (mrad)
Measurement resolution	Angular subtense at which SRF = 0.99	Measured (mrad)
Pixels	Number of detector elements or number of digital data points	A number (unitless)
Area weighted average resolution	A composite resolution averaged over the entire FOV	Calculated from measured values. (Same units as used for resolution)
Ground resolved distance	The smallest test target (1 cycle) that a photo interpreter can distinguish	Measured or calculated (feet or meters)

FOCUS and SYSTEM RESOLUTION   147

The IFOV, geometric IFOV, or detector angular subtense, DAS, is often used to describe the resolution of systems when the detector is the limiting subsystem. If the detector size, d, is much smaller than the effective focal length, then the small angle approximation provides

$$IFOV = \alpha = \frac{d}{fl_{system}} \qquad 5\text{-}2$$

The detector horizontal and vertical dimensions maybe different so that the IFOV in the two directions may be different. The detector MTF is

$$MTF = \frac{\sin(\pi \alpha f_x)}{\pi \alpha f_x} \qquad 5\text{-}3$$

The MTF is equal to zero when the spatial frequency $f_x = 1/\alpha$. Resolution can be defined when the MTF = 0. However, for most systems, the MTF does not abruptly reach zero but approaches zero asymptotically. The apparent IFOV, α_{app} can be estimated by fitting a $\sin(\pi\alpha_{app}f_x)/(\pi\alpha_{app}f_x)$ curve to the MTF (Figure 5-10).

If the $\sin(\pi\alpha_{app}f_x)/(\pi\alpha_{app}f_x)$ curve does not fit the MTF well, the limiting resolution can be defined as that spatial frequency where the MTF drops to 2% or 5% of its maximum value. For undersampled systems, the system MTF is defined only up to the Nyquist frequency and the MTF may not approach zero (Figure 5-11). The effective instantaneous-field-of-view, EIFOV, offers an alternate measure of resolution. For many systems, the EIFOV and IFOV are approximately equal.

TV limiting resolution is determined from the finest detail that can be discerned when viewing a star, a wedge or resolution patterns. The star is composed of concentric square waves whose spatial frequency increases toward the center of the star. If the system is out of focus, spurious resolution can reverse periodic images such that white bars become black and black bars become white. When spurious resolution occurs, the low spatial frequency bars are visible, become invisible (neutral gray) at mid-spatial frequencies, and reappear at higher spatial frequencies but with reversed contrast. Resolution is the region at which the bars first become invisible. The usable region for image fidelity is that region in which all the bars can be seen with certainty. TV limiting resolution is a subjective measure. The spatial frequency at which the pattern disappears is approximately the same as the measured limiting resolution. Besides TV limiting resolution, variety of other measures of resolution exits for

148 *TESTING & EVALUATION OF IR IMAGING SYSTEMS*

monitors[6] and a variety of measurement techniques is available[7].

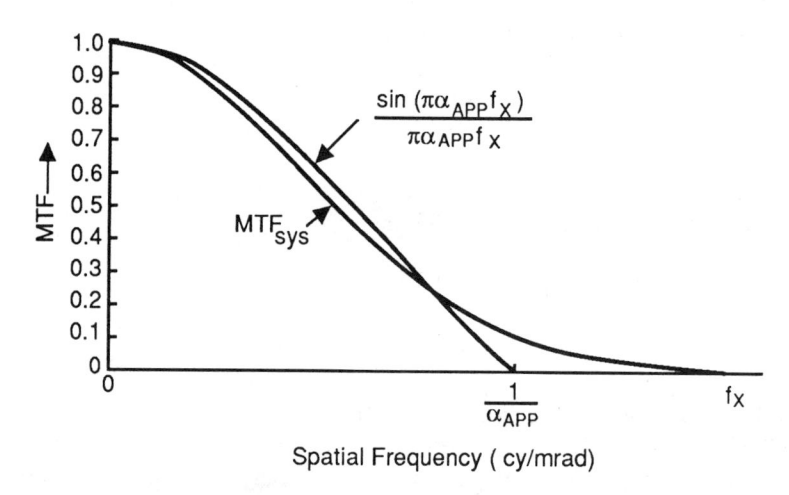

Figure 5-10: Fitting a detector MTF to the system MTF to obtain an effective system cutoff.

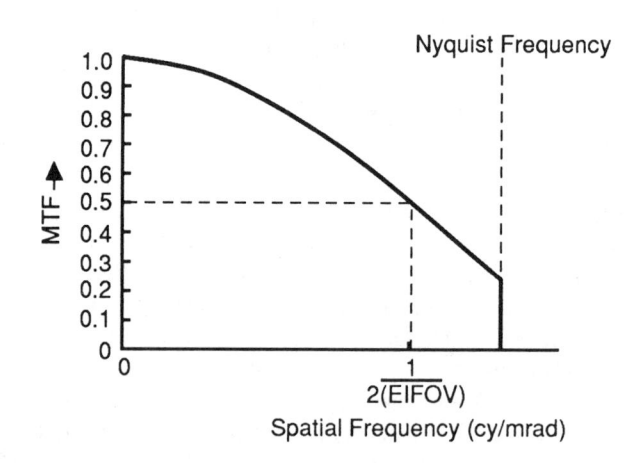

Figure 5-11: Definition of EIFOV for an undersampled system.

The slit response function, SRF, provides the imaging resolution. Methods to measure the SRF are given in Section 6.2. The imaging resolution is that target angular subtense that produces a 50% response in the SRF (Figure 5-12). The imaging resolution includes both optical and electronic responses and may be considered more representative of actual system response than the calculated IFOV. For an ideal system, the IFOV is twice the imaging resolution. The measurement resolution selected for SRF = 0.99 is approximately the smallest sized target that will be faithfully reproduced in intensity. It is the absolute minimum size that can be used for responsivity measurements and radiometric calibration.

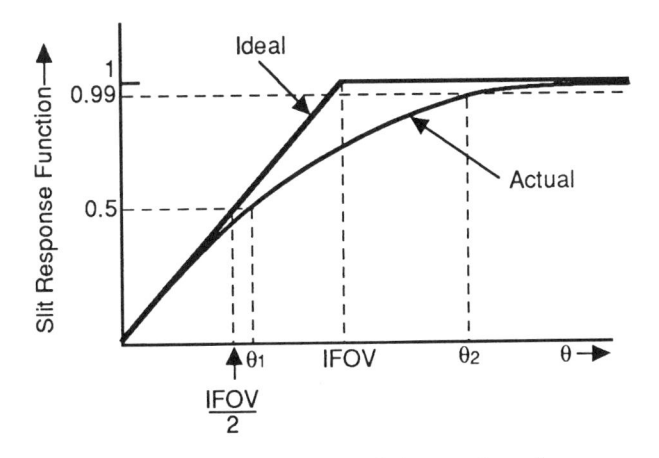

Figure 5-12: Slit response function. θ_1 is the imaging resolution and θ_2 is the measurement resolution.

An image may be broken into many discrete data points. Each datum represents a location and local image intensity. Each datum is considered a picture element or pixel. The number of pixels selected to represent the image depends upon the number of detectors and for scanning systems, the sampling rate. Each datum represents the output of a detector and represents an IFOV. In oversampled systems, the pixels are closely spaced and do not add any additional information to the image. Information is contained in a resolution element or resel. A resel is the smallest region in object space whose dimension is equal to the resolution in that dimension. The measures of resolution in Table 5-2 (page 146) are different ways of defining a resel. For staring systems, a pixel and resel are equal. In oversampled systems, the resel may consist of many pixels.

Example 5-2
RESELS AND PIXELS

An infrared imaging system has an IFOV of 0.1 mrad and a FOV of 51.2 mrad. What is the number of resels and pixels if the system is designed with a staring array and a scanning array? Assume that the scanning array is oversampled either at 1024 samples/line or 2048 samples/line and that staring array consists of 512 x 512 elements. For this example, the resel is equal to the IFOV.

For the staring array, there are 512 pixels per line. Each pixel represents an IFOV and the resel and pixel are equal. For the scanning array, there are 1024 pixels per line. Each pixel represents an IFOV but the IFOVs overlap in object space so that there are 2 pixels for each resel. If the sampling rate increases to 2048 samples/line, the number of pixels would increase to 2048 but the resel value would remain the same. Increasing the sampling rate does not change the system's resolution but increases the accuracy of all measurements by reducing phasing effects.

Example 5-3
MONITOR PIXELS VERSUS SYSTEM RESELS

The output of an infrared imaging system is displayed on a digital monitor which consists of 1024 x 1024 pixels. The focal plane array consists of 512 x 512 detectors. What is the system resolution?

Each infrared detector is considered a resel. Each resel is mapped onto 4 monitor pixels. The system resolution is determined by the imaging system, not the monitor. High quality monitors only insure that the image quality is not degraded.

FOCUS and SYSTEM RESOLUTION 151

The area weighted average resolution is determined by summing the resolution in concentric annular zones over the entire picture format (Figure 5-13):

$$AWAR = \sum_{i=1}^{N} R_i \frac{A_i}{A_T} \qquad 5\text{-}4$$

where R_i is the resolution in zone i which has an area A_i. A_T is the total area:

$$A_T = \sum_{i=1}^{N} A_i \qquad 5\text{-}5$$

The angles midway between successive test target angular locations determine the zone boundaries.

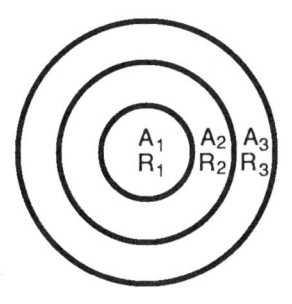

Figure 5-13: Area weighted average resolution. R_i is the resolution in zone i whose area is A_i.

☞─────────────────────────────

Example 5-4
AWAR

An infrared imaging system with a circular format picture has a FOV of 40 mrad. The TV limiting resolution was measured in 4 zones as shown in Figure 5-15 and Table 5-3. It is assumed that the resolution is constant within each zone or annulus. What is the AWAR?

The area of a zone i is $\pi(r_i^2 - r_{i-1}^2)$. The AWAR is given in Table 5-3.

152 TESTING & EVALUATION OF IR IMAGING SYSTEMS

Table 5-3
AWAR MEASUREMENTS

ZONE	Resolution measured at θ	Resolution lines/mm	r_{zone}	A_{zone}	A_{zone}/A_T	Weighted resolution
1	0 mrad	500	2.5	19.6	0.016	8
2	5 mrad	480	7.5	157.0	0.125	60
3	10 mrad	470	12.5	314.1	0.250	117
4	15 mrad	460	20.0	765.7	0.609	280
						AWAR =465

Since Zone 4 has the largest area, the resolution obtained for Zone 4 dominates the AWAR.

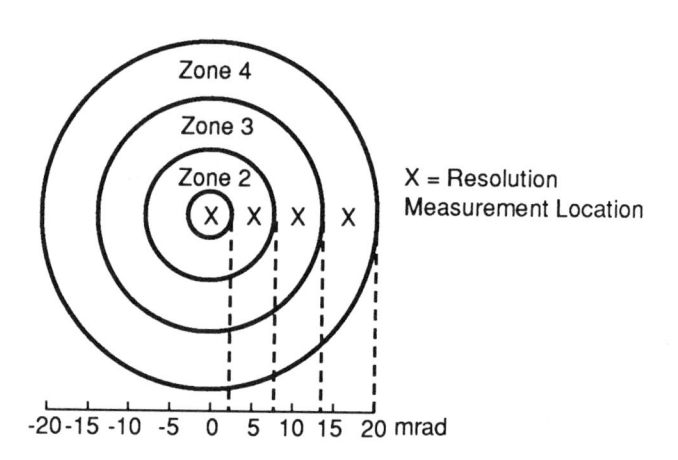

Figure 5-14: Regions of measurement.

For aerial reconnaissance and associated image interpretation, resolution is measured by the ground resolved distance[8,9] GRD. GRD is the minimum test target (one cycle) size that can be resolved on the ground by an experienced photo interpreter. The smallest detail that can be distinguished will have a physical width of GRD/2.

GRD is:

$$GRD = (Resolution) \cdot R_1 \quad \text{feet or meters} \quad 5\text{-}6$$

where R_1 is the slant path range to the target. The IFOV is often used as the resolution measure. Any resolution measure listed in Table 5-2 (page 146) can be used if it is converted to mrad/cy. GRD cannot be measured in the laboratory since it depends upon the distance to the target but may be calculated from an appropriate resolution measure.

5.3.2. RESOLUTION TARGETS

Resolution targets are created by placing lines or fiducial marks at various spatial frequencies of interest. The marks are found on most charts used for determining the limiting resolution of monitors. These charts typically have both horizontal and vertical wedges that allow resolution measurements in both the horizontal and vertical directions. The horizontal resolution is obtained from wedges that are vertical and the vertical resolution is obtained from the wedges that travel horizontally. The wedges require interpolation of the limiting resolution relative to the fiducial mark location. TV limiting resolution occurs where the stars or wedges blend into a neutral gray. If the resolution is rotationally symmetrical, the blurry area is circular (Figure 5-1: page 138). Stars or wedges can be thought of as concentric periodic bar targets. If the blurry arcs or circles are approximated by chords then:

$$Resolution = \frac{N_{sectors}}{\pi \cdot blur\ diameter} \quad \frac{cycles}{mm} \quad 5\text{-}7$$

where the blur diameter is measured in mm. A sector is one cycle or one bar plus one space and the star contains $N_{sectors}$. When placed in a collimator with focal length fl_{COL},

$$Resolution = \frac{N_{sectors}}{\pi \cdot blur\ diameter} fl_{COL} \quad \frac{cycles}{mrad} \quad 5\text{-}8$$

When viewing a star on the monitor, the angular size of the blur diameter is:

$$\theta = \frac{blur\ diamter\ on\ monitor}{monitor\ width} (HFOV) \quad mrad \quad 5\text{-}9$$

154 TESTING & EVALUATION OF IR IMAGING SYSTEMS

The angular blur diameter, when related to object space provides:

$$Resolution = \frac{N_{sectors}}{\pi}\left[\frac{1}{\theta}\right] \quad \frac{cycles}{mrad} \qquad 5\text{-}10$$

A sweep frequency target also can be calibrated by fiducial marks. Since the target bars are in discrete steps, it is easy to determine the system limiting resolution. For production testing purposes where the requirement may be to perceive a certain frequency, a fiducial mark could be placed at the specification value (Figure 5-15). If the bars can be seen above the mark, then the system passes whereas if the only bars below the mark can be seen, the system fails. Note that with this method, the system might have much greater resolution than the required specification (i.e., can see many bars smaller than the specification).

FAIL | PASS

notch

Figure 5-15: Calibrated linear sweep target used to determine acceptable focus during production.

The USAF 1951 tri-bar targets (Figure 5-4: page 140) consist of 5:1 aspect ratio bars whose perimeter form a square. The change in pattern size is a geometric progression based upon the sixth root of 2 or 1.1255. With this progression, the number of lines per unit distance doubles with every sixth target. Each set of six elements is known as a group.

Each group heading is the power of two to which the first element is raised:

$$\frac{\text{line-pairs}}{\text{mm}} = 2^{\left(group + \frac{element - 1}{6}\right)} \qquad 5\text{-}11$$

For example, the first element of group -2 has 2^{-2} or 0.25 lines/mm. The sizes of the bars are given in Table 5-4.

Table 5-4
USAF 1951 TRI-BAR RESOLUTION CHART
(LINES/MM)

Group Element	-2	-1	0	1	2	3
1	0.250	0.500	2.00	4.00	8.00	16.0
2	0.280	0.561	1.12	2.24	4.49	8.98
3	0.315	0.630	1.26	2.52	5.04	10.1
4	0.353	0.707	1.41	2.83	5.66	11.3
5	0.397	0.793	1.59	3.17	6.35	12.7
6	0.445	0.819	1.78	3.56	7.13	14.3

5.4. SUMMARY

Focus and resolution are metrics used to discern detail. Best focus cannot be ascertained by simply viewing an image. Focus adjustment is a dynamic test in which the focus lens-assembly must be moved to verify that the best focus has been obtained. Table 5-5 lists four focus techniques. The method selected is a matter of convenience and will depend upon the system application and availability of test targets. Resolution can have a variety of definitions depending upon the discipline and purpose of the metric (Table 5-6). Both focus and resolution can be affected by phasing effects. The target should be adjusted to achieve maximum modulation or maximum amplitude. For any target that has diagonal lines with respect to the detector array axes, phasing effects will produce Moire patterns. Resolution measures are not all interrelatable and cannot be used interchangeably. Each measure provides different information about the system. For example, two different systems can have different MTF curves but the same resolution (Figure 5-16). How are these two systems compared? The answer lies in the intended use of the system and without the intended use, it is impossible to rank the systems.

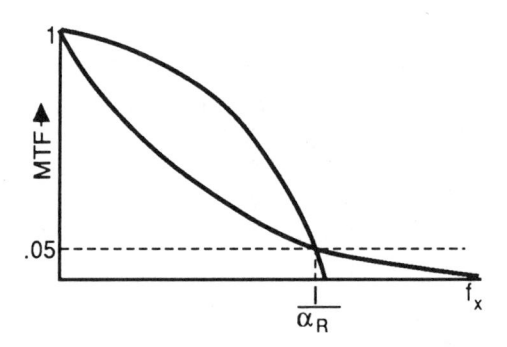

Figure 5-16: Two different systems with the same resolution when defined as that spatial frequency where the MTF is 0.05. The resolution is α_R mrad.

Resolution is a measure of the ability to resolve detail. It is determined by viewing high contrast targets whereas the MRT and MDT tasks are to detect and resolve low contrast targets embedded in noise. The three most popular resolution measures are the IFOV, imaging resolution and the number of detector elements. The system may not be able to produce the resolution implied by the IFOV due to sampling or if line-to-line interpolation is present. The imaging resolution overcomes this by being a measured value. In an ideal system the imaging resolution is ½ of the IFOV.

Table 5-5
FOCUS TECHNIQUES

FOCUS TECHNIQUE	TYPICAL TARGETS
Visual method	Stars, wedges, Sayce patterns, high spatial frequency bars, USAF tri-bar target
Analog video amplitude	Slits, high spatial frequency bars, point sources
Maximizing MTF	Slit or edge
Edge detection algorithms	Complex target with fine detail

Table 5-6
MEASURES OF RESOLUTION

DISCIPLINE	RESOLUTION METRIC
Optical designers	Rayleigh criterion Sparrow criterion Airy Disc diameter Blur diameter
Detector vendors	Number of detector elements
System analyst (geometric approach)	IFOV, DAS
System analyst (MTF approach)	Limiting resolution EIFOV
System calibration (SRF approach)	Imaging resolution Measurement resolution
Monitor designers	TV limiting resolution Number of addressable pixels
Photo reconnaissance and remote sensing	Ground resolved distance
Composite	Area weighted average resolution

158 *TESTING & EVALUATION OF IR IMAGING SYSTEMS*

Specifications must be both understandable and quantifiable. Since *in focus* is subjective, specifications can only place a lower limit on acceptability (Table 5-7). The system may have much better performance. Similarly, resolution specifications (Table 5-8) should not be at the theoretical value but should allow lee way for manufacturing tolerances.

Table 5-7
TYPICAL *IN FOCUS* SPECIFICATIONS

- The system shall be considered *in focus* if a 4 cy/mrad target is discernible.
- The system shall be considered *in focus* if group 3, element 4 of the USAF 1951 tri-bar resolution chart is discernible.
- The system shall be considered *in focus* if the MTF at 5 cy/mrad is greater than 0.3.
- The system shall be considered *in focus* if 450 lines/picture-height are discernible in the center of the field-of-view.

Table 5-8
TYPICAL RESOLUTION SPECIFICATIONS

- The blur diameter shall not be greater than 2 mrad.
- The point visibility factor shall be greater than 75% when measured in the center of the FOV.
- The IFOV shall be 1 mrad ± 0.01 mrad.
- The TV limiting resolution shall not be less than 400 lines/picture height.
- The EIFOV shall not be greater than 2 mrad.
- The imaging resolution shall not be greater than 0.5 mrad.
- The measurement resolution shall not be greater than 2 mrad.
- The AWAR shall be greater than 640 lines/mm when measured in 3 equi-area zones.
- The detector array shall consist of 240 by 320 elements.
- The ground resolved distance shall not be greater than 2 meters when measured from a height of 5000 meters.

5.5. REFERENCES

1. R. F. Hess, J. S. Pointer and R. J. Watt, "How are Spatial Filters Used in Fovea and Parafovea?", JOSA-A, Vol. 6 (2), pp. 329-339 (1989).
2. W. F. Schreiber, Fundamentals of Electronic Imaging Systems, Section 2.5: Springer-Verlag, New York (1986).
3. C. S. Williams and O. A. Becklund, Introduction to the Optical Transfer Function, Chap 9: John Wiley and Sons, New York, (1989).
4. J. G. Lourens, T. C. Du Tuit and J. B. Du Tuit, "Addressing the Focus Quality of Television Pictures" in Human Vision, Visual Processing and Digital Display, B. E. Rogowitz, ed.: SPIE Proceedings Vol. 1077, pp. 35-41 (1989).
5. L. M. Beyer, S. H. Cobb and L. C. Clune, "Ensquared Power for Obscured Circular Pupils With Off-Center Imaging", Applied Optics, Vol. 30(25), pp. 3569-3574 (1991).
6. L. M. Biberman, "Image Quality" in Perception of Displayed Information, L. M. Biberman, ed.: Plenum Press, New York (1973).
7. P. A. Keller, "Resolution Measurement Techniques for Data Display Cathode Ray Tubes", Displays, Vol. 7, pp. 17-29 (1986).
8. Air Standardization Agreement: "Minimum Ground Object Sizes for Imaging Interpretation", Air Standardization Co-ordinating Committee report AIR STD 101/11 dated 31 Dec. 1976.
9. Air Standardization Agreement: "Imagery Interpretability Rating Scale", Air Standardization Co-ordinating Committee report AIR STD 101/11 dated 10 July 1978.

EXERCISES

1. For each focus target given in this chapter, discuss the potential manufacturing difficulty and applicability to infrared imaging systems. Sketch typical experimental configurations for each target. Are there any other designs that would work?
2. An MRT 4-bar target will be used for focus adjustment, Sketch a line trace output for a slightly defocussed system when using a very low, mid and high spatial frequency target.
3. For the 3 targets given in Exercise 2, sketch the amplitude as a function focus lens-assembly position.
4. The GRD is 2 feet when the aircraft is at 5000 feet. What is the GRD when the aircraft is at 15000 feet? Plot the GRD as a function of altitude. Is the GRD a good measure of resolution?
5. A manufacturer specifies his resolution as 0.6 mrad when the SRF is 50%. The system has a focal length of $fl_{SYS} = 20$ inches and the detector size is 0.02 x 0.02 inches square. Sketch the theoretical and actual SRF.
6. An infrared imaging system with a circular format picture has a FOV of 40 mrad. The specification requires an AWAR of 450 line/mm. The TV limiting resolution was measured in 4 zones at 0, 5, 10, and 15 mrad. The resolution values are 350, 480, 470 and 460 lines/mm respectively. What is the AWAR? If the resolution was measured at 3 zones (5, 10 and 15 mrad), did the system pass the specification?. Would you consider this system usable?

6

SYSTEM RESPONSIVITY

The *responsivity function* is the input-to-output transformation in which the target size is fixed and the target intensity is varied. It provides information on gain, linearity, dynamic range and saturation. The signal transfer function (SiTF) or responsivity is the linear portion of the responsivity function and is a commonly used parameter to assess performance. Since a large target (compared to the IFOV) is used, the responsivity is sometimes called the low frequency response.

The output also can be obtained as a function of target area with the target intensity is fixed. This two-dimensional *responsivity* is the aperiodic transfer function (ATF). The ratio of system ATF to the ideal ATF provides the target transfer function and, for point sources, the point visibility factor or ensquared power value. If the target height is much greater then the vertical IFOV and the target width is varied, the one-dimensional slit response function (SRF) is obtained. The SRF provides the imaging resolution and measurement resolution. Measurement resolution is the smallest target that can be used for responsivity measurements and for radiometric calibration.

Dynamic range is the maximum measurable signal divided by the minimum measurable signal. For infrared imaging systems, the minimum value is taken as the noise level, NEDT. If the input is varied over 50 degrees and the NEDT is 0.2 degrees, then the dynamic range is 250:1 or 48 db. For non-noisy systems, the minimum signal depends upon the system design. For example, for digital circuitry, the minimum value is the least significant bit so that an 8 bit A/D converter has a dynamic range of 256:1.

6.1. SIGNAL TRANSFER FUNCTION

6.1.1. SYSTEM RESPONSE

The responsivity function is typically S-shaped. For scanning DC coupled systems (staring systems), the dark current (or noise floor) limits the minimum detectable signal and saturation limits the maximum detectable signal (Figure 6-1). For these systems it is customary to plot the output as a function of the absolute value of the input. For scanning AC coupled systems and DC coupled systems that have an AGC, the output is centered on an average value. As a result, it is convenient to plot the differential output as a function of the differential input. The word *difference* or *differential* is often omitted since it is understood that differential values are the main concern. Saturation for positive and negative differential inputs about this average value is typically limited electronically (Figure 6-2) by the dynamic range of an amplifier or A/D converter. The slope of the linear portion of the responsivity curve is the signal transfer function.

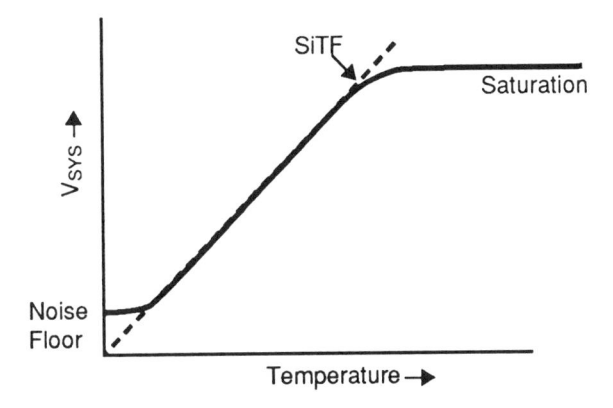

Figure 6-1: Responsivity function for a DC coupled system. The linear portion is the SiTF.

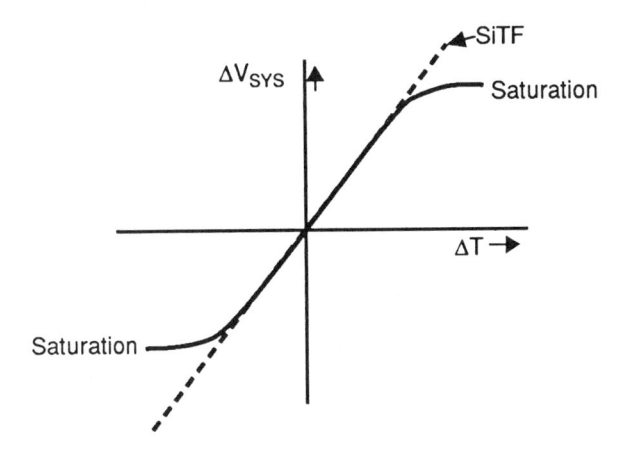

Figure 6-2: Responsivity function for an AC coupled system or a DC coupled system that employs an AGC.

162 TESTING & EVALUATION OF IR IMAGING SYSTEMS

Figure 6-3 illustrates the output of a staring array or scanning AC coupled system with DC restoration as a function of target intensity. As the input intensity increases, the signal increases but the background remains constant (Figure 6-3a). Either the absolute output (Figure 6-3b) or the voltage difference (Figure 6-3c) can be plotted.

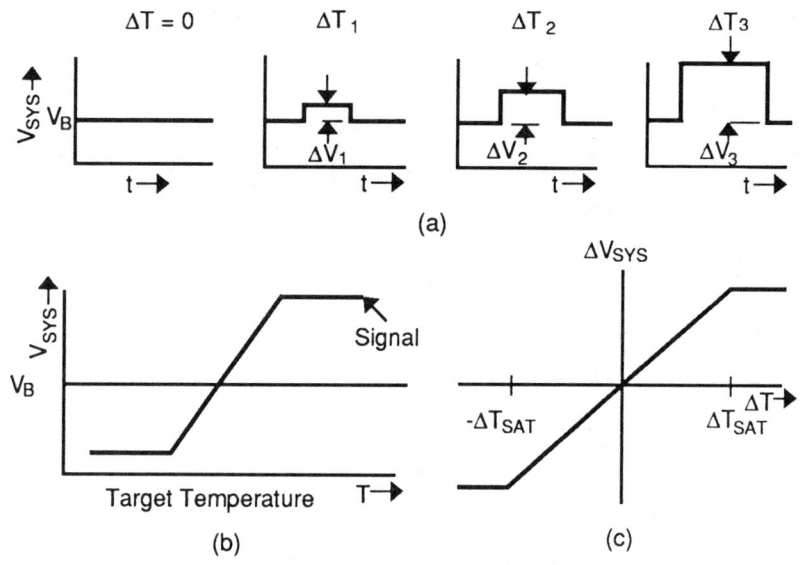

Figure 6-3: Output analog video voltages for a DC coupled system. (a) Line traces for three different input ΔTs: $\Delta T_3 > \Delta T_2 > \Delta T_1 > 0$, (b) V_{SYS} as a function of ΔT and (c) ΔV_{SYS} as a function of ΔT. ΔT_{SAT} is the differential input that saturates the output.

With AC coupling, the signal is coupled in the scan direction but not in the vertical direction. As a result there is a voltage shift (signal and background) in the scan direction that is not present in the vertical direction (Figure 6-4). The voltage difference must be measured in the scan direction. Figure 6-5 illustrates the effect of AC coupling on the output voltage for both large and small targets. Small targets reach saturation before large area targets do. The background voltage value decreases as the target ΔT increases (Figure 6-5a). The amount of decrease depends upon the target size (Figure 6-5b). AC coupling extends the dynamic range. The available output differential voltage can be equal to the full output available voltage when the target that covers 50% of the field-of-view (Figure 6-5c). If the target is any larger, the maximum available voltage starts to decrease. It approaches zero as the target fills the entire FOV. For a small target, the dynamic range is $\Delta T_{MAX2}/NEDT$ and for a large target, $\Delta T_{MAX1}/NEDT$.

SYSTEM RESPONSIVITY 163

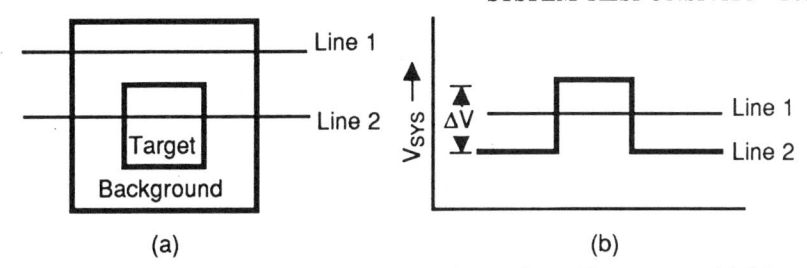

Figure 6-4: Effects of AC coupling on the analog video output. (a) Line 1 covers the background only and line 2 spans the background and target; and (b) the output analog voltages for lines 1 and 2. ΔV is the signal due to ΔT.

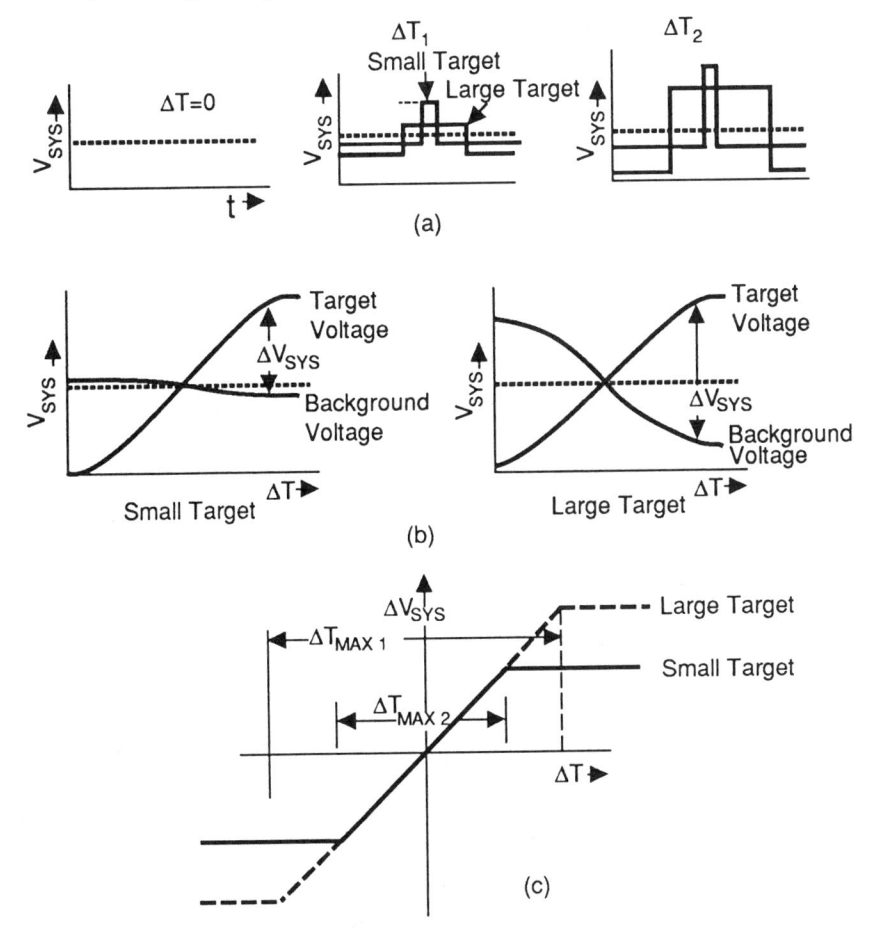

Figure 6-5: Target size effects when AC coupling is present. (a) ΔV_{SYS} for two different target sizes and with $\Delta T_2 > \Delta T_1 > 0$, (b) V_{SYS} as a function of ΔT; and (c) ΔV_{SYS} as a function of ΔT.

164 *TESTING & EVALUATION OF IR IMAGING SYSTEMS*

SiTF by itself is not a very good metric for comparing different systems because it depends upon system gain. The SiTF also can vary from system-to-system if the spectral response of each system is different. If the system SiTF is specified, it usually implies that the system is operating at maximum gain. The effect of gain change on the output voltage for DC coupled systems is shown in Figure 6-6a. Figure 6-6b illustrates gain change for AC coupled systems or DC coupled systems with an AGC.

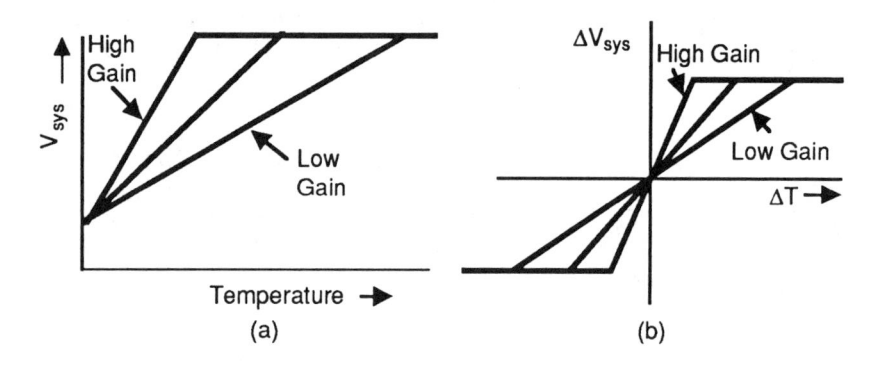

Figure 6-6: Gain affects the SiTF. (a) DC coupled systems and (b) AC coupled systems and DC coupled systems with an AGC.

Using the ΔT concept, for on-axis detectors, when a linear system is flood illuminated, the differential output voltage is:

$$\Delta V_{SYS} = \left[G \int_{\lambda_1}^{\lambda_2} R(\lambda) \frac{A_d}{4(f/\#)^2} \frac{\partial M_e(\lambda, T_B)}{\partial T} T_{SYS}(\lambda) T_{TEST}(\lambda) \, d\lambda \right] \Delta T \qquad 6\text{-}1$$

where $T_{TEST}(\lambda) = T_{COL}(\lambda) T_{ATM}(\lambda)$ and SiTF = $\Delta V_{SYS}/\Delta T$. Example 3-3 (page 78) illustrated a typical SiTF calculation. Variations in SiTF occur because most infrared imaging systems measure flux and not temperature. The SiTF is not strictly linear when plotted as a function of ΔT since the photon flux is not linear with ΔT (Figure 3-16: page 77). In principle, it is possible to calculate the SiTF temperature dependence by knowing the spectral response of the system. The SiTF should be specified at a background, say 293 K, and if measured at a different temperature, then that temperature must be recorded.

When performing an experiment, there will be slight variations in the measured values. Fitting a least-squares polynomial to the data set provides the best estimate of the SiTF:

$$\Delta V_{SYS} = m \Delta T + V_{OFFSET} \qquad 6\text{-}2$$

where

$$V_{OFFSET} = \Delta V_{AVE} - m \Delta T_{AVE} \qquad 6\text{-}3$$

If N data points are collected, then

$$\Delta T_{AVE} = \frac{\sum_{i=1}^{N} \Delta T_i}{N} \qquad \Delta V_{AVE} = \frac{\sum_{i=1}^{N} \Delta V_i}{N} \qquad 6\text{-}4$$

and the least-squares slope (which is the SiTF) is:

$$SiTF = m = \frac{N \sum_{i=1}^{N} \Delta V_i \Delta T_i - \sum_{i=1}^{N} \Delta V_i \sum_{i=1}^{N} \Delta T_i}{N \sum_{i=1}^{N} (\Delta T_i)^2 - \left(\sum_{i=1}^{N} \Delta T_i\right)^2} \qquad 6\text{-}5$$

Different setups at different laboratories can lead to different SiTFs due to different measuring conditions and different equipment. The offset, V_{OFFSET}, may be due to the inability to measure the background temperature accurately or due to different emittances between the target and its background. It also may be due to variations in ambient temperature that occur during the test (Figure 3-12: page 75). Background temperatures (and therefore ΔT) can be affected by air conditioning cycling, the location of air ducts and hot electronics. The offset should be noted when comparing system test techniques for it is an important comparison parameter but rarely recorded. The offset that exists in many test configurations can only be determined by tracing out the entire responsivity function. Unless it is known that no offset exists, the SiTF cannot be accurately determined by making a measurement at one input value only. When only one input is used, the underlying assumption is that the responsivity function passes through the origin. Furthermore, if the entire responsivity function was never established, the selected point may be outside the linear region. The responsivity also may be a function of target location within the FOV. For example, the off-axis SiTF may be modified by $\cos^N \theta$.

166 *TESTING & EVALUATION OF IR IMAGING SYSTEMS*

Example 6-1
SITF CALCULATION

The output of an infrared imaging system was measured for 10 different intensity inputs. The collimator transmittance is 0.90 and the atmospheric transmittance is assumed to be unity. What is the SiTF?

The data set is given in Table 6-1. The ΔT at the system's entrance aperture is the measured ΔT multiplied by the collimator transmittance.

Table 6-1
SiTF DATA SET

ΔT (measured)	ΔV (mv) (measured)	ΔT (at entrance aperture)	ΔV	$\Delta T \Delta V$	$(\Delta T)^2$
-5	-330				
-4	-320				
-3	-300	-2.7	-300	810	7.29
-2	-185	-1.8	-185	333	3.24
-1	-80	-0.9	-80	72	0.81
1	140	0.9	140	126	0.81
2	220	1.8	220	396	3.24
3	310	2.7	310	837	7.29
4	350				
5	355				
		sum = 0	sum = 105	sum = 2574	sum = 22.68

As shown in Figure 6-7, the input ΔT = -5, -4, 4 and 5 produced a nonlinear response. Therefore, these 4 values were omitted from the least-squares analysis.

The SiTF is:

$$SiTF = m = \frac{6 \cdot 2574 - 0}{6 \cdot 22.68 - 0} = 113.5 \ mv/^{\circ}C$$

The average voltage is $\Delta V_{AVE} = 105/6 = 17.5$, $\Delta T_{AVE} = 0$, and $V_{OFFSET} = 17.5 - 0 = 17.5$. The system output in mv is

$$\Delta V_{SYS} = 113.5 \cdot \Delta T + 17.5$$

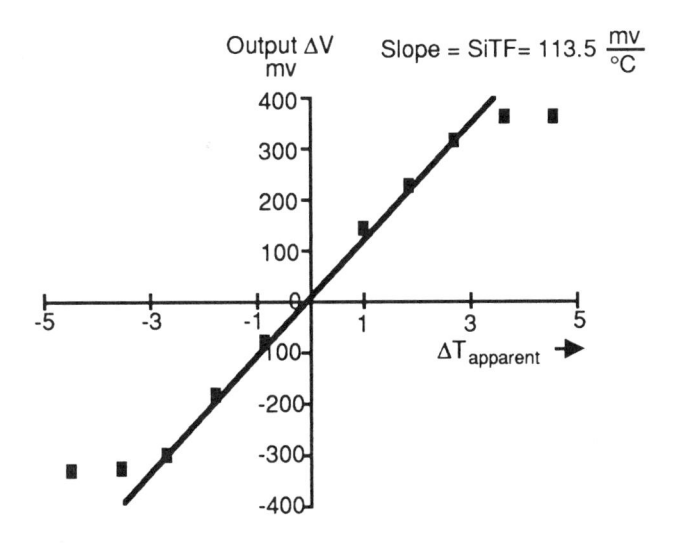

Figure 6-7: SiTF and data points for Example 6-1. $\Delta T_{apparent}$ is the ΔT at the entrance aperture. It is the target-background ΔT multiplied by the collimator transmittance and atmospheric transmittance.

6.1.2. RESPONSIVITY UNIFORMITY

The SiTF is a system transfer function. With scanning arrays, by selecting the output of a single analog video line, the output of a single detector is measured and the SiTF applies only to the detector being tested. If line-to-line interpolation is present, the analog video signal may have been created by several detectors. To obtain the average system SiTF, the SiTF should be measured on a variety of lines and averaged. Each detector/amplifier combination will have a different gain and different output. The variations in output will appear as intensity variation on the monitor.

168 *TESTING & EVALUATION OF IR IMAGING SYSTEMS*

Figure 6-8 illustrates a range of SiTFs that would be produced by a variety of individual detectors without gain/level normalization. For staring arrays a singe analog video line is produced by many detectors. Here, the SiTF is already an average of many detectors. The range of SiTFs is plotted as a histogram in Figure 6-9. The responsivity uniformity is a percentage of the average value: $U_R = 100\ \sigma_{SiTF}/SiTF_{ave}$ %. The range also can be used to specify the minimum and maximum allowable SiTFs for the detector array.

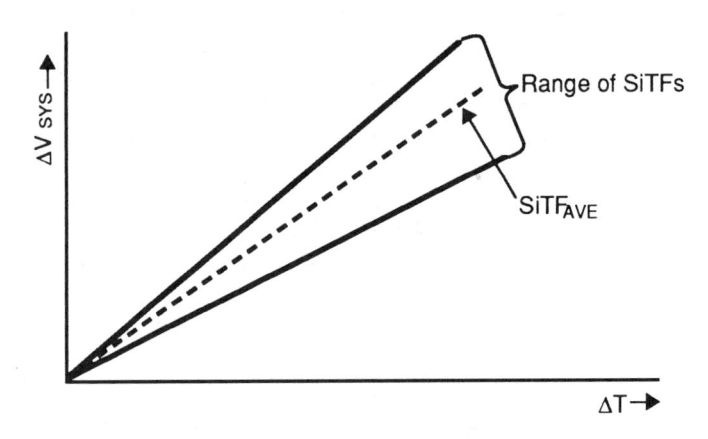

Figure 6-8: Different detector/amplifier combinations will have different SiTFs.

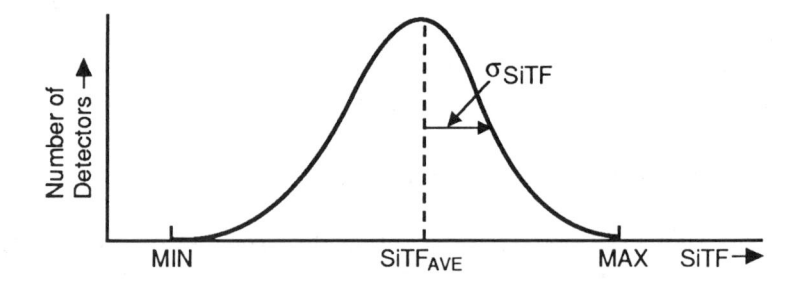

Figure 6-9: Histogram of SiTF variations due to different detector/amplifier gains. σ_{SiTF} is the standard deviation of all the SiTF values.

SYSTEM RESPONSIVITY 169

For staring arrays, the responsivity of each detector can be measured by flooding the array (flood illumination) with a source at temperature T_j and averaging the output, V_{ij} of each detector ij over N frames to reduce to noise (Figure 6-10):

$$\overline{V_{ij}(T_k)} = \frac{1}{N}\sum_{n=1}^{N} V_{ij}(n,T_k) \qquad 6\text{-}6$$

The process is repeated for various flood illumination intensities and then the responsivity function of each detector is plotted. Each detector gain and level can be used to determine the gain/level normalization constants or can be used to measure the deviation from linearity after gain/level normalization.

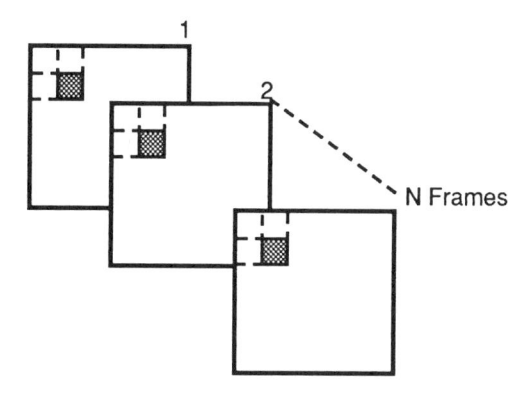

Figure 6-10: Methodology to measure the responsivity of individual detectors in an array. The output of detector ij is averaged over N frames.

6.1.3. SiTF TEST PROCEDURE

The test configuration for measuring the responsivity function is shown in Figure 6-11. ΔT is changed by increasing or decreasing the blackbody temperature. Although an emissive target is illustrated, a reflective target plate also could be used (Figure 4-8 or Figure 4-9: page 100). The target must be sufficiently large to insure that the output voltage reaches full value, i.e., the target size must be larger than the measurement resolution. MIL-STD-1859 recommends[1] that the target subtends 10% of the field-of-view but not less than

10 times the IFOV. The target selected depends upon the system operating characteristics (Figure 6-5: page 163). Ideally, the target plate or background should fill the entire field-of-view so that there are no competing effects from the surroundings. Data acquisition can be by any of the methods described in Section 4.6 (page 115).

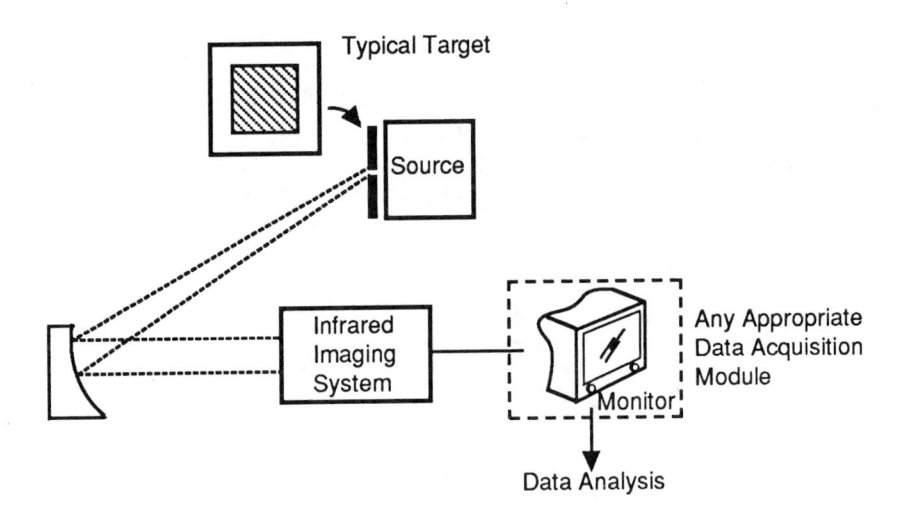

Figure 6-11: Responsivity test configuration. The blackbody, collimator and infrared imaging system should be placed on a vibration-stabilized optical table.

Figure 6-12 illustrates the correct ΔV_{SYS} when there is severe droop, ringing or trends present. Bell and Hoover[2] showed that the output signal difference can depend upon the system operation. An uncooled focal plane array that incorporates a translucent chopper[3] can provide signal differences that are a function of target size (Figure 6-13). For these systems, it is appropriate to select a target width that provides a maximum output. This will result in a maximum SiTF and a minimum NEDT.

SYSTEM RESPONSIVITY 171

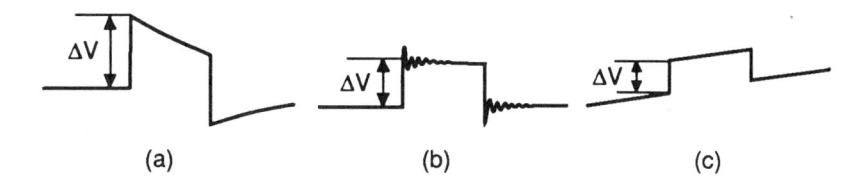

Figure 6-12: Possible analog video signals and the correct ΔV_{SYS}. (a) AC coupling exhibiting severe droop, (b) ringing and (c) trends or drifting baseline.

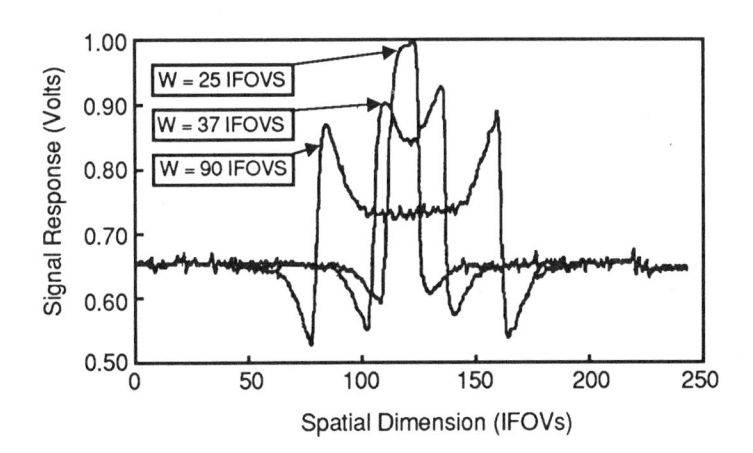

Figure 6-13: Output signal voltage as a function of target width for an uncooled system. The target width is measured in equivalent number of IFOVs. (From reference 2).

172 TESTING & EVALUATION OF IR IMAGING SYSTEMS

Noise (Figure 6-14) can introduce errors in the measured ΔV_{SYS}. Averaging pixels together reduces noise, but the target size limits the number of pixels that can be averaged. Signal averaging is best performed with a computer for it is difficult to estimate a signal value embedded in noise when it is displayed on an oscilloscope. The student t-test[*] provides the required number of pixels that must be averaged together to achieve a desired level of confidence. To insure that the signal has reached full value it is best to allow buffer zones for signal averaging (Figure 6-15).

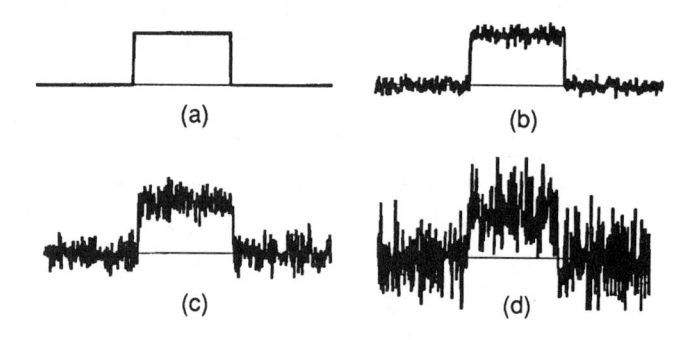

Figure 6-14: Effects of signal-to-noise ratio on target visibility. (a) No noise, (b) SNR = 10, (c) SNR = 5, and (d) SNR = 2.

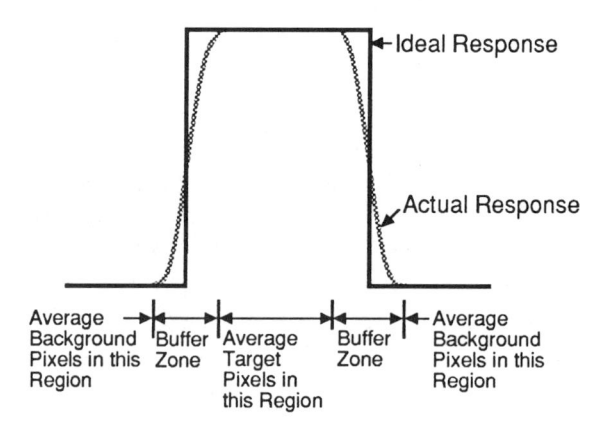

Figure 6-15: Selecting buffer zones for signal averaging. The buffer zones have been exaggerated for illustration.

[*]*The student t-test can be found in most statistics texts.*

Example 6-2
STUDENT T-TEST

Due to statistical variations, the measured value will not be exactly the true value. How many pixels should be average together so that the measured value is within 5% of the true value? The requirement is:

$$\left| \frac{\Delta V_{measured} - \Delta V_{true}}{\Delta V_{true}} \right| \leq 0.05$$

Using the student t-test, the variation in the measured value is

$$\Delta V_{measured} = \Delta V_{true} \pm \frac{ts}{\sqrt{N}}$$

where s is the rms value* calculated from the data set. Then

$$\frac{st}{\Delta V_{true}\sqrt{N}} \leq 0.05$$

At the 95% confidence level, as $N \to \infty$, $t \to 1.96$. As a first order approximation (not rigorously correct because t increases as N decreases):

$$N \geq \left(\frac{ts}{0.05 \Delta V_{true}}\right)^2 = \left(\frac{32.9 \, s}{\Delta V_{true}}\right)^2 = \left(\frac{32.9}{SNR}\right)^2$$

s and ΔV_{TRUE} can be referred to the input so that s is replaced with the NEDT and ΔV_{TRUE} is replaced by ΔT. If the sample data set has a standard deviation of NEDT = 0.1°C, the number of data points (rounded up to the next integer) required to know the value within 5% is given in Table 6-2. N is the number of data points that must be averaged together to obtain a 95% confidence level in the mean value.

*The rms value is identical to the data set standard deviation. This value is also called the 1-sigma value.

Table 6-2
REQUIRED NUMBER OF SAMPLES
(Assuming NEDT = 0.1° C)

ΔT °C	Signal-to-noise ratio	N
0.25	2.5	174
0.50	5.0	44
0.75	7.5	20
1.00	10	11
1.50	15	5
2.00	20	3

Table 6-2 is a first order approximation; the exact number of samples required must be recalculated using the correct t statistic for the data set of size N. As the SNR decreases, the number of data points required increases to maintain the desired levels of confidence. Figure 6-14 (page 172) illustrates this point.

Table 6-3 provides the SiTF test procedure. If the SiTF is different from that calculated, each input should be carefully examined (detector spectral responsivity, optical spectral transmittance, collimator spectral transmittance, source intensity/calibration, atmospheric spectral transmittance and system gain) as well as the possible SiTF modifiers given in Table 6-4.

Table 6-3
SiTF TEST PROCEDURE

1. Establish a test philosophy, criteria for success and write a thorough test plan (Section 1.3: page 12).
2. Verify that the test equipment is in good condition and that the test configuration is appropriate (Figure 6-11: page 170). Ask previous users if any problems were noticed.
3. Verify the spectral response of the system and its relationship to the source characteristics, collimator spectral transmittance and atmospheric spectral transmittance (Section 4-3: page 105 and Section 4-4: page 112).
4. Calculate the expected responsivity (Equation 6-1: page 164).
5. Turn off all automatic gain (AGC) circuits and fix gain at a predetermined value. If this is not possible, use an AGC target (Figure 4-15: page 105).
6. Insure that the infrared imaging system has reached operating equilibrium before proceeding.
7. Select a data acquisition method (Section 4.6: page 115) and collect data for a variety of source intensities.
8. Multiply all ΔT_i by the spectrally weighted collimator and atmospheric transmittances to obtain the effective ΔT at the infrared imaging system entrance window (Equation 3-50: page 81).
9. Calculate the SiTF using the least squares fit methodology.
10. Fully document any test abnormality and all results. The background temperature and the temperature offset should be recorded. The responsivity function should be shown graphically with the calculated least-squares regression line overlaying the data points. Data points also should be presented in tabular form. The target size used should be recorded.
11. As appropriate, repeat above steps for different locations in the field-of-view.

176 *TESTING & EVALUATION OF IR IMAGING SYSTEMS*

Table 6-4
CAUSES FOR VARIATIONS IN SiTF

- Low optical transmittance or dirty optics.
- Cosine$^N\theta$ effects (Figure 2-3: page 24).
- Vignetting or other shading.
- Low detector responsivity (bad detector or bad cooler) (Figure 2-11: page 31).
- Low emittance targets (Section 4.2: page 92).
- Detector temperature variation (Figure 2-11: page 31).
- Ambient temperature changing during test (Section 3.1.5: page 73).
- Active AGC.
- Gamma correction present (Section 2.4.3: page 48).
- Nonlinear response due to ΔT nonlinearity (Section 3.1.5: page 73).

6.2. APERIODIC TRANSFER FUNCTION & SLIT RESPONSE FUNCTION

Systems can sense radiation from targets whose angular subtense is smaller than the instantaneous-field-of-view. The aperiodic transfer function (ATF) is the transformation of input target area to output voltage for fixed target intensity. It is a two-dimensional response that depends upon the vertical and horizontal extent of the target. For convenience, a long slit that covers several detectors is used to obtain the one-dimensional slit response function (SRF). While the MTF describes the ability to discern detail, the ATF describes the ability to sense something and therefore is related to point source detection or hot spot detection.

The ideal and system ATFs are shown in Figure 6-16 where A_{IFOV} is the projected area of a detector element. As the source area approaches zero, the system point spread function is obtained. The point visibility factor (PVF) is that portion of the point spread function that is incident onto one detector. The ratio of the actual system ATF to the ideal ATF is the target transfer function (TTF) (Figure 6-17). As the A_S approaches zero, the TTF approaches the PVF. The PVF is also called the blur efficiency and ensquared energy.

SYSTEM RESPONSIVITY 177

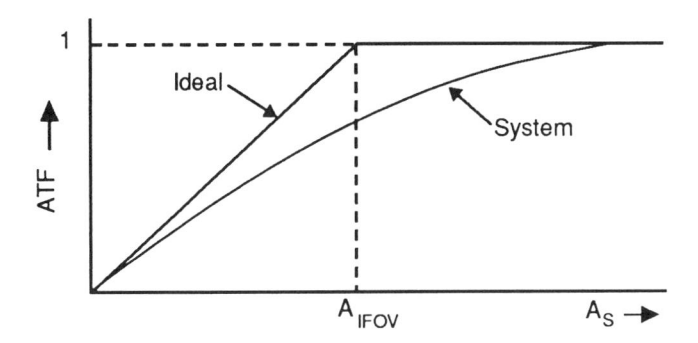

Figure 6-16: Aperiodic transfer function illustrating both the ideal and actual system responses.

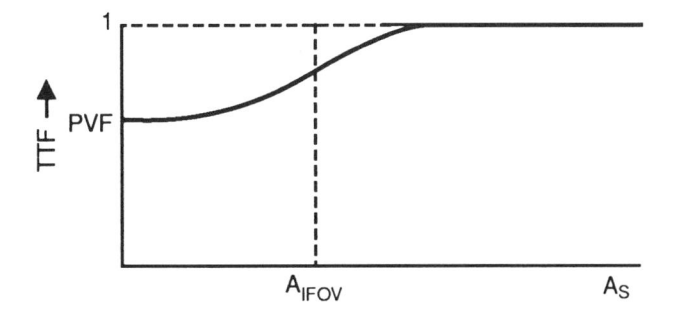

Figure 6-17: Target transfer function. The TTF is the ratio of system ATF to ideal ATF.

The SRF is illustrated in Figure 6-18. As the slit approaches zero, the system line spread function is obtained. Scanning systems tend to have different responses in the scan and cross scan directions and the SRF is different in the two directions. The SRF provides the measurement resolution and the imaging resolution.

178 TESTING & EVALUATION OF IR IMAGING SYSTEMS

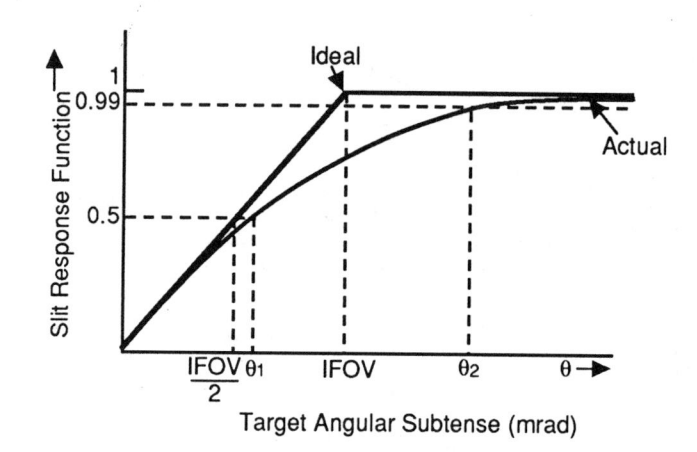

Figure 6-18: Slit response function. θ_1 is the imaging resolution and θ_2 is the measurement resolution.

When viewing a source in a collimator, the output ΔV_{SYS} is

$$\Delta V_{SYS} = \left[G \int_{\lambda_1}^{\lambda_2} R(\lambda) \frac{\Delta L_e(\lambda,T) A_o A_{IFOV}}{(fl_{COL})^2} T_{SYS}(\lambda) \, T_{TEST}(\lambda) \, d\lambda \right] ATF_{SYS} \qquad 6\text{-}7$$

where $\Delta L_e(\lambda,T) = L_e(\lambda,T_T) - L_e(\lambda,T_B)$ and $A_{IFOV} = A_d (fl_{COL}/fl_{SYS})^2$. The bracketed expression is the ATF normalization (i.e., normalized to flood illumination). For small temperature differences and when $A_S >> A_{IFOV}$ the bracketed expression is identical with Equation 6-1 (page 164). The target transfer function is:

$$TTF = \frac{ATF_{SYS}}{ATF_{IDEAL}} = ATF_{SYS} \left(\frac{A_{IFOV}}{A_S} \right) \qquad \text{when } A_s \leq A_{IFOV} \qquad 6\text{-}8a$$

$$= ATF_{SYS} \qquad \text{when } A_s > A_{IFOV} \qquad 6\text{-}8b$$

Substituting 6-8 into 6-7 yields:

$$\Delta V_{SYS} = \left[G \int_{\lambda_1}^{\lambda_2} R(\lambda) \frac{\Delta L_e(\lambda,T) A_o A_S}{(fl_{COL})^2} T_{SYS}(\lambda) T_{TEST}(\lambda) \, d\lambda \right] TTF \qquad 6\text{-}9$$

when $A_S \leq A_{IFOV}$

and

$$\Delta V_{SYS} = \left[G \int_{\lambda_1}^{\lambda_2} R(\lambda) \frac{\Delta L_e(\lambda,T) A_d}{4(f/\#)^2} T_{SYS}(\lambda) T_{TEST}(\lambda) \, d\lambda \right] TTF \qquad 6\text{-}10$$

when $A_S > A_{IFOV}$

As A_S approaches zero, it is convenient to represent $L_e A_S$ by the source radiant intensity I_e. Then Equation 6-9 becomes

$$\Delta V_{SYS} = \left[G \int_{\lambda_1}^{\lambda_2} R(\lambda) \frac{A_o \Delta I_e(\lambda,T)}{(fl_{COL})^2} T_{SYS}(\lambda) T_{TEST}(\lambda) \, d\lambda \right] PVF \qquad 6\text{-}11$$

when A_S is very small. This equation is the point source detection equation where fl_{COL} represents the distance from the source to the infrared imaging system.

Similar equations exist for the slit response function where the SRF is substituted for the ATF. When viewing a slit in a collimator, the output ΔV_{SYS} is

$$\Delta V_{SYS} = \left[G \int_{\lambda_1}^{\lambda_2} R(\lambda) \frac{\Delta L_e(\lambda,T) A_o A_{IFOV}}{(fl_{COL})^2} T_{SYS}(\lambda) T_{TEST}(\lambda) \, d\lambda \right] SRF \qquad 6\text{-}12$$

180 *TESTING & EVALUATION OF IR IMAGING SYSTEMS*

The test configuration for the ATF or SRF is shown in Figure 6-19. The peak output is recorded for each target size as shown in Figure 6-20.

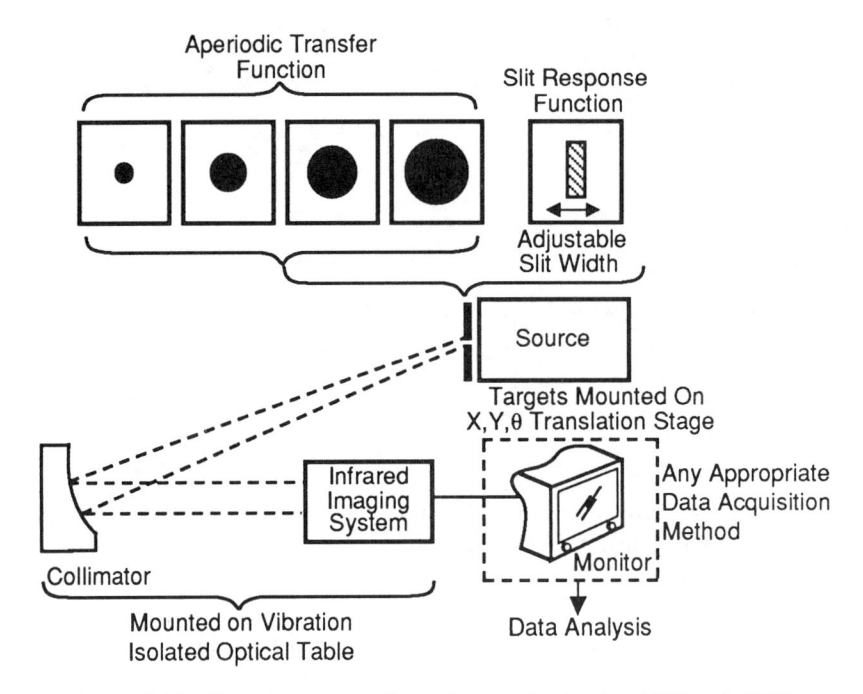

Figure 6-19: Generic test configuration to obtain the ATF and SRF. The blackbody, collimator and infrared imaging system should be placed on a vibration-stabilized optical table.

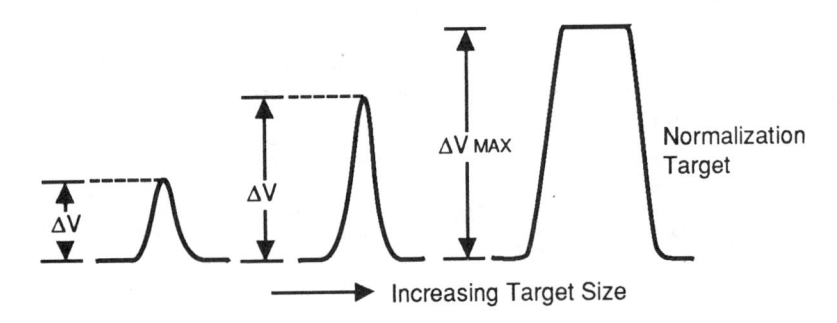

Figure 6-20. Output signal as a function of target size.

SYSTEM RESPONSIVITY 181

For the SRF, the target angular subtense should vary approximately from 0.1·IFOV to 5·IFOV. For the ATF, the target area should vary from 0.1 to 5 times the projected detector area. The small targets may be difficult to fabricate and this may limit the resultant data set. For the SRF, the slit height is unimportant but must be large enough so that there is no vertical edge effects. The slit height must cover more lines than that used in any line-to-line interpolation scheme. The slit must be aligned parallel to the array axis. Since this test depends on target size only, the source intensity is unimportant. It should be sufficiently high to provide a good signal-to-noise ratio without entering a nonlinear region.

However, as the target size decreases, the flux from the blackbody also decreases. A point is reached where the flux is below the detectable limit of the system. This represents a lower limit on the usable target size for a given source intensity. Then it is not possible to measure the ATF or SRF for any smaller targets. The ATF and SRF are normalized to unity by dividing all responses by the maximum output which is obtained by a large target. This value should be identical with the value obtained when performing the responsivity test with the same source temperature.

It is best to measure the responsivity function before measuring the ATF to determine the maximum level allowable without entering nonlinearity or saturation. To enhance the signal-to-noise ratio, the system noise can be reduced by reducing the system gain and the source intensity can then be increased. The visibility of detail is affected by both atmospheric turbulence and mechanical vibrations. Placing the test configuration in an enclosed chamber minimizes turbulence. Placing the configuration on a vibration-isolated optical table minimizes vibrational effects. How closely the system ATF follows the ideal ATF depends upon the relationship between the blur diameter and the detector size (Figure 3-8: page 70).

The ATF/SRF test procedure is given in Table 6-5. Possible causes for ATF/SRF variations are given in Table 6-6. The output voltage reaches a maximum value for a very large target. This value is used to normalize all other values so that $\Delta V/\Delta V_{MAX}$ is recorded as a function of target size (Figure 6-20). The most critical and most difficult part of the test is to center the target onto a detector. For staring arrays, alignment can be verified by viewing the output of adjoining detectors and verifying that the signals are equal (Figure 6-21). Alignment is more difficult when the fill factor is less than 100%. The shape of the output depends upon the detector array. With scanning systems, the detector electronics can produce an asymmetrical output (Figure 6-22a) whereas aring arrays usually produce a symmetrical output (Figure 6-22b).

Table 6-5
ATF/SRF TEST PROCEDURE

1. Establish a test philosophy, criteria for success and write a thorough test plan (Section 1.3: page 12).
2. Verify that the test equipment is in good condition and that the test configuration is appropriate (Figure 6-19: page 180). Ask previous users if any problems were noticed.
3. Verify the spectral response of the system and its relationship to the source characteristics, collimator spectral transmittance and atmospheric spectral transmittance (Section 4.3: page 105, and Section 4.4: page 112).
4. Turn off all automatic gain (AGC) circuits and fix gain at a predetermined value. If this is not possible, use an AGC target (Figure 4-15: page 105).
5. Insure that the infrared imaging system has reached operating equilibrium before proceeding.
6. Verify that the infrared imaging system is in focus (Section 5.2: page 137).
7. Select a source intensity that maximizes the signal-to-noise ratio without entering a nonlinear region or saturation. This can be obtained from the responsivity function (Table 6-3: page 175).
8. Critically align the target onto a detector to achieve maximum output.
9. Record the voltage difference between the target and its background.
10. Repeat steps 8 and 9 for each target size.
11. Divide all outputs by the maximum output. Maximum output is obtained when the largest target is used. This value should be the same as that obtained with the SiTF test target when the source is at the same temperature.
12. As appropriate, repeat steps 8-11 for different locations in the field-of-view.
13. Fully document any test abnormality and all results. The ATF/SRF should be presented graphically and in tabular form. As required, determine the PVF, imaging resolution or measurement resolution.

Table 6-6
CAUSES FOR VARIATIONS IN ATF/SRF

- Detector temperature variation (Figure 2-11: page 31).
- Ambient temperature changing during test (Section 3.1.5: page 73).
- Active AGC.
- Gamma correction present (Section 2.4.3 page 48).
- Errors in measuring the target size.
- Target not aligned onto a detector (Figure 6-21).

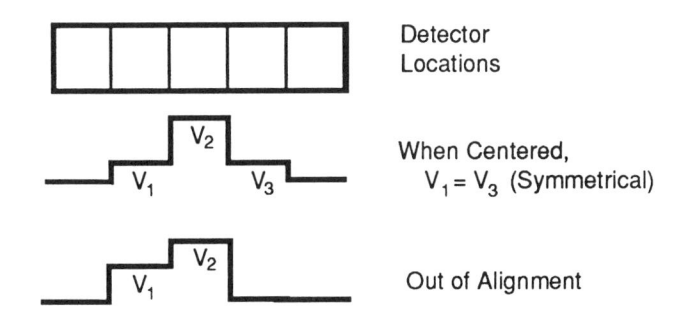

Figure 6-21: Correct alignment of a target for ATF and SRF measurements.

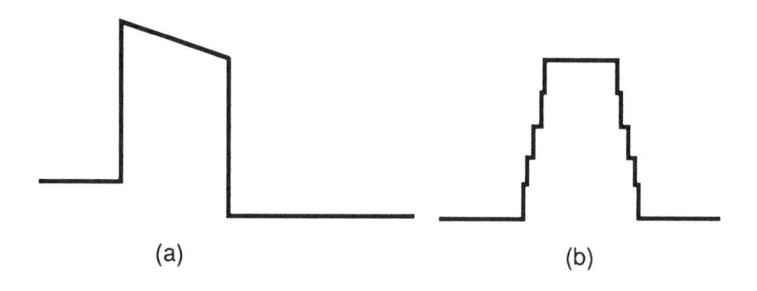

Figure 6-22: Possible video output signals. (a) Scanning system and (b) staring system.

6.3. DYNAMIC RANGE and LINEARITY

The responsivity function also provides dynamic range and linearity information. The system dynamic range is the maximum measurable input signal divided by the minimum measurable signal. The NEDT is assumed to be the minimum measurable signal. For AC coupled systems, the maximum available output depends on the target size and therefore the target size must be specified if dynamic range is a specification. Saturation is not well defined with systems that produce a S-shaped responsivity function. Depending upon the application, the maximum input value may be defined by one of four methods. In the first method, deviation from linearity is specified. A band that contains values within a desired percentage about the SiTF is defined. The range of data points that fall within the band is designated as the dynamic range. This band defines the acceptable linear region of the system responsivity. Figure 6-23 illustrates a 5% tolerance about the average value. When the signal level reaches some specified level, say as 90%, of full value is the second definition (Figure 6-24). The third is the intersection of the saturation value with the linear portion of the response function (Figure 6-24). The fourth is a specification on the minimum SiTF.

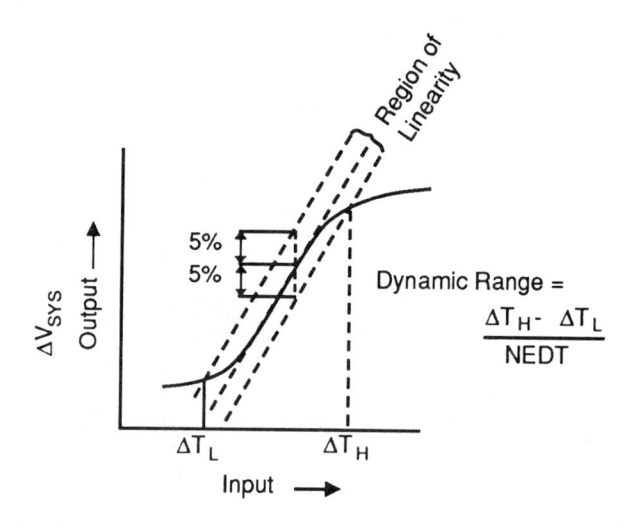

Figure 6-23: Dynamic range defined by linearity.

SYSTEM RESPONSIVITY   185

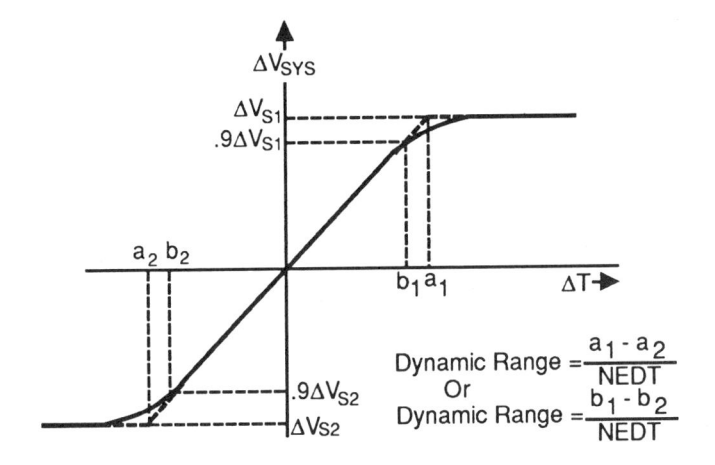

Figure 6-24: Dynamic range defined by saturation.

Example 6-3
DYNAMIC RANGE

The required dynamic range is 1000:1. If the NEDT is 0.05° C and the maximum output video voltage is 1 volt, what is the minimum SiTF required to achieve the required dynamic range?

The maximum signal before saturation is $(1000)(0.05) = 50°$ C. With a video output of 1 volt, the minimum SiTF is $1/50 = 20$ mv/°C. If the measured SiTF were greater, then the system would saturate with a smaller ΔT. However, it is assumed that the gain can be reduced so that the 1000:1 dynamic range can be met. The intersection of the saturation value with the linear portion of the response curve provides the same dynamic range as the minimum SITF approach. With this method, the entire responsivity curve need not be created but only the linear portion about ambient. From a practical view point, the slope method is easier to perform and it is independent of target size.

For most systems, the detector output is adjusted both in gain and level (offset) to maximize the dynamic range of the A/D converter. Figure 6-25 illustrates a system that contains an 8 bit A/D converter. The converter has an assumed input range between 0 and 1 volt and an output between 0 and 255 counts. By selecting the gain and offset, any detector voltage range can be

186 *TESTING & EVALUATION OF IR IMAGING SYSTEMS*

mapped into the digital output. Figure 6-26 illustrates 3 different gains and offsets. When the source flux level is less than ϕ_{min}, the source will not be seen (i.e., it will appear as 0 counts). When the flux level is greater than ϕ_{max}, the source will appear as 255 counts and the system is said to be saturated. ϕ_{min} and ϕ_{max} are redefined for each gain and level setting.

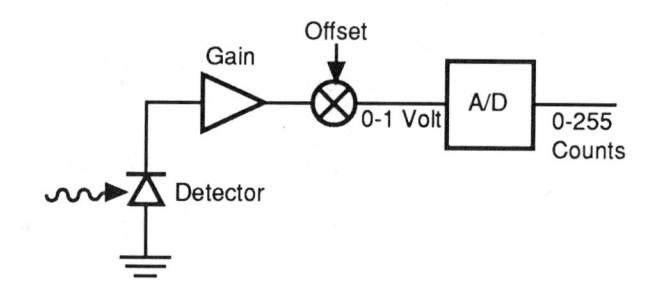

Figure 6-25: Generic block diagram. Depending upon the design, gain and level (offset) may be automatic or under manual control.

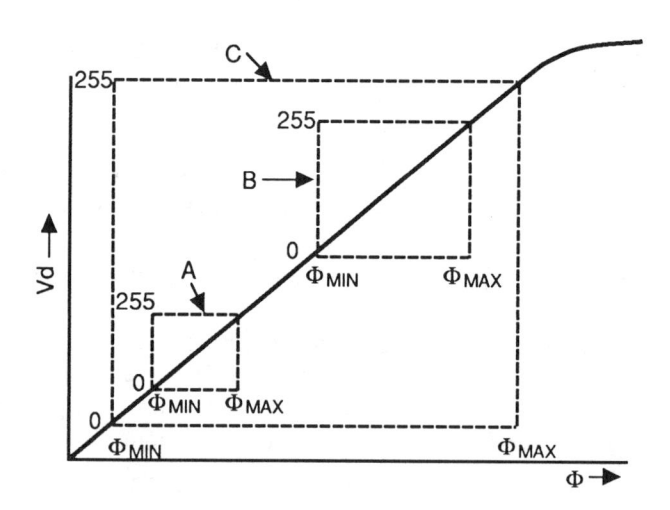

Figure 6-26: Different gain and levels affect the input-to-output transformation. Output A occurs with maximum gain. B occurs with moderate gain and C with minimum gain. For the 3 gains, the detector output is mapped into the full dynamic range of the A/D converter.

SYSTEM RESPONSIVITY 187

With a linear system, both the signal and noise increase with increasing gain. But the signal-to-noise and NEDT remain constant. However, saturation limits the maximum signal. When dynamic range is the input range divided by the NEDT, increasing the gain decreases the dynamic range (Figure 6-27). Dynamic range can be extended at the expense of the minimum measurable signal at any particular gain setting. For example, the NEDT is usually measured at maximum gain (range A in Figure 6-26). The system range can be changed and then the smallest measurable signal may be an LSB (range C in Figure 6-26). The NEDT may be measured on a different gain setting than that used to define the system total dynamic range. Figure 6-28 illustrates the difference between instantaneous dynamic range and system total dynamic range for a system with a fixed gain and variable offset.

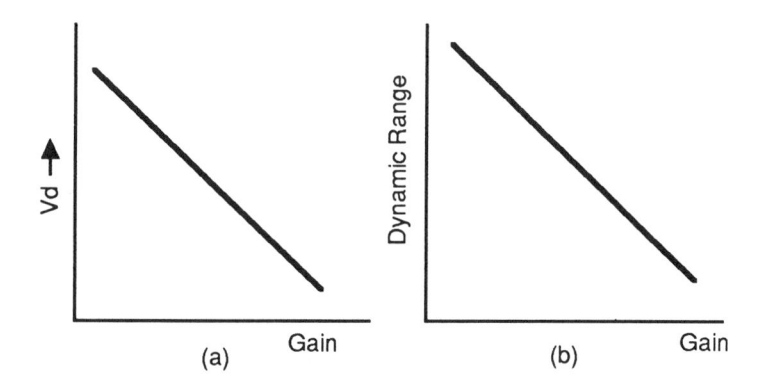

Figure 6-27: Gain reduces the dynamic range. (a) The A/D converter will saturate with lower input ΔTs as the gain increases and (b) the dynamic range as a function of gain. The gain must be specified when the dynamic range is a specification.

188 TESTING & EVALUATION OF IR IMAGING SYSTEMS

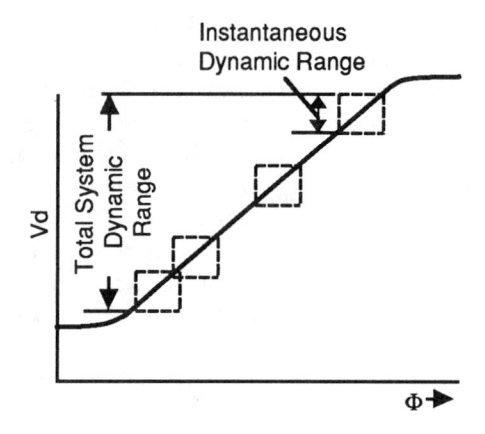

Figure 6-28: With fixed gain, the instantaneous dynamic range remains constant but the total system dynamic range can be extended by changing the level.

6.4. SUMMARY

The test engineer must thoroughly understand the infrared imaging system operation before performing system responsivity measurements. AC coupled, AC coupled with DC restoration and DC coupled systems require different data analysis methodologies. As new detector technologies emerge, it may be necessary to modify the test procedures.

The input-to-output transformations can be presented in a variety of ways (Table 6-7). When the target size is fixed and the target intensity is varied, the responsivity function is obtained. The slope of the linear portion is the SiTF. To increase the accuracy of the measurements, the student t-test can be used to determine the number of data points required. A least-squares fit to the data provides the best estimate of the SiTF. SiTF, by itself, is not a very good metric for comparing different systems because it changes with gain. The SiTF is

$$SiTF = \frac{\Delta V_{SYS}}{\Delta T} = G \int_{\lambda_1}^{\lambda_2} R(\lambda) \frac{A_d}{4(f/\#)^2} \frac{\partial M_e(\lambda, T_B)}{\partial T} T_{SYS}(\lambda) T_{TEST}(\lambda) d\lambda \qquad 6\text{-}13$$

Although the output is shown as volts, it can be any arbitrary unit such as ADU or monitor luminance.

Table 6-7
SYSTEM RESPONSIVITY TESTS

TEST	VARIABLE	TEST RESULTS
Responsivity function	Target intensity	SiTF Dynamic range Saturation
ATF	Target area	Target transfer function Point visibility function
SRF	Target width	Imaging resolution Measurement resolution

The ATF and SRF are obtained by varying the target size and keeping the target intensity constant. The point visibility factor is obtained from the ATF and the SRF provides the imaging resolution and measurement resolution. The PVF can be obtained experimentally by noting where the target transfer function asymptotes to a constant value or by the method illustrated in Example 5-1 (page 145). The TTF approach provides multiple data points and thereby increases the confidence that the PVF has been accurately obtained. For SRF and ATF measurements, the target must be critically aligned to the detector. The system response to point sources is a primary specification for IRST systems. The responsivity, $\Delta V_{SYS}/\Delta E_e$ is

$$\frac{\Delta V_{SYS}}{\Delta E_e} = \frac{\left[G \int_{\lambda_1}^{\lambda_2} R(\lambda) A_o \, \Delta I_e(\lambda,T) \, T_{SYS}(\lambda) T_{TEST}(\lambda) \, d\lambda \right] PVF}{\int_{\lambda_1}^{\lambda_2} \Delta I_e(\lambda,T) \, T_{TEST}(\lambda) \, d\lambda} \qquad 6\text{-}14$$

Although not often stated explicitly, the inputs and outputs are differential values.

Dynamic range is the maximum measurable value divided by the minimum measurable value. The NEDT is taken as the minimum value. The instantaneous dynamic range may be different from the total system dynamic range. The NEDT may be measured at a different gain setting than the total dynamic range.

Finally, specifications should be both understandable and testable (Table 6-8).

Table 6-8
TYPICAL SPECIFICATIONS

- The SiTF (in the center of the FOV) shall be greater than 0.2 V/deg when the background temperature is 20°C.
- For AC coupled scanning systems: The dynamic range shall be greater than 50 db when the target is equal to 20% of the field-of-view.
- For DC coupled and staring systems: The instantaneous dynamic range shall be 4000:1.
- The imaging resolution shall not be greater than 0.5 mrad in the center of the FOV.
- The measurement resolution shall not be greater than 2 mrad in the center of the FOV.
- The point visibility factor shall be greater than 0.80 (measured in the center of the FOV).
- The SiTF for each analog video line shall be within 5% of the average SiTF when operating at minimum gain and at maximum gain.

6.5. REFERENCES

1. MIL-STD-1859, "Thermal Imaging Devices, Performance Parameters of", 15 Sept. 1981.
2. P. A. Bell and C. W. Hoover, Jr., "Standard NETD Test Procedure for FLIR Systems with Video Outputs", in *Infrared Imaging Systems: Design, Analysis, Modeling and Testing IV*, G. C. Holst, ed.: SPIE Proceedings Vol. 1969, pp 194-205 (1993).
3. R. E. Flannery and J. E. Miller, "Status of Uncooled Infrared Imagers", in *Infrared Imaging Systems: Design, Analysis, Modeling and Testing III*, G. C. Holst, ed.: SPIE Proceedings Vol. 1689, pp. 379-395 (1992).

EXERCISES

1. The minimum detectable temperature (MDT) is MDT = K NEDT/ATF. Sketch the MDT assuming an ideal response and an actual ATF as a function of target *diameter*.
2. Sketch the output voltage (target and background) as the level is changed for both an AC and DC coupled system.
3. The following [ΔV_{SYS}, ΔT] data values were obtained during a responsivity function test: (-50,-6), (-50,-5), (-49,-4.5), (-48,-4), (-30,-3), (-20,-1.8), (18, 2), (36,4), (58,6), (62,7), (68,8), (68,10). The voltages are in units of mv and the temperature in Celsius. Plot the responsivity function and calculate the SiTF. Draw the SiTF on same graph. What are some causes for V_{OFFSET}? If the NEDT is 0.2 C, what is the dynamic range using the three definitions given in the text. What would cause the data values to be asymmetrical about the origin?
4. The maximum ATF is obtained when the target image is centered on a detector. Sketch ΔV_{SYS} and the ATF if the image is centered on 2 adjoining detectors. Sketch the output if the image is at the corner of a detector such that it is centered on 4 contiguous detectors. Compare these results with those when the source is centered on a detector.
5. ATF measurements are defined for the case where the source image is centered onto a detector. The equations imply that there is 100% fill factor. Sketch the output if the fill factor was 50% and 10%.
6. An error is made during a SiTF test. The system is AC coupled system and the measured ΔV was the peak voltage on line 2 and the background was on line 1 of Figure 6-4 (page 163). Sketch the output (target, background and difference) with the error and compare that to the correct output. Sketch the output for a very small target and a target that covers 50% of the field-of-view.
7. The entire ATF cannot be obtained because there is insufficient flux available to make the measurements. The test engineer increases the target temperature to increase the flux during the experiment. Discuss the problems with this approach. (Hint: consider normalization issues).

7

SYSTEM NOISE

The noise analysis approach taken follows the three-dimensional noise model developed by D'Agostino and Webb[1]. The noise is divided into a set of eight components that relate temporal and spatial noise to a three-dimensional coordinate system. The three-dimensional approach allows full characterization of all noise sources including random noise, fixed pattern noise, streaks, rain, 1/f noise, and any other artifact that may have been introduced. Analyzing the noise in this manner has the advantage of simplifying the understanding of a complex phenomenon by breaking it down into a manageable set of components. Furthermore, analysis provides insight to possible hardware and software factors that may be responsible for the noise. For the system performance analyst, the method simplifies the incorporation of complex noise factors into model formulations[2].

Usually, both the temporal and spatial noise data are collected simultaneously. The data analysis technique separates the noise into the eight components. Each component has its own noise power spectral density* (NSPD). The NSPD provides information on the spectral content of the noise. Spectral peaks in the noise will affect observer detection of targets whose spatial features are the same as the noise spectral features (See Chapter 10). By using the system SiTF the measured noise can be referred to the system input to obtain the noise equivalent differential temperature (NEDT). NEDT is that input that produces a signal-to-noise of unity for flood illumination. NEDT is taken as the smallest measurable signal and is used to define the dynamic range. NEDT is also called noise equivalent temperature (NET) and noise equivalent temperature differential (NETD). For point source detection, the lowest measurable signal is the noise equivalent flux density (NEFD) or, equivalently, the noise equivalent irradiance (NEI). The NEDT and NEFD are only two performance parameters of many and are a measure of system sensitivity only. They are excellent diagnostic tools for production testing to verify performance. They are poor system-to-system comparison parameters and should be used cautiously when comparing systems built to different designs because they are a function of spectral responsivity and the noise equivalent electronic bandwidth. These laboratory metrics are measured in *rms* units but the units are seldom used. Similarly, the terms *difference* or *differential* are often omitted since it is understood that the system noise is referred to a differential temperature.

* Although commonly called spectral density, *it is a function of spatial frequency.*

In principle, all noise sources, except random noise, can be eliminated or at least be reduced below a measurable value. When random noise is due solely to the random events associated with photon detection, the system is said to have background limited performance (BLIP). Assuming photon events can be described by Poisson statistics, when the system is background limited, the noise variance is equal to the number of photon present. As the background flux changes, the noise variance changes proportionally. The rms noise value is the square root of the variance. Therefore, as the background changes, the rms noise changes by the square root of the average value. The mean-variance technique exploits this fact. It provides additional information about the system such as dynamic gain, the presence of other noise sources and saturation.

7.1. NOISE STATISTICS

The underlying noise probability density function must be known so that the appropriate statistics are used. Subconsciously, assumptions are made about the noise data distribution when calculating the standard deviation (also called the 1-sigma value or rms value). It is these underlying assumptions that must be routinely verified. Figure 7-1 illustrates a randomly varying voltage whose distribution of values can be described by a Gaussian probability distribution:

$$P(v) = \frac{1}{\sqrt{2\pi}\sigma_o} e^{-\frac{1}{2}\left(\frac{v-\mu}{\sigma_o}\right)^2} \qquad 7\text{-}1$$

where σ_o^2 is the population variance and μ is the population mean.

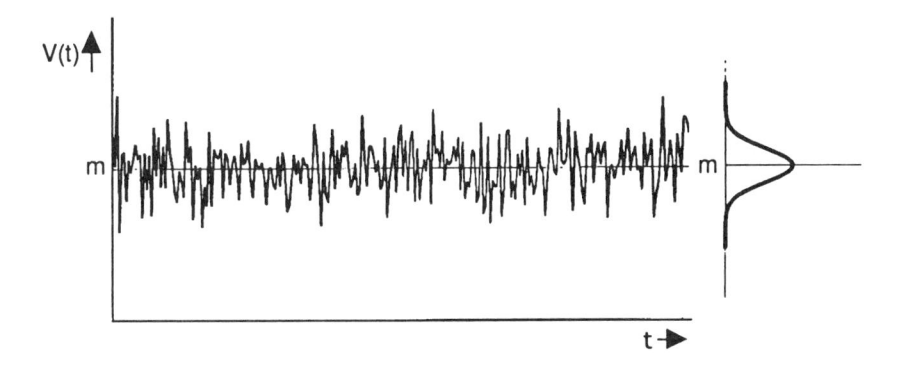

Figure 7-1: Gaussian noise. The probability density function describes the signal fluctuations about a mean value, m.

It is not possible to perform exhaustive testing, rather an inference is made about the population mean and variance based upon a finite data set. The measured variance, s^2, is an estimate the true variance, σ_o^2:

$$s^2 = \frac{N_e \sum_{i=1}^{N_e} v_i^2 - \left(\sum_{i=1}^{N_e} v_i\right)^2}{N_e(N_e-1)} \qquad 7\text{-}2$$

where there are N_e elements in the data set. The measured mean, m, is an estimate of the true mean, μ:

$$m = \frac{\sum_{i=1}^{N_e} v_i}{N_e} \qquad 7\text{-}3$$

Gaussian (or normal) distributions appear as straight lines when plotted on normal-probability graph paper[3]. The chi-squared goodness of fit methodology also determines if a sample data set follows a Gaussian distribution[4]. While the normal-probability graph or chi-squared test provides a rigorous test for normality, a *quick look* may be adequate: fitting a Gaussian curve to the data values. Figure 7-2 illustrates the probability density function of a typical noise trace. Overlaid is the Gaussian distribution where the mean and standard deviation were calculated from Equations 7-2 and 7-3. If the distribution is not Gaussian, further description is necessary for a complete characterization of the noise. Furthermore, if the data is not Gaussian distributed, the three-dimensional noise model must be reevaluated for the specific noise statistics present.

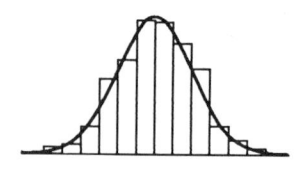

Figure 7-2: Histogram of noise data with an overlaid Gaussian curve. The data appears to be Gaussian in nature.

SYSTEM NOISE 195

From a statistical point of view, if the variance is measured k times, each measurement will provide a different result because a finite data set was analyzed each time. The best estimate of the variance is the average of the measured variances:

$$s_{ave}^2 = \frac{s_1^2 + s_2^2 + \ldots + s_k^2}{k} \qquad 7\text{-}4$$

where it is assumed that the same number of data points was used to calculate each s_i. The best estimate of the standard deviation is:

$$\sigma = \sqrt{s_{ave}^2} = \sqrt{\frac{s_1^2 + \ldots + s_k^2}{k}} \qquad 7\text{-}5$$

The average rms noise is the square root of the average variance.

7.2. THREE-DIMENSIONAL NOISE MODEL

D'Agostino and Webb[1] developed a three-dimensional noise model (Figure 7-3). They created the D_i directional averaging operators that allow the mathematical derivation of eight noise components from the noise data set. The operators simply average the data in the direction indicated by their subscripts. The D-operators will not be used in this text but the results will be used to describe the various noise sources.

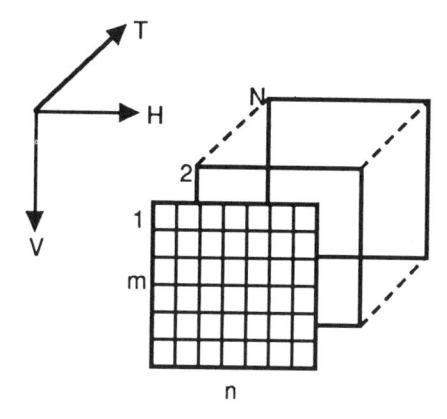

Figure 7-3: Three-dimensional noise model coordinate system illustrating data set N_{TVH}.

196 *TESTING & EVALUATION OF IR IMAGING SYSTEMS*

The T-dimension is the temporal dimension representing the framing sequence. The other two dimensions provide spatial information. However, depending upon the infrared imaging system design, the horizontal dimension may represent time for a scanning system or may represent space for a staring system. For a staring array, m and n indicate detector locations. For parallel scanning systems, m indicates detector locations and n is the digitized analog signal. m is the number of raster lines and n is the digitized analog signal for serial scanning systems. The number of data elements for each component is given in Table 7-1 and illustrated in Figure 7-4. The standard deviation, σ_i, of the data set N_i is the rms noise value for that data set. The noise components are *bidirectional* if the calculated variance is independent of the direction chosen to perform the calculation. That is, the variances will be the same whether measured on a line, column, globally or frame-to-frame. Systems which exhibit this behavior are said to be ergodic[5].

Table 7-1
NUMBER OF ELEMENTS IN EACH NOISE COMPONENT

3-D NOISE COMPONENT	NUMBER OF ELEMENTS, N_e
N_{TVH}	m x n x N
N_{VH}	m x n
N_{TH}	n x N
N_{TV}	m x N
N_V	m
N_H	n
N_T	N
S	1

SYSTEM NOISE 197

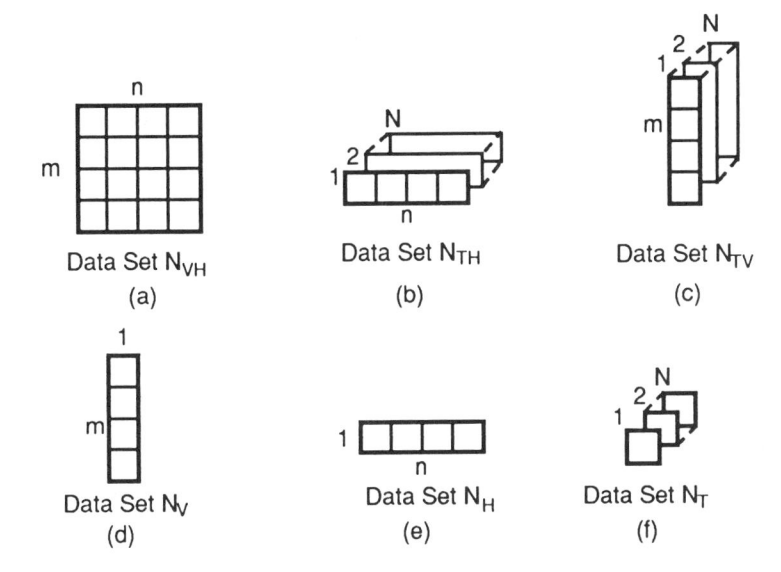

Figure 7-4: Data sets for the three-dimensional noise model.
(a) Data set N_{VH}: each pixel is averaged over N frames
(b) Data set N_{TH}: each column is averaged over m pixels
(c) Data set N_{TV}: each row is averaged over n pixels
(d) Data set N_V: each row is averaged over n pixels and N frames
(e) Data set N_H: each column is averaged over m pixels and N frames
(f) Data set N_T: each frame is averaged over m x n pixels.

Table 7-2 lists seven noise components and some possible contributors to the components for serial scanning, parallel scanning and staring array imaging systems. For mathematical completeness, the noise model has eight components with the eighth being the global average value, S. Depending upon the system design and operation, any one of these noise components could dominate. The origin of these components is significantly different and the existence and manifestation depend upon the specific design of the infrared imaging system. Not all of the components may be present in every infrared imaging system. Certain noise sources such as microphonics are more difficult to describe since they may appear in variety of forms. "Readout electronics" is a catchall phrase for possible staring array artifacts. The 3-D noise model provides the basic framework for analyzing the various noise sources. Depending upon the system design and operation, the same noise source may appear in different noise components. The magnitude of each measured noise value depends, in part, on the size of the 3-dimensional volume chosen.

Table 7-2
SEVEN NOISE COMPONENTS OF THE 3-D NOISE MODEL

3-D NOISE COMPONENT	DESCRIPTION	SERIAL SCAN	PARALLEL SCAN	STARING ARRAY
N_{TVH}	Random 3-D noise	Random and 1/f noise	Random and 1/f noise	Random
N_{VH}	Spatial noise that does not change from frame-to-frame	nonuniformity	nonuniformity	FPN, nonuniformity
N_{TH}	Variations in column averages that change from frame-to-frame (rain)	-	EMI*	Read out electronics
N_{TV}	Variations in row averages that change from frame-to-frame (streaking)	EMI*	Transients (flashing detectors), 1/f noise	Read out electronics
N_V	Variations in row averages that are fixed in time (horizontal lines or bands)	Line-to-line interpolation	Detector gain/level variations, line-to-line interpolation	Read out electronics, line-to-line interpolation
N_H	Variations in column averages that are fixed in time (vertical lines)	-	-	Read out electronics
N_T	Frame-to-frame intensity variations (flicker)	Frame processing	Frame processing	Frame processing

*electromagnetic interference

The most widely used noise metrics are the NEDT, nonuniformity and FPN. NEDT is the high frequency component of random noise and 1/f noise as the low frequency noise component of N_{TVH}. Fixed pattern noise (high frequency spatial noise) and nonuniformity (low frequency spatial noise) are components of N_{VH} (Table 7-3). "High" and "low" will be defined later. The low frequency components tend to be annoying whereas the high frequency components can significantly affect the ability to resolve detail. High frequency noise interferes with the basic operation of the system and is the factor most often called the system sensitivity. Many performance models predict the high frequency noise component rms value and call this value the *system* NEDT.

Table 7-3
NEDT, FPN AND NONUNIFORMITY

3-D NOISE COMPONENT	FREQUENCY COMPONENT	SERIAL SCAN	PARALLEL SCAN	STARING ARRAY
N_{TVH}	high	NEDT	NEDT	NEDT
	low	1/f	1/f	-
N_{VH}	high	-	-	FPN
	low	non-uniformity	non-uniformity	non-uniformity

There are four known sources of spatial noise: detector responsivity nonlinearities, variations in detector spectral response, detector 1/f noise, and array 1/f noise[6,7]. *Fixed pattern* noise refers to any pattern that does not change significantly from frame-to-frame. Detector responsivity and spectral response usually do not change with time and the manifestation of the variation will be a pattern that rarely changes. However, if the detector temperature changes, the responsivity also changes (Figure 2-11: page 31). 1/f noise is a low frequency phenomenon that changes slowly with time. Since each detector has different 1/f noise characteristics, these low frequency components appear as different DC offsets. The different offsets appear as fixed pattern noise. Since each detector is independent, the 1/f drift will be different and the *fixed pattern* noise will change with time. A completely random nonuniform spatial pattern changes slowly with time and this *spatial* noise increases monotonically as a function of time after normalization (Figure 2-30: page 48). Although 1/f noise is low frequency temporal noise, its appearance depends upon the system design and operation. 1/f noise may appear as streaks in scanning systems[8,9]. The line-to-line variations or streaks may be considered unacceptable for cosmetic reasons or may interfere with system usage. Since each line may have a different average value, it appears in N_{TV}. FPN is not normally associated with scanning systems, but the variations in the vertical direction are a form of FPN. For staring arrays, the measured noise depends on the time of the last gain/level calibration. After random noise, the next prevalent[2] noise components for parallel scan systems are σ_{TV} and σ_V. For staring arrays, the next most important noise source is σ_{VH}.

Although nonuniformity was originally a measure of optical defects (shading, mottling, and blemishes), it now includes all effects from any cause that produce a nonuniform output (e.g., it includes narcissus and scan noise). Nonuniformity is sometimes considered a cosmetic defect. It may be more apparent visually for some gain settings or background intensities than others. It may not be apparent at laboratory ambient temperatures but may become obvious in the field when viewing bland scenes or the cold sky. Narcissus may be diminished by matching the blackbody temperature to the equivalent value of the narcissus signal. Narcissus and shading are always present and will be seen if the system gain is sufficiently high and if the appropriate background temperature is selected.

7.3. NOISE MEASUREMENTS

It is the data analysis techniques that allow separation of the various temporal and spatial components. Scanning and staring systems require different approaches. The measurement technique determines the lowest frequency that can be obtained. Usually a line of video is stripped out. The active line time for RS 170 is approximately 52 μsec (Figure 7-5).

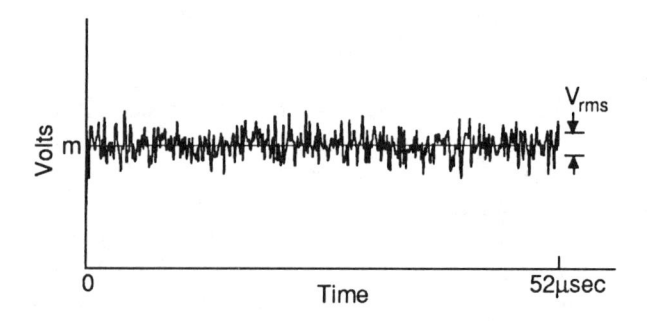

Figure 7-5: Single RS 170 analog video line illustrating high frequency temporal noise.

SYSTEM NOISE 201

For scanning systems, the lowest frequency that can be accurately measured is f = 1/(2·52μs) ≈ 10 kHz. Thus a frequency much lower than this appears as DC offset (Figure 7-6) and any frequency somewhat less appears as monotonically increasing or decreasing levels (Figure 7-7). This occurs because the selected 3-dimensional noise volume is finite. The measured frequency response is limited by the finite data set. The NEDT is the "high" frequency noise and is V_{rms} in Figure 7-5. The inclusion of low frequency components increases the calculated variance.

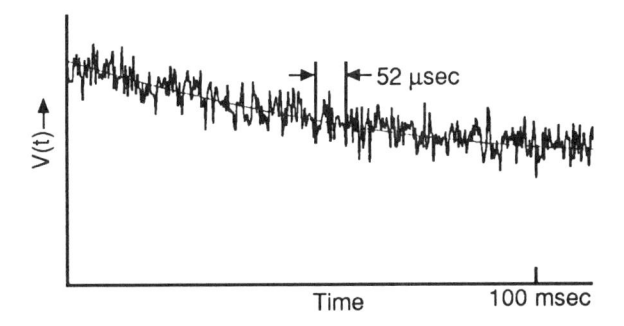

Figure 7-6: Output of a single detector with 1/f noise. During an active RS 170 line, the 1/f noise appears as a DC offset.

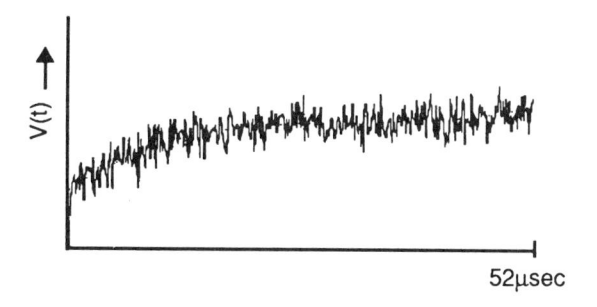

Figure 7-7: Single RS 170 analog video line illustrating mid spatial frequency 1/f noise.

202 TESTING & EVALUATION OF IR IMAGING SYSTEMS

Figure 7-8 illustrates 2 noise traces with excessive low frequency components. The total system noise, which includes all noise components, is significantly larger than the NEDT and portrays a different picture of the system sensitivity. That is, the calculated variance is significantly affected by any large low frequency excursions and can be affected by large nonuniformities.

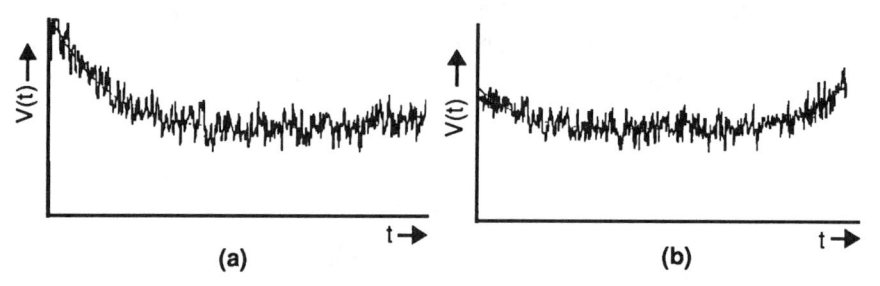

Figure 7-8: Different noise traces that have the same high frequency components but different low frequency components due to 1/f noise and nonuniformity. The high frequency noise is the same for both traces but the overall noise is different.

The removal of low frequency components from the high frequency signals is called trend removal[10]. With appropriate trend removal, the traces in Figure 7-8 will yield the same high frequency noise value as Figure 7-5. Trends can be removed by passing one- or two-dimensional filter over the data set. Trends usually can be described by a 2nd order polynomial[11] ($y = a_o + a_1 x + a_2 x^2$). Subtraction of the underlying polynomial (Figure 7-9) provides the high frequency data set. The polynomial is applied first to each line (horizontal trend removal) and then vertically to each pixel column (vertical trend removal). The constant, a_o, removes the average value in each line and each column and provides the same result as the D_i directional averaging operators.

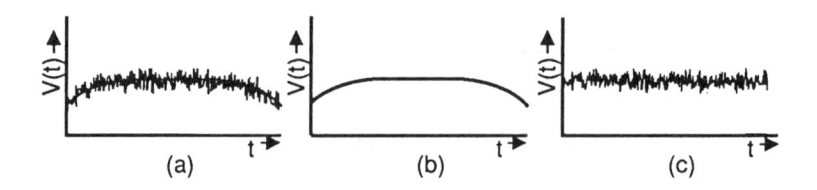

Figure 7-9: Removal of low frequency components with a 2nd order polynomial approximation. (a) Total system noise, (b) 2nd order polynomial that fits the low frequency components, and (c) subtraction of (b) from (a) establishes the high frequency components.

SYSTEM NOISE 203

For staring arrays, the output of each detector will vary with time and this variation is seen frame-to-frame. In one frame, the output can vary as shown in Figure 7-10. Fixed pattern noise occurs when the average value of each detector is different. Gain/level correction can, in principle, remove FPN. However, due to 1/f noise, the average value of each detector will drift. The amount of FPN depends, in part, on how often correction is applied (i.e., every frame or only once when the system is turned on).

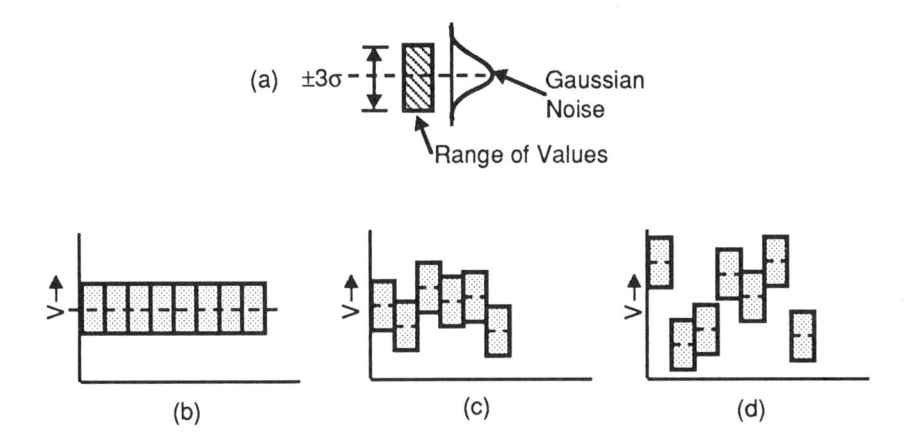

Figure 7-10: Noise in a staring array. (a) Single detector range of output values based upon Gaussian statistics, (b) output range of values for 8 detectors when no FPN is present, (c) output range of values when FPN is barely perceptible, and (d) output range of values when FPN is about 5 times NEDT.

Low frequency noise is any noise that has frequency components less than 150 kHz. High frequency noise is noise that has components above 150 kHz when referred to the standard RS 170 video format[12]. Figure 7-11 illustrates the ideal electronic filters required to separate the noise sources. These filters can either be implemented in hardware or simulated in software[13]. Since RS 170 has a bandwidth of approximately 5 MHz, the low frequency cutoff represents about 3% of the noise bandwidth. For other video formats, the low frequency cutoff should be 3% of the bandwidth. The high pass filter will remove any DC offset or trends in the data. Since the rms noise is proportional to the square root of the noise bandwidth, the measured high frequency noise rms value will be 98.5% of the total system rms value when using an ideal filter. As an alternative, a 2nd order polynomial can be subtracted from the data to obtain the high frequency components.

```
              │‾‾‾‾‾‾‾‾|                                      _____
              │        |                                      |
_____|        |_____              _____ |
              150 KHz          Hz →                  150 KHz         Hz →
                 (a)                                     (b)
```

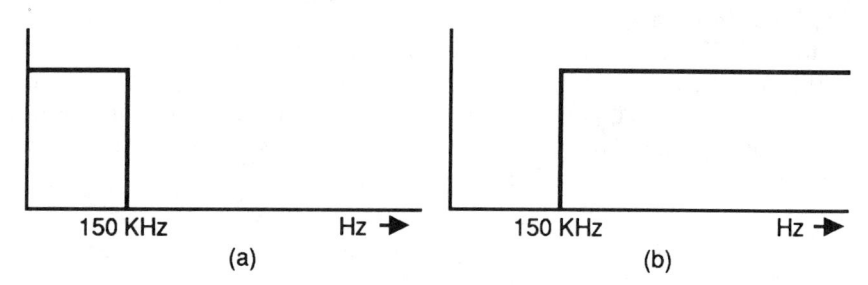

Figure 7-11: Ideal filters for separating the high frequency components from the low frequency components. (a) Filter for nonuniformity and (b) filter for NEDT and FPN. The 150 kHz filter is appropriate for RS 170 analog video.

Figure 7-12 illustrates the generic test configuration used to measure all noise sources. The data can be collected by any of the methods shown in Section 4.6 (page 115). The detectors should be uniformly illuminated (flood illumination) by either placing a nonreflective cloth over the infrared imaging system or having the system view a large blackbody source. The source or cloth must cover the entire area that the detector can sense. To insure that any nonuniformity seen is not due to the source, the source should be moved to verify that the image variations do not move. If the image does move, then it is due to imperfections of the source. The blackbody source permits noise measurements for a variety of background temperatures. This provides additional information about the system characteristics and permits the determination of FPN variations. The same system gain must be used for the responsivity and the noise measurements. If the system has an active AGC, it may be necessary to use an AGC target (Figure 4-15: page 105)

Many frames of data must be collected to measure all eight noise components. This requires image storage capability. If the system does not exhibit any fixed pattern noise and the noise is truly bidirectional, a transient digitizer may be used to examine each analog video line sequentially: the test time (i.e., the time to scroll through the image) does not become an issue.

Noise on a single analog video line can be *crudely* estimated by examining the line trace on an oscilloscope (Figure 4-28: page 118). The standard deviation is estimated as (peak-to-peak)/6. The value of 6 comes from Gaussian distribution considerations in which 99.74% of the data falls within $\pm 3\sigma$ of the mean. It is difficult to visually estimate the high frequency values when trends are present.

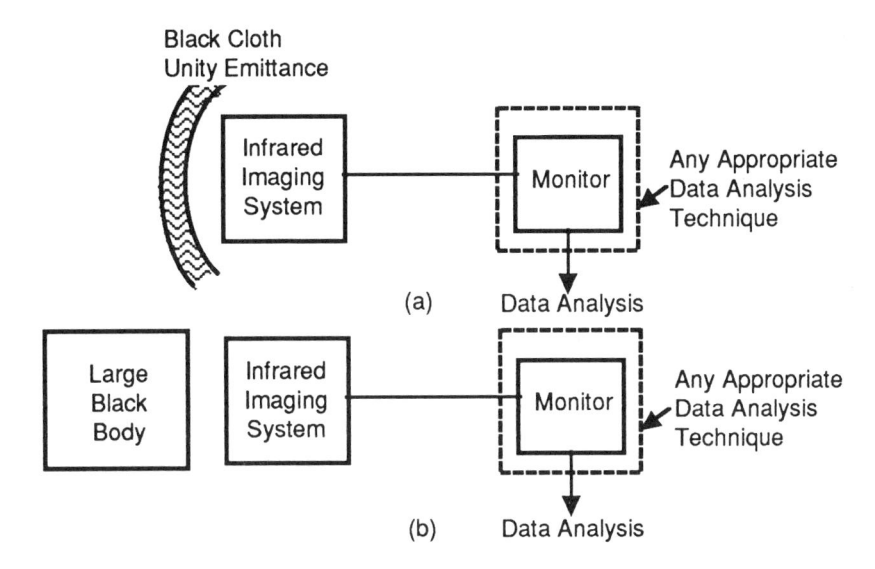

Figure 7-12: Typical configurations for measuring noise components. (a) Flood illumination produced by an opaque cloth with unity emittance and (b) large blackbody that fills the field-of-view.

7.3.1. NOISE EQUIVALENT DIFFERENTIAL TEMPERATURE

Historically[14], NETD was a measure of detector performance and was defined at the output of the common module post amplifiers. For this classical measurement, a simple single pole filter was added whose 3 db cutoff frequency is equal to the reciprocal of twice the detector dwell time (Figure 7-13). The filter, purportedly, approximated the response of subsequent electronics. NEDT has grown into a metric of *system* sensitivity. For system measurements, the NETD is defined at the analog video or the output of the monitor. When measuring the system NETD, the external 3 db filter should *not* be used. FLIR92 maintains the historical definition[2] and calls the system noise σ_{TVH} to indicate that the system noise is more complex than that originally perceived when the simple RC circuit was introduced. However, in keeping with popular parlance, from now on, NEDT will refer to the high frequency component of σ_{TVH}.

206 TESTING & EVALUATION OF IR IMAGING SYSTEMS

$f_{3db} = \dfrac{1}{2\tau_d}$ τ_d = Detector Dwell Time

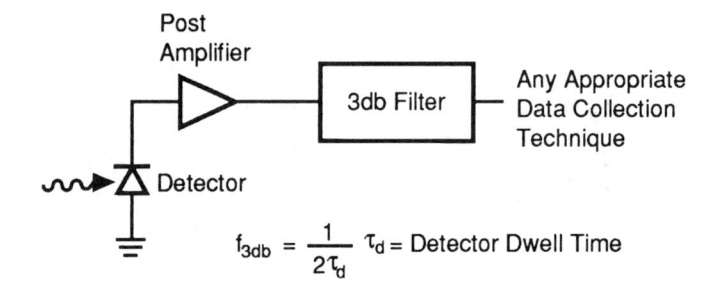

Figure 7-13: The 3 db filter. The filter should *NOT* be used for system measurements.

It is the data collection and analysis techniques that separates the various temporal and spatial components. By placing the 150 kHz filter on the analog RS 170 output, the data is filtered in the horizontal direction only. The high frequency temporal components can be separated from the spatial components via a frame subtraction technique (Figure 7-14). The temporal components are reduced by the square root of the number of frames averaged whereas the fixed pattern components are not. By averaging 100 frames, the temporal noise is reduced 10 fold. This averaged frame, containing fixed pattern noise, is then subtracted from a single frame to leave only the temporal variations. This subtraction method produces one frame of the data set, N_{TVH}. The complete data set contains N frames. The trends in the vertical direction can be removed via a 2nd order polynomial fit to the data set (column by column). The rms value can then be calculated for each line, each column or globally.

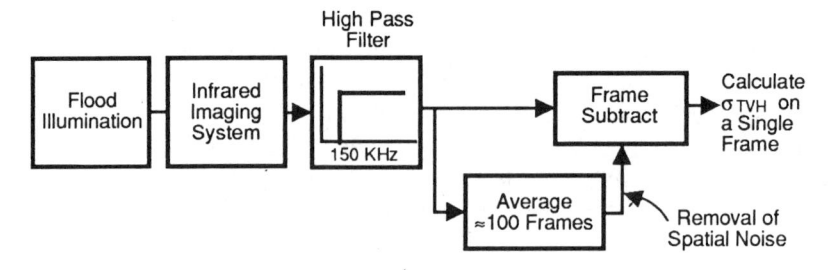

Figure 7-14: Experimental configuration to determine σ_{TVH}. Averaging 100 frames reduces the temporal noise and leaves only the spatial noise. The frame subtraction removes the spatial noise from the current frame.

SYSTEM NOISE 207

If the noise is truly bidirectional (i.e., $\sigma_{TH} = \sigma_{VH} = \sigma_{TV} = \sigma_V = \sigma_H = \sigma_T = 0$), then a line grabber or transient recorder can be used to measure σ_{TVH} line-to-line (Figure 7-15).

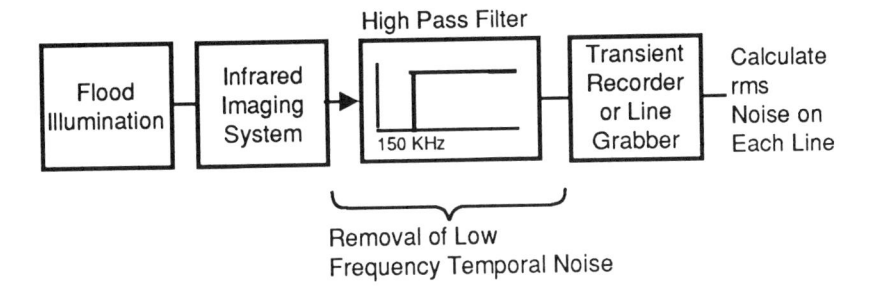

Figure 7-15: Experimental configuration to determine σ_{TVH} when $\sigma_{TH} = \sigma_{VH} = \sigma_{TV} = \sigma_V = \sigma_H = \sigma_T = 0$. Instead of using the high pass filter, a 2nd order polynomial can be used to remove the low frequency components.

The NEDT is obtained from the measured noise and the SiTF:

$$NEDT = \frac{\overline{V_{RMS}}}{\left(\frac{\Delta V_{SYS}}{\Delta T}\right)} = \left(\frac{\overline{\sigma_{TVH}}}{SiTF}\right)_{trends\ removed} \qquad 7\text{-}6$$

Substituting the theoretical value for the SiTF (Equation 6-1: page 164) into Equation 7-6 and assuming that the system gain, G, applies to both V_{rms} and SiTF yields:

$$NEDT = \frac{\pi V_{RMS}}{\alpha \beta A_o \int_{\lambda_1}^{\lambda_2} R(\lambda) \frac{\partial M_e(\lambda,T)}{\partial T} T_{SYS}(\lambda)\, d\lambda} \qquad 7\text{-}7$$

where the detector angular subtenses α and β are equal to $\alpha\beta = A_d/(fl_{SYS})^2$.

208 TESTING & EVALUATION OF IR IMAGING SYSTEMS

As the background temperature increases, the thermal derivative increases (Figure 3-11: page 74). For background limited systems, V_{rms} also will increase, but not as fast the thermal derivative. Therefore, as the background temperature increases, the NEDT will decrease. To avoid inconsistencies, the NEDT must be specified for a specific background temperature. If another temperature is selected, that temperature must be stated. The NEDT test procedure is given in Table 7-4 and possible causes for variation in the NEDT are given in Table 7-5.

Table 7-4
NEDT TEST PROCEDURE

1. Establish a test philosophy, criteria for success and write a thorough test plan (Section 1.3: page 12).
2. Verify that the test equipment is in good condition and that the test configuration is appropriate (Figure 7-12: page 205). Ask previous users if any problems were noticed.
3. Set system gain at the same level used for responsivity measurements (Table 6-3: page 175).
4. Move the source around to verify that any nonuniformity seen is due to the system and not the source.
5. Insure that the infrared imaging system has reached operating equilibrium before proceeding.
6. Select a data acquisition method (Section 4.6: page 115) and, as appropriate, collect data for a variety of background temperatures.
7. Separate noise components (Figures 7-14: page 206 or 7-15: page 207).
8. Plot each line of data to insure that no trends are present.
9. Plot the data in a histogram format and plot the calculated Gaussian distribution approximation to the data (Figure 7-2: page 194).
10. Calculate NEDT for each flood intensity: NEDT = σ_{TVH}/SiTF
11. Correct the NEDT according to the high pass filter used. If an ideal filter were used, the system NEDT is the measured NEDT divided by 0.985. If the system is ergodic, no filter is necessary.
12. Fully document any test abnormality and all test results. Record the background temperature.

Table 7-5
REASONS FOR POOR or VARYING NEDT

- Poor grounding to the test setup.
- SiTF too low (Table 6.4: page 176).
- Additional noise sources not expected.
- Quantization error introduced by test equipment.
- Bad detector/cooler (Figure 2-11: page 31).
- Detector temperature not optimum (Figure 2-11: page 31).
- Detector temperature changing (Figure 2-11: page 31).
- Ambient temperature changing (Section 3.1.5: page 73).

Example 7-1
SCANNING ARRAY AVERAGE NEDT

The SiTF of a scanning system is 0.3 v/deg. The following rms noise voltages have been measured on one line (same detector): 33, 35, 36, 37, and 40 mv. What is the average NEDT?

If the same number of data points was used to calculate each standard deviation, the best estimate of the variance is

$$s_{ave}^2 = \frac{33^2 + 35^2 + 36^2 + 37^2 + 40^2}{5} = 1315.8$$

The best estimate of the noise is

$$\overline{\sigma} = \sqrt{s_{ave}^2} = 36.3 \quad mv$$

and the average NEDT is

$$NEDT = \frac{\overline{\sigma}}{SiTF} = 0.121 \ °C$$

210 *TESTING & EVALUATION OF IR IMAGING SYSTEMS*

☞

Example 7-2
STARING ARRAY AVERAGE NEDT

16 individual detectors have the following rms noise values: 0.24, 0.25, 0.25, 0.26, 0.265, 0.265, 0.27, 0.27, 0.27, 0.27, 0.275, 0.28, 0.28 0.285, 0.29 and 0.29 v. The SiTF is 2 v/deg for each detector. What is the average NEDT?

In Example 7-1, the same detector output was measured 5 times. If the system is ergodic, the same average rms noise values will be obtained whether the average is taken over 16 different detectors or if the same detector is measured 16 times (Figure 4-38: page 128). Assuming an ergodic system, Equation 7-5 (page 195) describes the best estimate:

$$\sigma = \sqrt{s^2_{AVE}} = 0.270 \quad volts \; rms$$

and the average NEDT = 0.135 °C.

If the system is not ergodic, each detector is considered an independent source. Each output is statistically different. For example, the 1st detector will always produce an average rms value of 0.24, the 2nd detector will always produce an average rms value of 0.25 and so on. A population mean and variance of the rms values are calculated from Equations 7-2 and 7-3 (page 194):

$$m = 0.269 \; volts$$

$$s = 0.014 \; v \; rms$$

Yielding an average NEDT = 0.1345 °C and a NEDT variation of s/m = 5.2%. The histogram of values is shown in Figure 7-16 with the Gaussian approximation overlaying the histogram.

SYSTEM NOISE 211

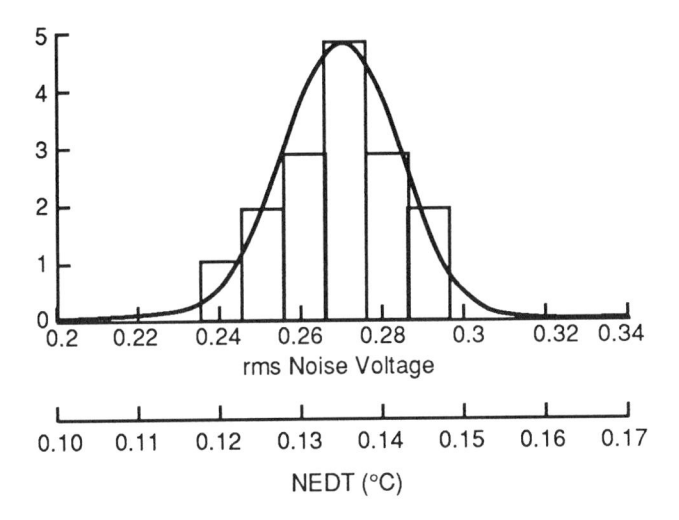

Figure 7-16: Noise histogram for 16 statistically independent detectors.

7.3.2. FIXED PATTERN NOISE

Fixed pattern noise is measured similarly as the NEDT with the notable distinction that the temporal component is removed instead. FPN can be measured after averaging many frames together and then passing the data through a two-dimensional high pass filter to remove the low frequency components (Figure 7-17). FPN is the standard deviation, σ_{VH}, of the resulting data set.

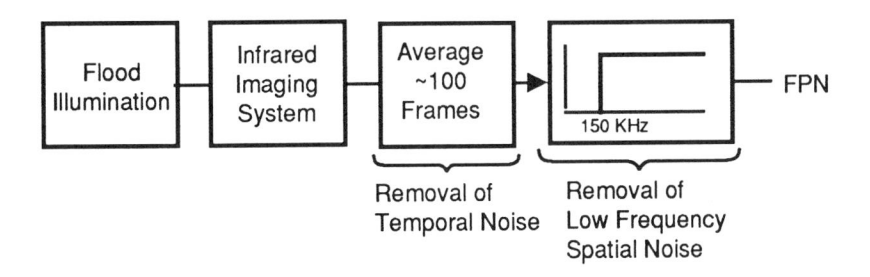

Figure 7-17: Experimental configuration to determine FPN. The low frequency noise can also be removed with a 2nd order polynomial.

212 TESTING & EVALUATION OF IR IMAGING SYSTEMS

When the data set N_{VH} is placed in a histogram format (Figure 7-18), FPN can be expressed as a percentage of an average value.

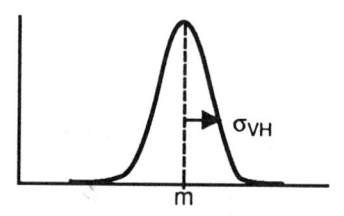

Figure 7-18: Assumed Gaussian probability density function of FPN data.

For DC coupled systems, FPN is a percentage of the average signal. By referencing it to the average value, the FPN is independent of the system gain.

$$FPN = 100\% \left(\frac{\overline{\sigma_{VH}}}{m}\right)_{trends\ removed} \qquad 7\text{-}8$$

For AC coupled systems, and DC systems that float the average value, FPN is defined as a percentage of the NEDT. By referencing it to the NEDT, FPN is independent of the system gain. When using this definition, it is important that the NEDT and FPN are measured with the same system gain.

$$FPN = 100\% \left(\frac{\overline{\sigma_{VH}/SiTF}}{NEDT}\right) = 100\% \left(\frac{\overline{\sigma_{VH}}}{\sigma_{TVH}}\right)_{trends\ removed} \qquad 7\text{-}9$$

The above definitions (Equation 7-8 or 7-9) are appropriate for systems that have a linear variation in SiTF (Figure 7-19). As the input changes, or as the system gain changes, the output changes linearly so that

$$FPN = \frac{\sigma_1}{m_1} = \frac{\sigma_2}{m_2}$$

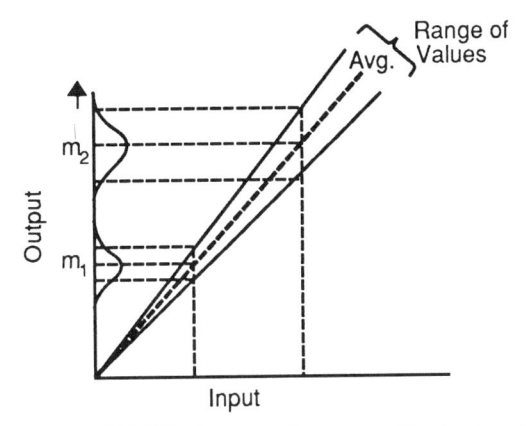

Figure 7-19: Range of SiTFs for a staring array illustrating the range of output for two inputs. m_1 and m_2 are the average values obtained for the 2 different inputs.

Theoretically, FPN can be totally removed for any particular input. However, due to variations in hardware and truncation errors imposed by the correction circuitry, the fixed pattern noise will not be totally removed. FPN is a minimum at the calibration points. At other background inputs, the FPN may be significant (Figure 7-20). It may be appropriate to determine FPN as a function of background intensity. After correction, Equations 7-8 or 7-9 are no longer universally valid as suggested by Figure 7-19. FPN is now a function of the input intensity. With two-point correction, the FPN will be a minimum at the correction points and a maximum mid way between the correction points. It is appropriate to specify FPN at T_1, T_2 and $(T_1+T_2)/2$. The generic test procedure for measuring the FPN is given in Table 7-6. Incomplete gain/level normalization, nonlinear detector response and 1/f drift cause excessive FPN.

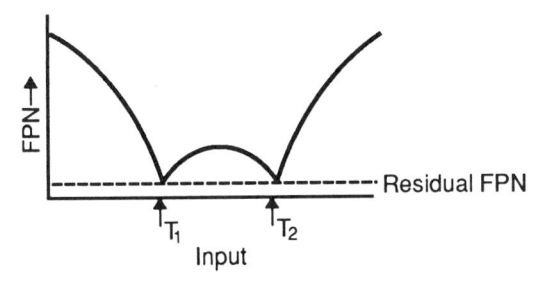

Figure 7-20: FPN as a function of input intensity for a system with two-point correction. FPN is usually negligible at the correction points.

214 *TESTING & EVALUATION OF IR IMAGING SYSTEMS*

Table 7-6
FPN TEST PROCEDURE

1. Establish a test philosophy, criteria for success and write a thorough test plan (Section 1.3: page 12).
2. Verify that the test equipment is in good condition and that the test configuration is appropriate (Figure 7-12: page 205). Ask previous users if any problems were noticed.
3. Set system gain at the same level used for responsivity measurements (Table 6.3: page 175).
4. Move the source around to verify that any nonuniformity seen is due to the system and not the source.
5. Insure that the infrared imaging system has reached operating equilibrium before proceeding.
6. Select a data acquisition method and collect data for a variety of background temperatures.
7. Separate noise components (Figure 7-17: page 211).
8. Plot the data in a histogram format (Figure 7-18: page 212) and plot the calculated Gaussian approximation to the data.
9. As appropriate, calculate σ_{VH}/m or $(\sigma_{VH}/SiTF)/NEDT$.
10. Determine FPN as a function of flood intensities and calibration intensities.
11. Fully document any test abnormality and all results. This includes the raw data, the background intensity used. For staring arrays that have FPN correction, record the calibration temperatures.

7.3.3. NONUNIFORMITY

Nonuniformity is a measure of large area blemishes, blotches and shading effects that may be distracting to the observer. Nonuniformity is the existence of blemishes whereas uniformity is the inverse. High uniformity is synonymous with low nonuniformity. Nonuniformity must be defined for a specific input flood intensity and the measurement conditions must be clearly stated. Nonuniformity requirements may vary depending upon the region of interest (Figure 7-21). Generally, the lowest nonuniformity is required in the center of the FOV ("sweet spot") with allowable imperfections increasing toward the periphery of the FOV (Figure 7-21c).

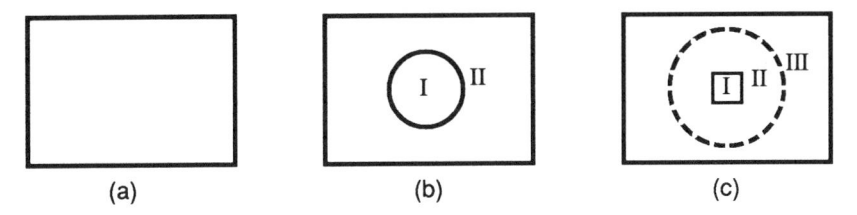

Figure 7-21: Nonuniformity requirements for an imaging system with 4:3 aspect ratio FOV. (a) Same requirement over entire field-of-view, (b) two different requirements with the nonuniformity being lower in region I, and (c) three requirements. In region I (sweet spot), the nonuniformity is very low with allowable nonuniformity increasing in region II and more so in region III.

Nonuniformity is any low frequency signal that is less than 150 kHz in a RS 170 video line. Nonuniformity appears as a change in level across the field-of-view that is present in every frame (Figure 7-22). It is the low frequency component of N_{VH}. The test configuration used to measure nonuniformity is similar to that used for FPN measurements with the exception that a low pass filter is used (Figure 7-23). Ideally, the low pass filter is a two-dimensional filter that removes the high frequency components both horizontally and vertically. Alternatively, the 2nd order polynomial directly provides the low frequency components of data set N_{VH}. If the system is truly bidirectional (i.e., $\sigma_{TH} = \sigma_{TV} = \sigma_V = \sigma_H = \sigma_T = 0$), then a transient digitizer can be used to evaluate each line (Figure 7-24).

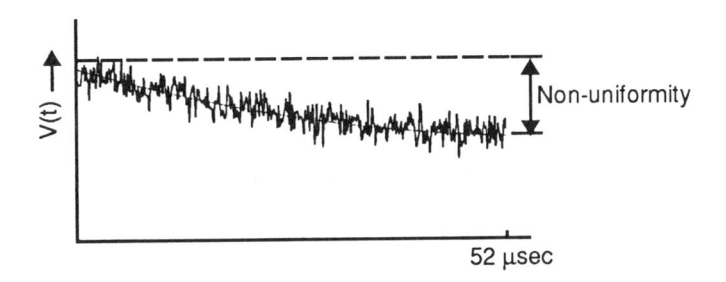

Figure 7-22: Typical RS 170 line trace illustrating nonuniformity. The nonuniformity will be present in every frame.

216 TESTING & EVALUATION OF IR IMAGING SYSTEMS

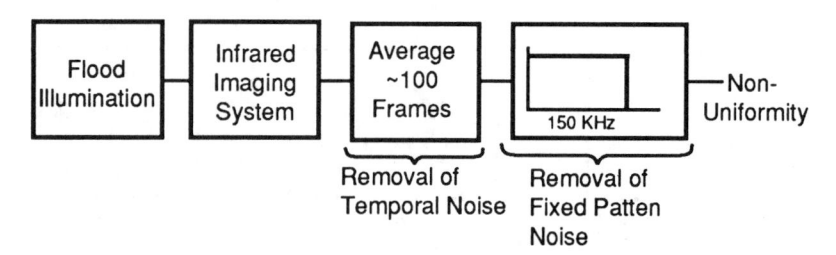

Figure 7-23: Experimental configuration to determine nonuniformity. A 2nd order polynomial also provides the low frequency components.

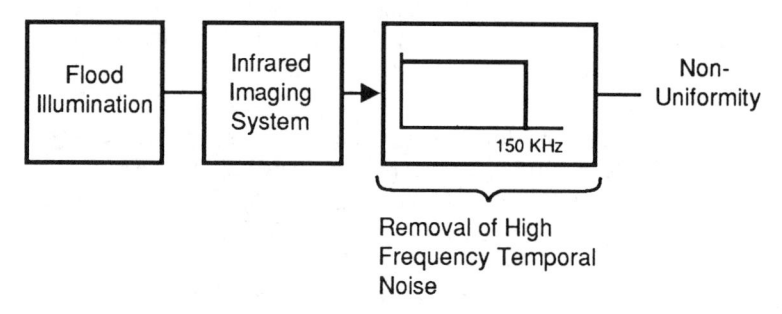

Figure 7-24: Experimental configuration to determine nonuniformity assuming $\sigma_{TH} = \sigma_{TV} = \sigma_V = \sigma_H = \sigma_T = 0$. The noise can be measured with a transient digitizer.

Nonuniformity is the standard deviation σ_{VH} of the resultant low frequency data set. Figure 7-25 illustrates a histogram of pixel values of the low frequency data set. For DC coupled systems, nonuniformity is a percentage of the average signal. By referencing it to the average value, the nonuniformity is independent of the system gain.

$$N_U = 100\% \left(\frac{\overline{\sigma_{VH}}}{m} \right)_{f \leq 150 kHz} \quad \text{7-10}$$

For AC coupled systems, and DC systems that float the average value, nonuniformity is a percentage of the NEDT. By referencing it to the NEDT, nonuniformity is independent of the system gain.

$$N_U = 100\% \left(\frac{\overline{\sigma_{VH}} / SiTF}{NEDT} \right)_{f \le 150 kHz} = \left(\frac{\overline{\sigma_{VH}}|_{f \le 150 kHz}}{\sigma_{TVH}} \right) \quad 7\text{-}11$$

It is inappropriate to define nonuniformity as a percentage of the dynamic range since the system gain affects dynamic range (Section 6.3: page 184).

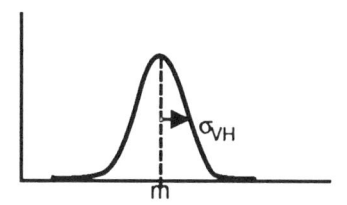

Figure 7-25: Assumed Gaussian probability density function of nonuniformity data.

With AC coupled scanning systems, nonuniformity may be a function of the background intensity. During the inactive scan time, the detectors are viewing a cold shield, the housing, or part of the scene. If the scene temperature is much different from what the detector sees during the inactive time, then the AC coupling effect will mix the inactive time output with the active time output. As a result, nonuniformity may depend upon the scene (background) intensity[15] (Figure 2-15: page 35). It may be appropriate to measure nonuniformity as a function of background intensity.

The methodology to measure nonuniformity is given in Table 7-7. There is a myriad of reasons for failing the nonuniformity test. These include shading, AC coupling, poor gain/level correction, 1/f noise, and narcissus. Reviewing Chapter 2 helps to determine the probable cause.

Table 7-7
NONUNIFORMITY TEST PROCEDURES

1. Establish a test philosophy, criteria for success and write a thorough test plan (Section 1.3: page 12).
2. Verify that the test equipment is in good condition and that the test configuration is appropriate (Figure 7-12: page 205). Ask previous users if any problems were noticed.
3. Set system gain at the same level used for responsivity measurements (Table 6.3: page 175).
4. Move the source around to verify that nonuniformities seen are due to the system and not the source.
5. Insure that the infrared imaging system has reached operating equilibrium before proceeding.
6. Select a data acquisition method and collect data for a variety of source intensities.
7. Separate noise components (Figure 7-23 or 7-24: page 216).
8. Plot the data in a histogram format (Figure 7-25) and plot the calculated Gaussian distribution approximation to the data.
9. As appropriate, calculate σ_{VH}/m or $(\sigma_{VH}/SiTF)/NEDT$.
10. Fully document any test abnormality and all results.

Nonuniformity, as defined by Equation 7-10 or 7-11 (pages 216-217), represents a method of quantifying the distribution of low frequency spatial noise. It does not suggest if the variation in intensity is perceptible. Nonuniformity can be redefined as a measure of perceptible changes in intensity that may be considered annoying. Small intensity changes that may be noticeable in adjacent pixels may not be visible when separated a large distance with a gradient between them (Figure 7-26). This occurs because the eye is more sensitive to edges than gradients[16]. The perception of any noise pattern depends upon the difference between the signal and background intensities and the gradient associated with the edge[17]. Generally, a discrete step of about 0.3·NEDT is discernible. However, as the slope of the edge decreases, the signal must increase for perception. When the edge gradient spans one-half of the FOV, the signal amplitude must be greater than about one NEDT to be seen. Subjective evaluation of image quality is simply: "Can you see any gradients in the FOV and where are they located?" The Cooper-Harper methodology (Section 1.2.2: page 9) is useful for subjective evaluation of nonuniformity.

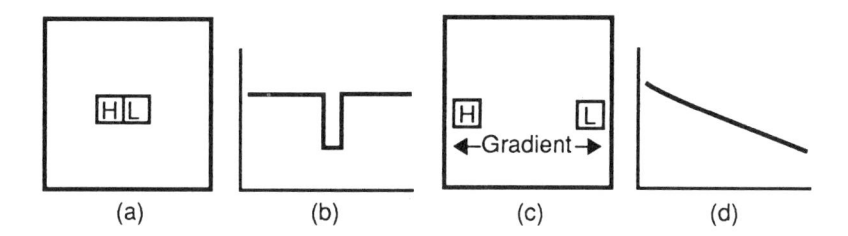

Figure 7-26: Visibility of discrete steps and gradients. (a) Discrete step in monitor brightness that is perceptible, (b) line trace through (a), and (c) same brightness difference as (a) but separated by a gradient which is not perceptible. (d) Line trace though (c).

7.3.4. NOISE EQUIVALENT FLUX DENSITY

The noise equivalent flux density is used to describe an infrared imaging system's response to point source targets. It is that flux density at the entrance aperture that produces a signal-to-noise ratio of unity. It is:

$$NEFD = \frac{V_{RMS}}{\left(\frac{\Delta V_{SYS}}{\Delta E_e}\right)} \qquad 7\text{-}12$$

Using Equation 6-14 (page 189) and noting that A_S appears in the numerator and denominator,

$$NEFD = \frac{V_{RMS} \int_{\lambda_1}^{\lambda_2}[L_e(\lambda,T_T) - L_e(\lambda,T_B)]T_{TEST}(\lambda)\,d\lambda}{PVF \int_{\lambda_1}^{\lambda_2} A_o R(\lambda)[L_e(\lambda,T_T) - L_e(\lambda,T_B)]T_{SYS}(\lambda)\,T_{TEST}(\lambda)\,d\lambda} \qquad 7\text{-}13$$

The system response to any sized target can be described by substituting the TTF for the PVF in Equation 7-13. As illustrated in Figure 7-27, the NEFD increases as the source area approaches zero is due to the point visibility factor. If all the irradiance were incident onto a detector, the PVF would be unity and the NEFD would be independent of target size. With independency, any size target can be used[18].

220 TESTING & EVALUATION OF IR IMAGING SYSTEMS

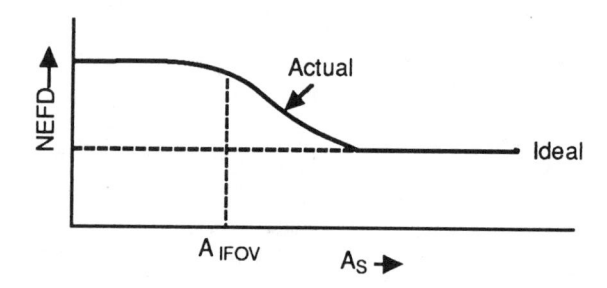

Figure 7-27: NEFD as a function of target area.

To measure the NEFD, a pinhole is placed in the focal plane of a collimator. The pinhole area, A_s, should be sufficiently small to insure that the TTF has reached its constant value, PVF. The NEFD is calculated from the measured noise voltage, V_{RMS}, measured signal difference, ΔV_{SYS}, and the calculated radiant exitance difference, ΔE_e:

$$\Delta E_e = \frac{A_{source}}{\pi (fl_{COL})^2} \int_{\lambda_1}^{\lambda_2} [M_e(\lambda, T_T) - M_e(\lambda, T_B)] \, T_{TEST}(\lambda) \, d\lambda \qquad 7\text{-}14$$

Figures 6-19 and 6-20 (page 180) illustrate the technique to measure the signal. V_{rms} is the high frequency noise and is measured in the same manner as the NEDT. The PVF should be determined (Section 6.2: page 176) before measuring the NEFD. The largest errors in this measurement are due to pinhole size uncertainty (uncertainty in flux density) and the inability to center the pinhole image onto a detector element. An alternate approach is to measure the NEDT with flood illumination and mathematically correct for the PVF and to convert ΔT to flux:

$$NEFD = \frac{V_{RMS}}{SiTF} \frac{SiTF}{\left(\dfrac{\Delta V_{SYS}}{\Delta E_e}\right)} \qquad 7\text{-}15$$

SYSTEM NOISE 221

or

$$NEFD = NEDT \frac{\alpha\beta \int_{\lambda_1}^{\lambda_2} R(\lambda) \frac{\partial M_e(\lambda,T)}{\partial T} T_{SYS}(\lambda) T_{TEST}(\lambda) d\lambda \int_{\lambda_1}^{\lambda_2} \Delta L_e(\lambda,T) T_{TEST}(\lambda) d\lambda}{\pi \, PVF \int_{\lambda_1}^{\lambda_2} R(\lambda) \Delta L_e(\lambda,T) T_{TEST}(\lambda) T_{SYS}(\lambda) d\lambda} \qquad 7\text{-}16$$

If the spectral bandwidth, $\Delta\lambda = \lambda_2 - \lambda_1$, is small so that the integral can be evaluated at the center of the interval $\lambda_c = (\lambda_2 + \lambda_1)/2$ then

$$NEFD \approx NEDT \; \frac{\alpha\beta}{\pi \, PVF} \; \frac{\partial M_e(\lambda_c, T_B)}{\partial T} T_{TEST}(\lambda_c) \, \Delta\lambda \qquad 7\text{-}17$$

For linear systems, ΔV_{SYS} is proportional to ΔE_e so that the ratio, $\Delta V_{SYS}/\Delta E_e$ is independent of the background intensity. For background limited systems, V_{rms} will increase as the background temperature increases. Therefore, as the background temperature increases, the NEFD increases. Figure 7-28 illustrates a typical response as a function of temperature. The actual response depends upon the specific system design.

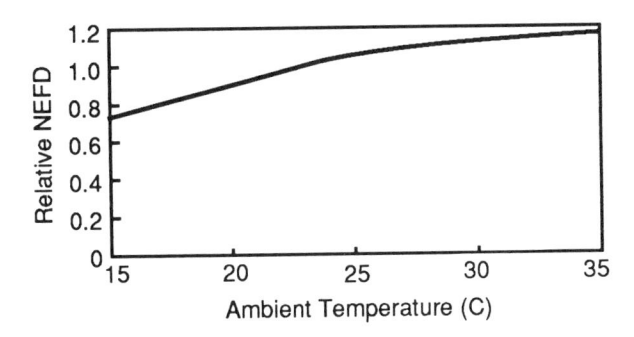

Figure 7-28: Effect of background temperature on NEFD.

7.3.5. NOISE POWER SPECTRAL DENSITY

The spectral nature of the noise was illustrated with the high pass and low pass filters used to identify NEDT, FPN and nonuniformity. The noise power spectral density provides additional information on the noise characteristics. Each noise component of the 3-dimensional model has its own NPSD. The total noise power is the integral of the power spectrum and the square root of this value is the system rms noise:

$$\sigma = \left[\int_0^\infty NPSD(f_x) \, df_x \right]^{0.5} \qquad 7\text{-}18$$

The noise equivalent bandwidth (NEBW) is a convenient mathematical construct used to calculate noise voltages. Normalizing the NSPD provides the NEBW:

$$NEBW = \frac{\int_0^\infty NPSD(f_x) \, df_x}{NSPD(f_x = 0)}$$

then

$$\sigma = \left[NPSD(f_x = 0) \right]^{\frac{1}{2}} (\Delta f)^{\frac{1}{2}}$$

where Δf is the NEBW. The relationship between the actual system response and the NEBW is illustrated in Figure 7-29. The areas under both curves are equal.

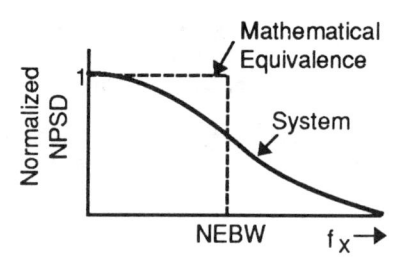

Figure 7-29: Noise equivalent bandwidth. The area under both curves is equal. The system's frequency response may be much greater than that implied by the NEBW.

Example 7-3
TEST EQUIPMENT BANDWIDTH

An infrared imaging system has a noise equivalent bandwidth of 6 Mhz when measured on the analog video (RS 170). Can a frame grabber that collects 512 samples horizontally correctly measure the total noise?

The frame grabber's Nyquist frequency is $f_N = 512/(2 \cdot 52 \mu sec) = 4.9$ MHz. Any frequency above 4.9 Mhz will be aliased to a lower frequency. For example, a bar pattern that exists at 5.5 MHz will appear as a bar pattern with a frequency of $(2)(4.9) - 5.5 = 4.3$ MHz, an unacceptable condition when measuring image fidelity. However, the higher frequency noise components will be aliased to lower frequencies and will be appropriately included in the noise calculations (Equation 7-2: page 194, or Equation 7-18: page 222). With aliasing, the measured NPSD will contain the aliased components and will not be the true NPSD.

A word of caution: the correct noise rms value will be measured if aliasing takes place. Some test equipment have internal anti-aliasing filters that limit the bandwidth. This equipment will not provide the correct rms noise values if the highest noise frequencies exceed the anti-aliasing filter cutoff. The frequencies present may be much higher than that suggested by the noise equivalent bandwidth (Figure 7-29). The highest spatial frequency present must be considered when selecting test equipment.

The NPSD is important when subjectively determining image quality (See Chapter 10). If the noise has any spectral peaks then a target whose spatial frequency is near these peaks will be more difficult to perceive. Figure 7-30 illustrates several features of the NPSD. The electronic bandwidth typically limits the high frequency response (Figure 7-30b). Both 1/f and nonuniformity produce excess low frequency noise (Figures 7-30c and 7-30d). If AC coupling is present (Figure 7-30e), it will result in a zero response at zero spatial frequency. This creates a normalization problem (see Section 8.2.11). This low frequency feature may not be seen since the frequency resolution of the discrete Fourier transform is often greater than the AC coupling cut-on frequency. Fixed pattern noise also occurs at the pixel clock frequency due to the noise associated with the switching circuitry[19]. Power supply regulation, microphonics and electromagnetic interference can introduce spectral peaks in the NPSD (Figure 7-30f). These fixed pattern noises may not be noticeable in a time trace but may become obvious in the noise power spectral density.

224 TESTING & EVALUATION OF IR IMAGING SYSTEMS

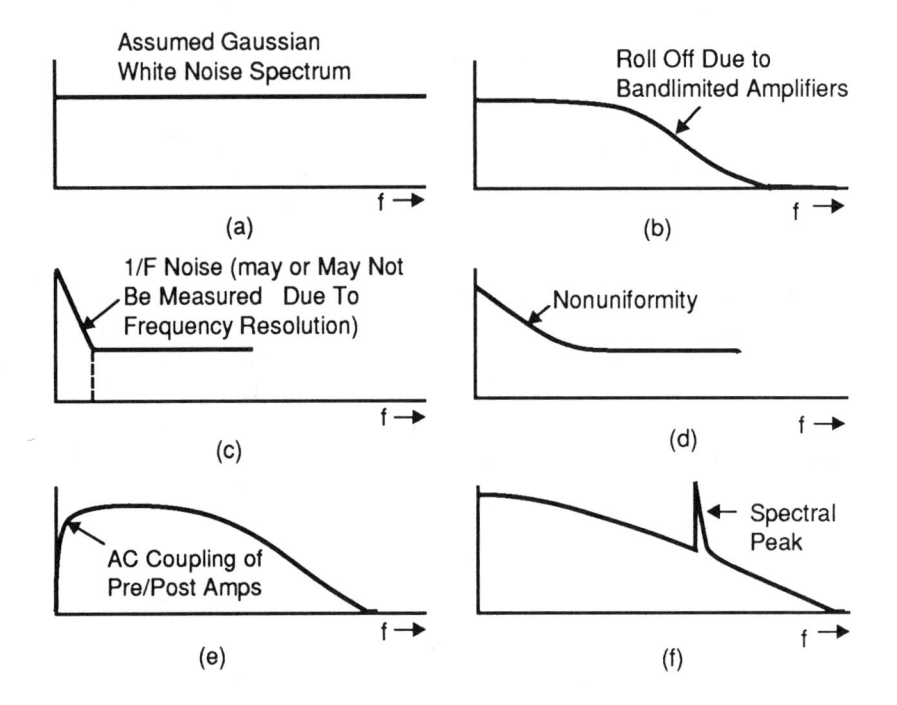

Figure 7-30: Features of the NPSD. (a) Assumed spectrum with white noise, (b) high frequency roll-off due to band limited electronics, (c) peak at very low frequencies due to 1/f noise, (d) peak at low to mid frequencies due to nonuniformity, (e) zero response at DC due to electronic AC coupling and (f) spectral peak due switching circuits, power supply regulation etc.

Methods to obtain the NPSD are identical with those used to obtain the MTF discussed in Chapter 8. The only difference is that the Fourier transform is performed on a noise trace rather than a line spread function. When the Fourier transform operates on a time trace it provides electrical frequency, Hertz. The power is $| MTF^2 + PTF^2 |$ and is scaled in volts^2/Hz. These units are obtained by dividing the power by the frequency bandwidth. This is determined from the test equipment sampling frequency. Equivalently, the power is multiplied 2 T_s where T_s is the sample-to-sample time. Most software programs provide volts^2/Hz units. When operating on 2N samples, the discrete Fourier transform provides N independent NSPD data points.

The system total noise (before scaling) is

$$\sigma = \sqrt{\sum_{n=0}^{N-1} NPSD(n)}$$

or after scaling

$$\sigma = \sqrt{\sum_{f=0}^{f_{max}} NPSD(f)\,\Delta f}$$

Where σ is the σ_i component of the three-dimensional noise model.

7.3.6. MEAN-VARIANCE TECHNIQUE

With linear systems, the NEDT, FPN, nonuniformity and NEFD are independent of system gain. Some noise sources and saturation are gain dependent. The mean-variance technique[20,21] provides insight into the gain dependency and also determines system dynamic range, saturation and conversion efficiency for DC coupled systems only. Figure 4-36 (page 124) illustrated the reduced noise voltage as the input signal approaches saturation. The mean-variance technique exploits this feature. For this test, the infrared imaging system is flood illuminated with a blackbody source (Figure 7-12b: page 205). The source temperature is slowly increased and the output average value and noise variance is measured.

Figure 7-31 illustrates a simplified infrared imaging system noise model where noise is added to the ideal detector and to the ideal amplifier. $\sigma_{DETECTOR}$ is the noise generated within the detector and consists of shot noise, σ_{SHOT}, and other noise sources, σ_{OTHER}. Additional amplifier noise, σ_{AMP}, is added by the electronics. Since noise powers are additive the total system noise is

$$\sigma_{SYS}^2 = G^2\left(\sigma_{SHOT}^2 + \sigma_{OTHER}^2\right) + \sigma_{AMP}^2 \qquad 7\text{-}19$$

where G is the gain of the system.

226 TESTING & EVALUATION OF IR IMAGING SYSTEMS

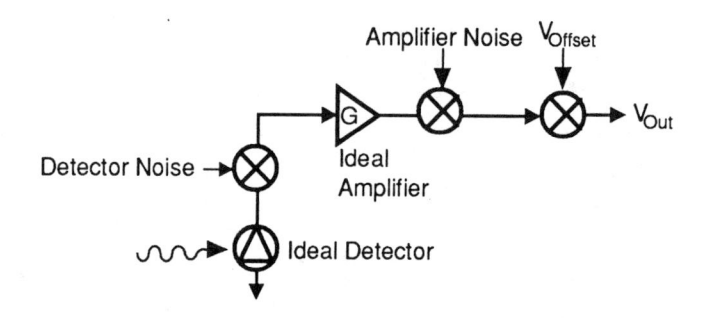

Figure 7-31: Idealized detector/amplifier circuit with added noise sources.

The signal generated by ϕ_P photons is:

$$V_{DC} = GKR_e \phi_P \qquad 7\text{-}20$$

where R_e is the detector responsivity in volts/photon and K relates the number of incident photons onto the detector from the blackbody source. The test configuration geometry determines K. It depends upon the source area, distance to the source, system f/#, and system transmittance. Equation 7-20 is a convenient representation for the calculated output voltage. With broad band spectral response systems, Equation 3-60 (page 86) should be used with the exception that the photon flux, $\Delta M_q(\lambda,T)$, is used (defined later).

Since shot noise is associated with the random arrival and detection of photons, it can be described by Poisson statistics where the variance is equal to the average value:

$$\sigma_{SHOT}^2 = KR_e \phi_P \qquad 7\text{-}21$$

Substituting Equations 7-20 and 7-21 into Equation 7-19 yields

$$\sigma_{SYS}^2 = GV_{DC} + G^2 \sigma_{OTHER}^2 + \sigma_{AMP}^2 \qquad 7\text{-}22$$

The measured output may contain an offset:

$$\sigma_{SYS}^2 = G(V_{OUT} + V_{OFFSET}) + G^2 \sigma_{OTHER}^2 + \sigma_{AMP}^2 \qquad 7\text{-}23$$

When σ_{SYS}^2 is plotted versus V_{OUT} (Figure 7-32a), the relationship should be linear. Departures from linearity indicate that other noise sources are present. The finite variance at $V_{OUT} = 0$ is caused either by the offset in the A/D converter or by the presence of additional noise sources. As the system starts to reach saturation, the mean-variance curve starts to change slope.

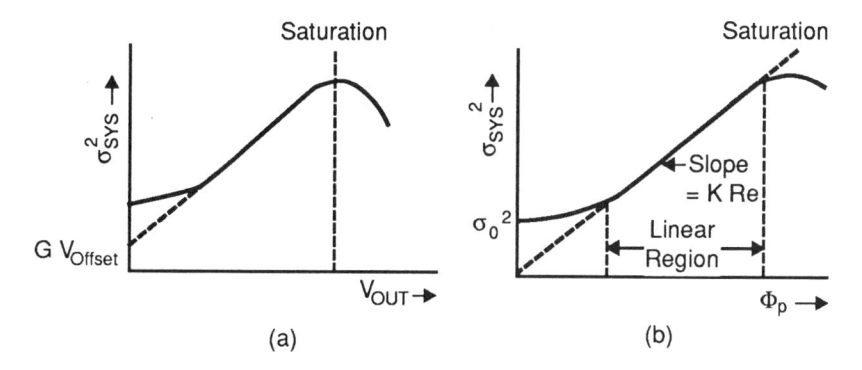

Figure 7-32: Mean-variance plot. As the system approaches saturation, the slope changes. (a) σ_{SYS}^2 as a function of V_{OUT} and (b) σ_{SYS}^2 as a function on input photon flux.

The slope of σ_{SYS}^2 versus ϕ_P is the conversion efficiency KR_e. For broad band spectral response systems,

$$\sigma_{SHOT}^2 = K \int_{\lambda_1}^{\lambda_2} R_e(\lambda) \phi_P(\lambda) \, d\lambda = K \int_{\lambda_1}^{\lambda_2} R_e(\lambda) M_q(\lambda, T) \, d\lambda \qquad 7\text{-}24$$

where $M_q(\lambda,T)$ is the photon flux from an ideal blackbody source and is equal to:

$$M_q(\lambda,T) = \frac{c_3}{\lambda^4 \left[e^{(c_2/\lambda T)} - 1\right]} \qquad 7\text{-}25$$

where $c_2 = 1.4388 \times 10^4$ μm-K and $c_3 = 1.8837 \times 10^{23}$ μm³/sec-cm². The total system noise referred to the input is

$$N_{SYSTEM} = \frac{\sigma_{SYS}}{SiTF_{SYS}} = \frac{\sqrt{G^2(\sigma^2_{SHOT} + \sigma^2_{OTHER}) + \sigma^2_{AMP}}}{SiTF_{SYS}} \qquad 7\text{-}26$$

Defining the system SiTF at unity gain as $SiTF_{G=1}$, then

$$SiTF_{SYS} = G \cdot SiTF_{G=1} \qquad 7\text{-}27$$

and the noise variance is

$$N^2_{SYSTEM} = \frac{(\sigma^2_{SHOT} + \sigma^2_{OTHER})}{SiTF^2_{G=1}} + \frac{\sigma^2_{AMP}}{G^2 SiTF^2_{G=1}} \qquad 7\text{-}28$$

As the gain increases, the total system noise becomes independent of gain (Figure 7-33). Most systems are designed such that the total system noise, which includes all of the components of the three-dimensional noise model, is independent of gain setting.

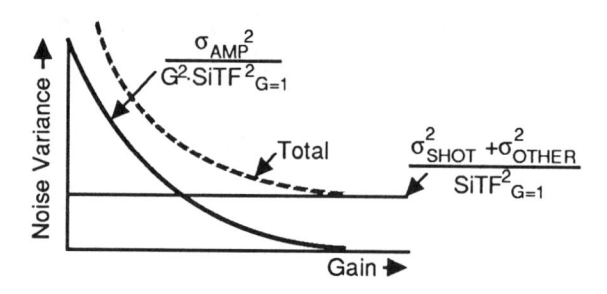

Figure 7-33: Effect of system gain on the total noise variance. For most systems, the system noise will be independent of gain.

7.4. SUMMARY

Noise is defined in the broadest sense as any unwanted signal components that arise from a variety of sources within an infrared imaging system. Noise may appear in a variety of ways such as random noise, fixed pattern noise, line-to-line nonuniformity, rain, moving bands, and flashing channels; any one of which may be the dominate noise source. Some of these effects may be difficult to quantify due to their transitory nature. Others may be easy to perceive but difficult to measure. For example, the eye is very sensitive to frame-to-frame intensity variations or flicker. Flicker may not be obvious on a single analog video line trace or in a single frame of data.

Table 7-8 highlights the 8 noise elements of the three-dimensional noise model. This model is useful for describing very complex phenomena and provides a convenient method of describing both temporal and spatial noise. Each noise element has its own noise power spectral density. By separating the NPSD into low and high frequency components, the most widely used noise parameters, NEDT, FPN and nonuniformity, are defined. The NEDT and NEFD are generally taken as the lowest measurable levels and are used to define system sensitivity. As other noise sources become dominant, these additional noise elements also should be measured. The NEDT is the high frequency component of N_{TVH}. The next prevalent noise components for parallel scanning systems are N_{TV} and N_V. For staring systems, fixed pattern noise is a major contributor to the total system noise. Both the temporal and spatial noise components are usually collected simultaneously and it is the data methodology that separates them. The test engineer must understand the sources of noise so that he can choose the appropriate test configuration and data analyses to separate the noise components. Because NEDT is a temperature difference, NEDT decreases as the background temperature increases (Figure 3-15: page 76). On the other hand, the NEFD increases as the background temperature increases (Figure 7-28: page 221).

Table 7-8
THREE-DIMENSIONAL NOISE DESCRIPTORS

NOISE COMPONENT	PIXEL VARIATIONS	ROW VARIATIONS	COLUMN VARIATIONS	FRAME VARIATIONS
TEMPORAL	N_{TVH}	N_{TV}	N_{TH}	N_T
SPATIAL	N_{VH}	N_V	N_H	S

NEDT, FPN and nonuniformity are measured values whereas NEFD is derived from both calculated and measured values. For DC coupled systems, it is convenient to express FPN and nonuniformity as a percentage of the average value. For AC systems or those systems that float the DC value, the metrics can be described as a percentage of the NEDT. Because of the widespread usage of Gaussian statistics, it is usually assumed that the data set is Gaussian distributed. This should be routinely confirmed by plotting a histogram of the noise data and fitting a Gaussian curve to the data. When the measured noise greatly exceeds the anticipated value, the NPSD can provide insight into the cause of the excess noise. For DC coupled systems, the mean-variance technique provides additional information about the system noise and dynamic range.

Specification should be both understandable and quantifiable (Table 7-9). The system total noise is a measure of the total noise from any source and includes all the noise sources described by the three-dimensional noise model. Detector array responsivity metrics are sometimes included in system performance specifications. These include the number of defective elements and the location of the defective elements.

Defining fixed pattern noise and nonuniformity as a percentage of NEDT is a valid method for insuring that the noise is barely visible. For specifications, however, the amount of FPN or N_U should a percentage of the NEDT specification, $NEDT_{SPEC}$, instead of the measured NEDT. This avoids the potential problem that selected components may provide an extremely low NEDT and yet maintain the same σ_{VH}. Here, the fixed pattern noise and nonuniformity may be visible because the random noise is so low. Usually, the specifications assume that the NEDT is the dominant noise component of the 3-D noise model. As the NEDT is reduced, other noise components may become important. While this represents a challenge to system designers, it must not affect specifications.

Table 7-9
TYPICAL SPECIFICATIONS

SYSTEM NOISE:
- The system total noise (from any source including fixed pattern, random and nonuniformity) shall not be greater then 0.35° C when the background is 23° C.

NEDT (High frequency random noise):
- The NEDT shall not be greater than 0.2° C on a background of 23° C.

FPN (high frequency spatial noise):
- The FPN shall not be greater then 2% of the average value when the background temperature is 23° C.
- Over the entire operational range of -20° C to 40° C, the FPN shall not be greater than 1% of the $NEDT_{SPEC}$.
- The FPN shall not be greater than 0.5% of the $NEDT_{SPEC}$ at the calibration points and not greater than 1% any where between the calibration points.

Nonuniformity (low frequency spatial noise):
- Nonuniformity shall not be greater than 2% of the average value when the background temperature is 23° C.
- Nonuniformity shall not be greater than 2% of the $NEDT_{SPEC}$ when the background temperature is 23° C when measured over the central 50%.
- Nonuniformity shall not be greater than 5% of the $NEDT_{SPEC}$ when the background temperature is 23° C (over the entire FOV).
- There shall be no defective detector elements in the central 50% of the FOV.
- There shall be no adjacent defective detector elements vertically, horizontally or diagonally.
- There shall be no perceptible nonuniformity (rating of 8 or greater on the Cooper-Harper scale: page 9) when the system is viewing a uniform background at 20° C.

NEFD
- The NEFD shall not be greater than 10^{-13} w/cm^2 for an unresolved source when the background temperature is 20° C.

7.5. REFERENCES

1. J. D'Agostino and C. Webb, "3-D Analysis Framework and Measurement Methodology for Imaging System Noise", in Infrared Imaging Systems: Design, Analysis, Modeling and Testing II, G. C. Holst, ed.: SPIE Proceedings Vol. 1488, pp. 110-121 (1991).
2. L. Scott and J. D'Agostino, "NVEOD FLIR92 Thermal Imaging Systems Performance Model" in Infrared Imaging Systems: Design, Analysis, Modeling and Testing III, G. C. Holst, ed.: SPIE Proceedings Vol. 1689, pp. 194-203 (1992).
3. W. J. Dixon and F. J. Massey, Introduction to Statistical Analysis, pp. 55-57: McGraw-Hill, New York (1957).
4. J. S. Bendat and A. G. Piersol, Random Data: Analysis and Measurement Procedures, 2nd edition, pp. 91-94: John Wiley and Son, New York (1986).
5. M. Schwartz, Information Transmission, Modulation, and Noise, 2nd edition, pp. 391-395: McGraw-Hill, New York (1970).
6. J. M. Mooney, F. D. Shepherd, W. S. Ewing, J. E. Murguia and J. Silverman, "Responsivity Nonuniformity Limited Performance of Infrared Staring Cameras", Optical Engineering Vol. 28(11), pp. 1151-1161 (1989).
7. J. M. Mooney, "Effect of Spatial Noise on the Minimum Resolvable Temperature of a Staring Array", Applied Optics, Vol. 30(23), pp. 3324-3332, (1991).
8. M. D. Nelson, J.F. Johnson and T. S. Lomheim, "General Noise Processes in Hybrid Infrared Focal Plane Arrays" Optical Engineering, Vol. 30(11), pp. 1682-1700 (1991).
9. H. V. Kennedy, "Modeling Noise in Thermal Imaging Systems", in Infrared Imaging Systems: Design, Analysis, Modeling and Testing IV, G. C. Holst, ed.: SPIE Proceedings Vol. 1969, pp 66-77 (1993).
10. J. S. Bendat and A. G. Piersol, Random Data: Analysis and Measurement Procedures, 2nd edition, pp. 362-365: John Wiley and Son, New York (1986).
11. C. Webb, P. A. Bell and G. P, Mayott, "Laboratory Procedure for the Characterization of 3-D Noise in Thermal Imaging Systems" in IRIS Passive Sensors Symposium, March 1991.
12. E. F. Cross and T. M. Reese, "Figures of Merit to Characterize Integrating Image Sensors: a Ten Year Update", in Infrared Technology XIV, I. Spiro, ed.: SPIE Proceedings Vol. 972, pp. 195-206 (1988).
13. A. V. Oppenheimer and R. W. Schafer, Digital Signal Processing, pp. 195-271: Prentice Hall, Englewood Cliffs, NJ (1985).
14. J. Ratches, W. R. Lawson, L. P. Obert, R. J. Bergemann, T. W. Cassidy and J. M. Swenson, Night Vision Laboratory Static Performance Model for Thermal Viewing Systems, US Army Electronics Command Report ECOM Report 7043, Ft. Monmouth NJ (1975).
15. L. O. Vroombout and B. J. Yasuda, "Laboratory Characterization of Thermal Imagers", in Thermal Imaging, I. R. Abel, ed.: SPIE Proceedings Vol. 636, pp. 36-39 (1986).
16. R. J. Farrell and J. M. Booth, "Design Handbook for Imagery Interpretation Equipment", pp. 3.1-3.29, Boeing Aerospace Co. Report D180-19063-1, Reprinted with corrections Feb. 1984.
17. G. D. Tener, "Perception of Unwanted Signals in Displayed Imagery" in "Infrared Imaging Systems: Design, Analysis, Modeling and Testing III, G. C. Holst, ed.: SPIE Proceeding Vol. 1689, pp. 304-318 (1992).
18. R. D. Hudson, Infrared System Engineering, pp. 202, 249, 268, 420: Wiley Interscience, New York, NY, (1968).
19. G. D. Boreman, "Fourier Spectrum Techniques for Characterization of Spatial Noise in Imaging Arrays", Optical Engineering, Vol. 26(10), pp. 985-991 (1987).
20. J. E. Murguia, J. M. Mooney and W. S. Ewing, "Diagnostics on a PtSi Infrared Imaging Array" in Infrared Technology XIV, I. Spiro, ed.: SPIE Proceedings Vol. 972, pp. 15-25 (1988).

21. B. Stark, B. Nolting, H. Jahn and K. Andert, "Method for Determining the Electron Number in Charge-coupled Measurement Devices", Optical Engineering, Vol. 31(4), pp. 852-856 (1992).

EXERCISES

1. Draw a histogram for the noise traces shown in Figure 7-7 (page 201) and Figure 7-8 (page 202).
2. How would excessive N_{TH} affect system performance?
3. How would excessive N_{TV} affect system performance?
4. How would you measure N_T?
5. The 3 db filter was inadvertently used during system testing. If the filter reduced the noise bandwidth by ½, by what fraction would the rms noise be reduced?
6. A staring array consists of 1024 x 1024 detector elements. The analog video is a modified RS 170 format. Can a frame grabber that has only 480 x 640 resolution, be used to measure the noise? If so, why? If not, why not?
7. Describe the effect of system gain on nonuniformity when nonuniformity is described as a percentage of dynamic range.
8. Modify the Cooper-Harper methodology (Figure 1-7: page 10) to include perceptible nonuniformity and fixed pattern noise.
9. Write a test procedure to measure the NEFD.
10. Discuss the advantages and disadvantages of calculating the NEFD from the measured NEDT.

8

CONTRAST, MODULATION and PHASE TRANSFER FUNCTIONS

The optical transfer function (OTF) plays a key role in the theoretical evaluation and optimization of an optical system. The modulation transfer function (MTF) is the magnitude and the phase transfer function (PTF) is the phase of the complex-valued OTF. When an ideal system is viewing incoherent illumination, the OTF is real-valued and positive so that the OTF and MTF are equal. When focus errors or aberrations are present, the OTF may become negative or even complex valued. Electronic circuitry also can be described by an MTF and PTF. The combination of the optical MTF and the electronic MTF creates the infrared imaging system MTF. The MTF is the primary parameter used for system design, analysis and specifications.

Theoretically, the slit response function (SRF) and contrast transfer function (CTF) can be calculated from these. When coupled with the three-dimensional system noise parameters, the MTF and PTF uniquely define system performance. If the eye's detection threshold is accurately modeled, then the minimum resolvable temperature and minimum detectable temperature also can be calculated. However, to verify system performance, it is desirable to measure these parameters directly.

The MTF, CTF, and PTF are measures of how the system responds to spatial frequencies. They do not contain any signal intensity information. The MTF is a measure of how well the system will faithfully reproduce the scene. The highest spatial frequency that can be faithfully reproduced is the system cutoff frequency. For oversampled systems, it is where the MTF approaches zero. For undersampled systems, it is the Nyquist frequency. Systems can *detect* signals whose spatial frequencies are above cutoff but cannot faithfully reproduce them. For example, a very high-frequency (above system cutoff) 4-bar pattern may appear as one low contrast blob in an oversampled system. For an undersampled system, patterns above the Nyquist frequency are aliased to a frequency below Nyquist and a 4-bar pattern may appear as a distorted 3-bar pattern. From a design point of view, the MTF should be "high" over the spatial frequencies of interest; this range of frequencies is application-specific.

Because of its frequency dependence, the MTF is more descriptive of system performance than a single value such as limiting resolution. The area under the MTF curve[1,2] appears to be a measure of image quality: slight variations in the system spatial frequency cutoff generally do not affect this definition of image quality. But the MTF may be incomplete when presented a single curve. That is, the MTF may be different for different portions of the field-of-view and for different orientations. In general, the vertical and horizontal MTFs are different.

The system MTF and PTF alter the image as it passes through the circuitry. For linear-phase-shift systems, the PTF is of no special interest since it will only indicate a spatial shift with respect to an arbitrarily selected origin. An image in which the MTF is drastically altered is still recognizable whereas large nonlinearities in the PTF can destroy recognizability. Modest PTF nonlinearity may not be noticed visually except those applications where target geometric properties must be preserved (i.e., mapping or photogrammetry). Generally, PTF nonlinearity usually increases at high spatial frequencies. Since the MTF is small at high spatial frequencies, the nonlinear phase shift effect is diminished.

Modulation is the variation of a sinusoidal signal about its average value (Figure 8-1). It can be considered as the AC amplitude divided by the DC level. The modulation is:

$$MODULATION = M = \frac{B_{max} - B_{min}}{B_{max} + B_{min}} = \frac{AC}{DC} \qquad 8\text{-}1$$

where B_{max} and B_{min} are the maximum and minimum signal levels. The modulation transfer function is the output modulation produced by the system divided by the input signal modulation at that spatial frequency:

$$MTF = \frac{OUTPUT\ MODULATION}{INPUT\ MODULATION} \qquad 8\text{-}2$$

The concept is presented in Figure 8-2. Three input and output signals are plotted in Figures 8-2a and 8-2b and the resultant MTF is shown in Figure 8-2c. The MTF and PTF are usually measured on the analog video signal, but, by convention, are presented as a function of spatial frequency.

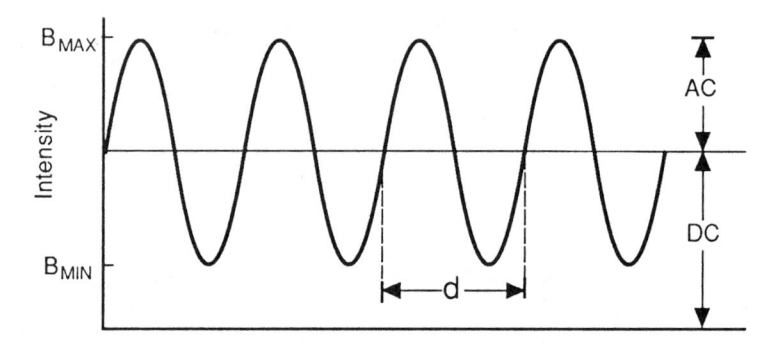

Figure 8-1: Definition of Target Modulation. d is the extent of one cycle. For optical systems, d is measured in space. If the target is placed in a collimator of focal length fl$_{COL}$, the spatial frequency is f$_x$ = fl$_{COL}$/d. For electronic circuitry, d is measured in time and the electrical frequency f$_{Hz}$ = 1/d.

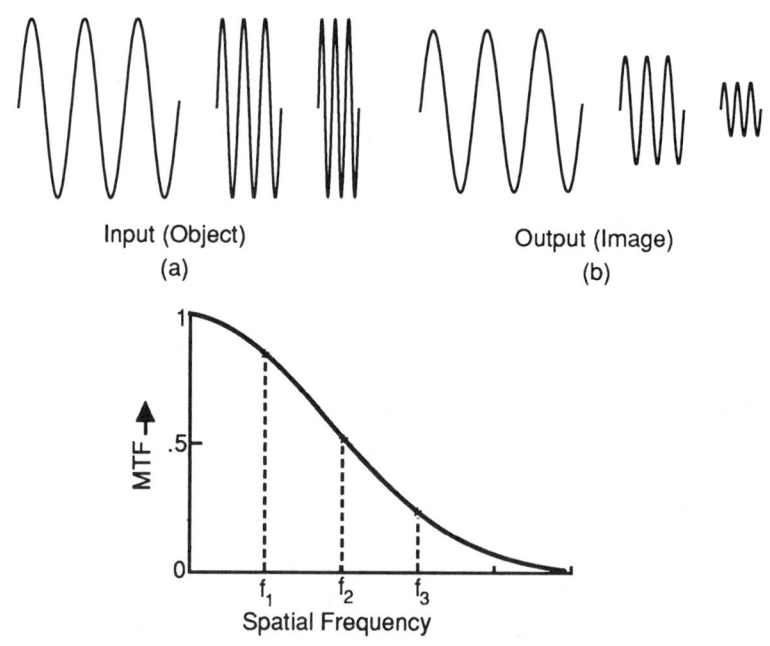

Figure 8-2: Modulation transfer function. (a) Input signal for three different spatial frequencies, (b) output for the three frequencies, and (c) MTF is the ratio of output-to-input modulation.

For infrared imaging systems, the system can sense radiation from both the target and its background. The input modulation depends on the target and background temperatures and emittances. Assuming blackbody radiators, as the emittances depart from unity, the input (target) modulation changes[3]:

$$M = \frac{B_{max} - B_{min}}{B_{max} + B_{min}} = \frac{\varepsilon_T M_e(\lambda, T_T) - \varepsilon_B M_e(\lambda, T_B)}{\varepsilon_T M_e(\lambda, T_T) + \varepsilon_B M_e(\lambda, T_B)} \qquad 8\text{-}3$$

The output B_{max} and B_{min} levels may be either voltage levels on the analog video, ADUs or luminance differences on the monitor. If the display luminance is measured, then it must be so stated since the measured MTF then includes the display MTF whereas the analog video MTF does not. Implicit in the MTF is the conversion from input infrared flux to output voltage or luminance. As a result, methods to measure the MTF depend upon both optical and electronic signal considerations.

There are two general methods for determining the MTF: the direct method which is based on measuring the response to sinusoidal or bar targets; and the indirect method, which is based upon computing the Fourier transform of the measured line spread function. There are benefits and shortcomings to both methods. Sinusoidal targets are available in the visible but are not easy to fabricate for the infrared. One could use square (bar) targets to obtain the CTF and mathematically convert to the sinusoidal response (MTF) using a series approximation. The CTF is not a replacement for the MTF, but represents a convenient measurement technique.

MTF and CTF measurements, while they appear straight forward, may be difficult to perform due to electronic nonlinearity, digitization effects, sample-scene phase effects, background removal, stray light, jitter, noise, and improper normalization. The magnitude of each of these effects depends upon the infrared imaging system design and measurement techniques.

The MTF also can be obtained from Young's fringes that are created with a coherent laser beam[4,5] or by a laser speckle approach[6]. However, the laser is a single wavelength and, therefore, the resultant MTF may not typify a polychromatic MTF. The laser approach is useful for comparative analyses.

8.1. CONTRAST TRANSFER FUNCTION

The system response to a square-wave target is the CTF, or square-wave response (SWR). The CTF is a convenient measure because of the availability of square wave targets. It is not a transfer function in the same sense that the MTF is and, as such, subsystem CTFs cannot be cascaded. Figure 8-3 illustrates the relationship between square wave and sinusoidal amplitudes. The CTF is typically higher than the MTF at all spatial frequencies. The relationship between the CTF and MTF was developed for analog optical systems. For sampled data systems, the mathematical relationship between CTF and MTF has not yet been developed. Therefore the relationship should be used cautiously for sampled data systems, if at all[7]. CTF measurements are appropriate tests for system performance verification. It is the conversion to MTF that has not been validated.

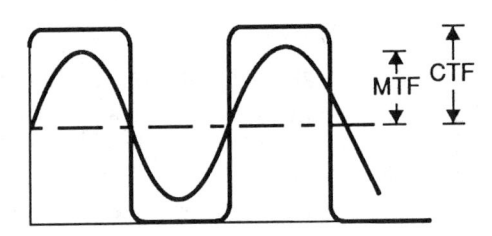

Figure 8-3: AC components of the CTF and MTF. The CTF is usually equal to or greater than the MTF.

Following Coltman's derivation[8], the CTF is expressed as an infinite series of OTFs. The OTF and MTF are equal for linear-phase-shift systems. A square wave can be expressed as a Fourier cosine series. The output amplitude of the square wave at frequency f_x is an infinite sum of the input cosine amplitudes modified by the system's MTF:

$$CTF(f_x) = \frac{4}{\pi} \left| MTF(f_x) - \frac{MTF(3f_x)}{3} + \frac{MTF(5f_x)}{5} - \ldots \right| \qquad 8\text{-}4$$

or

$$CTF(f_x) = \frac{4}{\pi} \left| \sum_{k=0}^{\infty} (-1)^k \frac{MTF[(2k+1)f_x]}{2k+1} \right| \qquad 8\text{-}5$$

Conversely, the OTF at frequency f_x can be expressed as an infinite sum of CTFs. Again, assuming a linear-phase-shift system, the MTF is[8,9]:

$$MTF(f_x) = \frac{\pi}{4} \left| CTF(f_x) + \frac{CTF(3f_x)}{3} - \frac{CTF(5f_x)}{5} + \frac{CTF(7f_x)}{7} + \frac{CTF(11f_x)}{11} + irregular\ terms \right| \qquad 8\text{-}6$$

or

$$MTF(f_x) = \frac{\pi}{4} \left| \sum_{k=0}^{\infty} B_k \frac{CTF(kf_x)}{k} \right| \qquad 8\text{-}7$$

where k takes on odd values only: 1, 3, 5, ..., etc. B_k is -1 or 1 according to:

$$B_k = (-1)^m (-1)^{\frac{k-1}{2}} \qquad for\ r = m \qquad 8\text{-}8$$

and $B_k = 0$ for $r < m$ where m is the total number of primes into which k can be factored and r is the number of different prime factors in k.

Theoretically, to obtain the MTF at frequency f_x, an infinite number of square wave responses must be measured. However, the number required is limited by the spatial cutoff frequency, $f_{MTF=0}$, where the MTF approaches zero and remains zero there after. For bar targets whose spatial frequency is above ⅓ $f_{MTF=0}$, the MTF is equal to $\pi/4$ times the measured CTF. That is, above ⅓ $f_{MTF=0}$, only one target is necessary to compute the MTF. It is difficult to determine exactly where $f_{MTF=0}$ is and, as a result, difficult to estimate ⅓ $f_{MTF=0}$. Unless there is evidence otherwise, the optical spatial frequency cutoff should be used. For unobscured, circular, diffraction limited optics the optical cutoff is D_o/λ_{AVE} where D_o is the aperture diameter and λ_{AVE} is the system's spectral responsivity average wavelength.

240 TESTING & EVALUATION OF IR IMAGING SYSTEMS

System resolution can be defined by the system's cutoff frequency. The detector MTF first zero occurs at $f_x = 1/\text{IFOV}$. The detector responds to higher spatial frequencies, but the output is distorted due to aliasing. On the other hand, the optical subsystem cannot respond to spatial frequencies above the optical cutoff because the MTF remains zero above the optical cutoff.

Example 8-1
MTF CALCULATIONS FROM SQUARE WAVE RESPONSES

An EO multiplexed infrared imaging system has an optical cutoff at 11 cy/mrad. The CTF is measured at 1, 3, 5, 7, and 9 cy/mrad in the horizontal direction (Table 8-1). What is the corresponding MTF at these frequencies?

$$MTF(1) = \frac{\pi}{4}\left[CTF(1) + \frac{CTF(3)}{3} - \frac{CTF(5)}{5} + \frac{CTF(7)}{7}\right]$$

$$MTF(3) = \frac{\pi}{4}\left[CTF(3) + \frac{CTF(9)}{3}\right]$$

$$MTF(5) = \frac{\pi}{4}CTF(5)$$

$$MTF(7) = \frac{\pi}{4}CTF(7)$$

$$MTF(9) = \frac{\pi}{4}CTF(9)$$

HISTORICAL NOTE: Early thermal imaging systems were designed for a single observer. He viewed the output of light emitting diodes, LEDs. Their intensities were directly proportional to the scene irradiance levels. For multiple observers, the LED outputs were converted into a TV format with a vidicon and were called EO multiplexed or EO mux systems. These were truly analog systems in the horizontal direction and therefore the conversion from CTF to MTF is valid. The vertical direction had a raster pattern and therefore the vertical direction was sampled.

CONTRAST, MODULATION & PHASE TRANSFER FUNCTIONS 241

The calculations are given in Table 8-1 and the results are plotted in Figure 8-4.

Table 8-1
MEASURED CTF AND CALCULATED MTF

SPATIAL FREQUENCY	CTF	MTF(1)	MTF(3)	MTF(5)	MTF(7)	MTF(9)
1	0.940	CTF(1) =0.940	-	-	-	-
3	0.691	CTF(3)/3 =0.230	CTF(3) =0.691	-	-	-
5	0.315	CTF(5)/5 =-0.063	-	CTF(5) =0.315	-	-
7	0.099	CTF(7)/7 =0.014	-	-	CTF(7) =0.099	-
9	0.018	-	CTF(9)/3 =0.006	-	-	CTF(9) =0.018
SUM		1.121	0.696	0.315	0.099	0.018
MTF= $\pi/4 \cdot$SUM		0.880	0.547	0.247	0.078	0.014

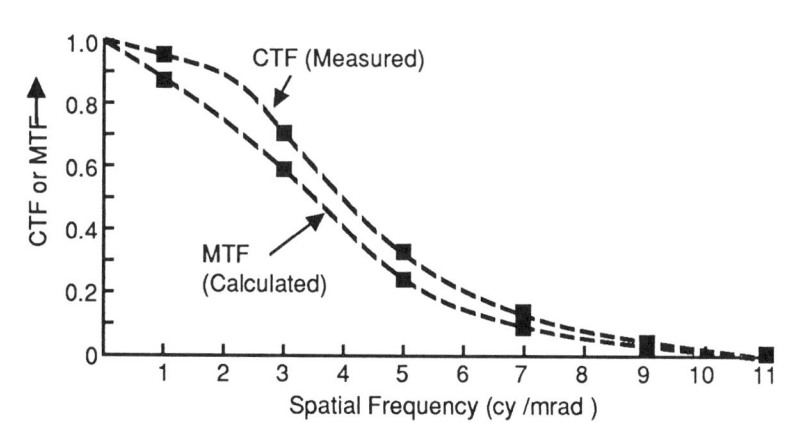

Figure 8-4: CTF versus MTF. The MTF can be calculated from the CTF for linear systems.

242 TESTING & EVALUATION OF IR IMAGING SYSTEMS

The generic CTF test configuration is shown in Figure 8-5. The source, target, collimator and infrared imaging system should be placed on a vibration-isolated optical table. The collimator aperture should be larger than the infrared imaging system aperture and be appropriately placed (Figure 4-20: page 110). No correction is needed for the collimator transmittance since the CTF is a relative measurement. If the data is captured by a digital recording device (frame grabber or transient recorder) its sampling rate should be high compared to the infrared imaging system sampling rate to avoid any signal degradation (Figure 4-34: page 122).

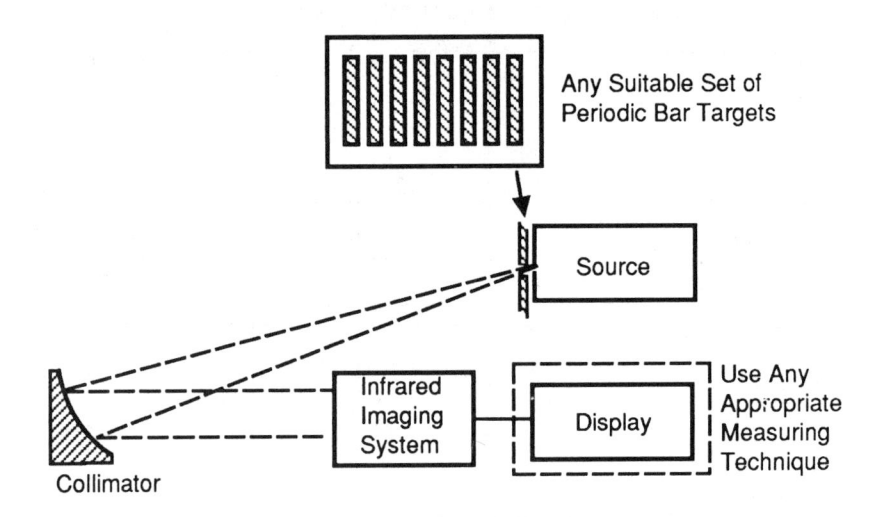

Figure 8-5: Generic CTF test configuration. The entire set up should be placed on a vibration-isolated optical table.

Ideally, the modulation should be measured for each target (Equation 8-3: page 237). This may be extremely difficult. For convenience, the modulation of each target is usually considered constant: $\epsilon_T M_e(\lambda, T_T)$ and $\epsilon_B M_e(\lambda, T_B)$ are the same for each target. If all the targets are identical, the input modulation may be considered as a fixed value (although unknown) and then only the output AC component need be measured. For convenience, the peak-to-peak values, which are twice the AC values, can be measured. Since the CTF is normalized to unity at zero spatial frequency, the absolute values of the DC components are unimportant.

CONTRAST, MODULATION & PHASE TRANSFER FUNCTIONS 243

A large target can be used to determine the low frequency AC response and this output value is used to normalize the other target outputs:

$$CTF = \frac{AC_{OUT}}{AC_{IN}} = \frac{AC_{OUT}}{AC_{OUT\ (large\ target)}} \qquad 8\text{-}9$$

The measurement methodology is shown in Figure 8-6. For low spatial frequencies, the output nearly replicates the input (Figure 8-6a). For frequencies near ⅓ cutoff, the output starts to appear sinusoidal (Figure 8-6b). Above ⅓ cutoff (Figure 8-6c), the CTF amplitude is $4/\pi$ larger than the MTF amplitude.

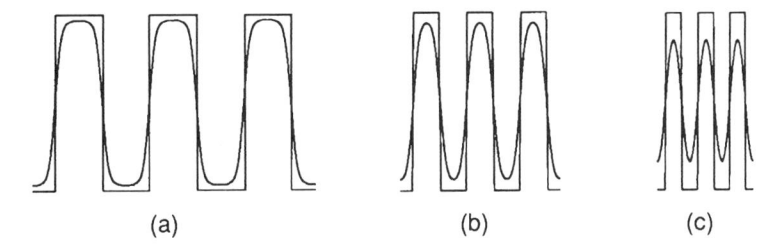

Figure 8-6: Input and output wave forms for a linear system. (a) Very low frequency signals are faithfully reproduced; (b) mid-spatial frequencies tend to look like sinusoids; and (c) input square waves whose spatial frequencies are above ⅓ $f_{MTF=0}$ appear as sinusoidal outputs.

Phasing effects between the infrared imaging system's internal sampler, detector and the location of the target introduce problems at nearly all spatial frequencies. This has been called[10] *sample-scene phase*. The interaction of the bar frequency and the system sampling frequency produces sum and difference frequencies, which, in turn, produce variations in amplitude and bar width (Section 2.3: page 36).

While the traditional CTF test procedure does not dictate any particular number of bars to use, phasing effects suggest that a target with many bars be used to insure that both the maximum and minimum signals are recorded[11,12]. The CTF is the difference between the maximum and minimum signal (Figure 8-7). Due to phasing, these values may not be next to each other. Sampling is present in all systems. For staring arrays, the discrete location of the detectors samples the scene. For scanning systems, the scene is sampled in the cross scan

244 TESTING & EVALUATION OF IR IMAGING SYSTEMS

direction by the discrete location of the detectors and by the A/D converter in the scan direction. Figure 8-7 represents the output of each detector for a staring array. For scanning systems, Figure 8-7 represents the output of the A/D converter. After the D/A converter, band-limited analog electronics will soften the edges.

(a) ⎵⎴⎵⎴⎵⎴⎵⎴⎵⎴⎵⎴⎵⎴⎵⎴⎵⎴⎵⎴⎵⎴

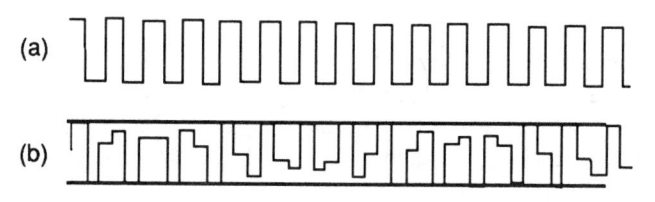

(b) (waveform with varying amplitudes)

Figure 8-7: Variation due to phasing. (a) Input and (b) system output. Local minimum and local maximum values vary with target phase. The absolute maximum and absolute minimum values define the CTF (solid lines in b).

For staring arrays, the CTF is well behaved, in the sense that the bar width remains constant, when the bar spatial frequency is proportional to the system Nyquist frequency:

$$f_x = \frac{f_N}{k} \qquad 8\text{-}10$$

where k is an integer. For staring arrays, $f_N = fl_{SYS}/(2\,d_{cc})$ where d_{cc} is the detector center-to-center spacing (detector pitch). f_N is the highest spatial frequency that can be faithfully reproduced. When using these specific targets, the phase should be selected for maximum response to obtain the in-phase CTF (IPCTF). The IPCTF approach provides only a few data points and these are insufficient to reconstruct the entire CTF curve. Even if the bar frequency is above ⅓ $f_{MTF=0}$, the conversion to MTF should not be performed. The IPCTF can be used to assess system performance at the spatial frequencies selected. Due to possible variations in focal length, f_N should be carefully determined before manufacturing the bar target. If there is a mismatch between the system Nyquist frequency and the bars, sum and difference frequencies will appear (Figure 2-20: page 39).

CONTRAST, MODULATION & PHASE TRANSFER FUNCTIONS

The CTF is not normally presented. As a result, it is very easy to confuse the MTF with the CTF. The CTF data must be clearly labeled as the square wave response and, for linear systems, the calculated MTF also should be shown on the same graph and appropriately labeled. The mathematical relationship between the CTF and MTF has not ben developed for sampled data systems[7]. The CTF test procedure is given in Table 8-2 for an analog system and the IPCTF for a sampled data system is given in Table 8-3. Possible causes of a poor CTF are given in Table 8-4.

Table 8-2
CTF TEST PROCEDURES
ANALOG SYSTEM

1. Establish a test philosophy, criteria for success and write a thorough test plan (Section 1.3: page 12).
2. Verify that the test equipment is in good condition and that the test configuration is appropriate (Figure 8-5: page 242). Ask previous users if any problems were noticed.
3. Verify infrared imaging system is in focus (Section 5.2: page 137).
4. Using the optical cutoff frequency as a guide, insure that all the targets required are available.
5. Insure that the infrared imaging system has reached operating equilibrium before proceeding.
6. Set the source intensity at a sufficiently high intensity so that the signal-to-noise ratio is high.
7. Verify that the system is operating in a linear region. This can be estimated from the responsivity curve (Table 6.3: page 175).
8. Using appropriate instrumentation (Section 4.6: page 115), select a line across the target and measure the output peak-to-peak values.
9. Normalize all data so that the CTF is unity at zero spatial frequency. The normalization value is obtained from a very large target.
10. Fully document any test abnormality and all results. The recorded data should include, at a minimum: the raw data, assumed system cutoff, source temperature, ambient temperature and any other pertinent measuring data. The measured CTF and calculated MTF data should be shown both graphically and in tabular form.

Table 8-3
IN-PHASE CTF TEST PROCEDURES
STARING ARRAYS AND SAMPLED DATA SYSTEMS

1. Establish a test philosophy, criteria for success and write a thorough test plan (Section 1.3: page 12).
2. Verify that the test equipment is in good condition and that the test configuration is appropriate (Figure 8-5: page 242). Ask previous users if any problems were noticed.
3. Verify infrared imaging system is in focus (Section 5.2: page 137).
4. Select targets according to $f_x = f_N/k$.
5. Insure that the infrared imaging system has reached operating equilibrium before proceeding.
6. Set the source intensity at a sufficiently high intensity so that the signal-to-noise ratio is high.
7. Verify that the system is operating in a linear region. This can be estimated from the responsivity curve (Table 6.3: page 175).
8. Adjust the target phase to obtain maximum modulation.
9. Using appropriate instrumentation (Section 4.6: page 115), select a line across the target and measure the output peak-to-peak values.
10. Normalize all data so that the CTF is unity at zero spatial frequency. The normalization value can be measured with a very large target.
11. Fully document any test abnormality and all results. The recorded data should include as a minimum: the raw data, optical cutoff, Nyquist frequency, source temperature, ambient temperature and any other pertinent measuring data.

Table 8-4
REASONS FOR POOR CTF

- System cannot provide optimum focus.
- Not normalized to zero spatial frequency properly.
- Signal out of linear region (Figure 4-36: page 124).
- Blackbody or ambient temperature changes during the test resulting in varying target modulation.
- Phasing effects.

8.2. MODULATION TRANSFER FUNCTION

8.2.1. INTRODUCTION

In an infrared imaging system, the infrared flux is focussed onto the detector(s), converted to a voltage, amplified, processed and then displayed on a monitor to create an image. Traditional MTF measurements assume that the system is linear and shift invariant. Optical systems tend to be linear and shift invariant; a condition called isoplanatism. These requirements are well known to the optical design community. MTF theory is also well known to the electronic circuit designer and the theory is routinely applied to analog circuitry. It is the combination of the optical and electrical responses that produce the system MTF and PTF.

Sampling (digitization) is an inherent feature of all infrared imaging systems. The scene is spatially sampled in either one or both directions depending upon the scanning scheme and the discrete nature of the detector elements. Sampled data systems are not shift invariant and do not have a unique MTF[10-16]. In most scanning systems, the detector output is a continuous analog signal which is then digitized (sampled and quantized) by the system's internal A/D converter. If the sampling rate is sufficiently high and the analog signal is band limited, the resultant digital signal will replicate the analog signal in frequency, amplitude and pulse width. However, to conserve on memory requirements and minimize data rates, infrared scanning imaging systems tend to operate at marginal sampling (clock) rates. Here, the output signal amplitude and pulse shape will be modified.

The two-dimensional MTF is the magnitude of the complex-valued two-dimensional Fourier transform of the point spread function (PSF). The PSF is the system response when viewing an ideal point source. For convenience, the MTF is measured in two orthogonal axes (usually coincident with the array axes) to obtain two one-dimensional MTFs. The one-dimensional MTF is the magnitude of the Fourier transform of the line spread function, LSF. The LSF is the resultant *image* produced by the infrared imaging system when it is viewing an ideal line. The *image* may be the analog video waveform or the intensity variation on the display.

The generic MTF test configuration is shown in Figure 8-8 with the data processing methodology shown in Figure 8-9. The targets should be placed in a holder similar to that illustrated in Figure 4-13 (page 104) so that the target can be appropriately aligned. The source, target, collimator and infrared imaging system should be placed on a vibration-isolated optical table. Air currents should

248 *TESTING & EVALUATION OF IR IMAGING SYSTEMS*

be minimized to avoid atmospheric turbulence effects (Section 4.4: page 112).

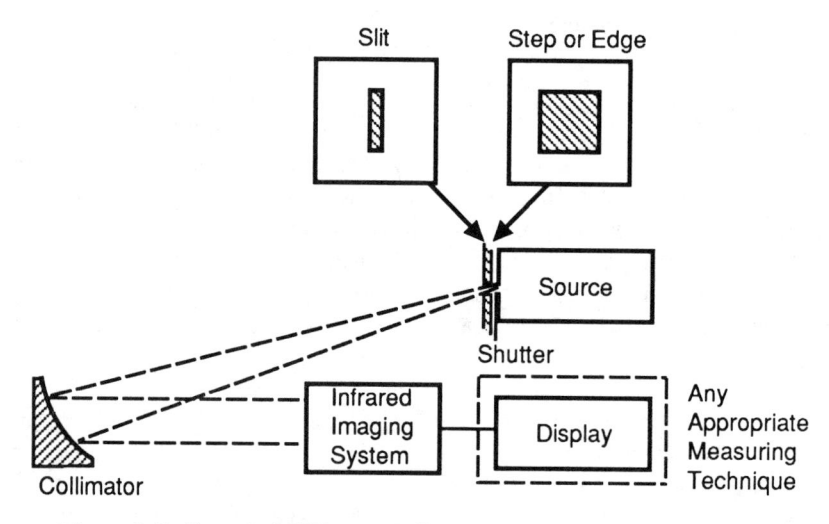

Figure 8-8: Generic MTF test configuration. The entire set up should be placed on a vibration-isolated optical table. A narrow slit produces the line spread function. A large target produces two edge responses.

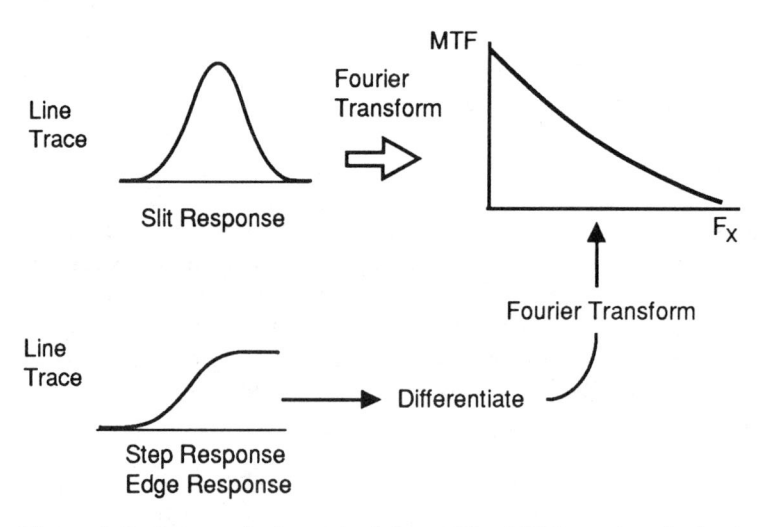

Figure 8-9: Data analysis methodology. The MTF can be obtained from the Fourier transform of the LSF. The LSF can be measured directly or can be obtained by differentiating the edge response.

A slit target is a practical implementation of the idealized line. The slit angular subtense must be smaller then the IFOV with a value of 0.1 · IFOV recommended. Ideally, the slit width should be even smaller, but as the slit becomes narrower, the flux passing through it diminishes. This produces a signal-to-noise ratio that is below a usable value. The slit width must be known accurately and should be measured in several places to ensure that it is constant. A heated wire may be used if the wire width is much smaller than the IFOV since it is difficult to measure the *heated* wire width. Since metal wires tend to elongate when heated, it may be necessary to hold the wire taught with springs (Figure 4-12: page 103).

The MTF also can be obtained from the edge spread response (ESF). It is also called the edge response, knife edge response, or step response. There are two advantages in using a knife edge target: (1) the target is simpler to construct than a narrow slit, and (2) there is no MTF correction as the slit requires. The edge is differentiated to obtain the line spread function and then Fourier transformed. However for noisy systems, differentiation accentuates the noise and this corrupts the resultant MTF. If an edge is used, it should be examined to insure that it is straight with no raggedness. There is no specific requirement on the slit or edge length.

Using the responsivity curve as a guideline, the slit or edge intensity should be as great as possible to enhance the signal-to-noise ratio without entering a nonlinear response region. If possible, the system gain should be reduced to reduce noise and the target signal should be increased to increase the signal-to-noise ratio. The absolute value of the source intensity is unimportant and the system gain is unimportant for MTF measurements.

For scanning arrays, the system views a slit with the resultant response collected from the analog video resulting in a time trace that is then sampled (by the test equipment) and Fourier transformed. A scale factor is applied to the discrete indices to convert electrical frequency into spatial frequency. While the same process can be used for staring arrays, it is more natural to express the LSF data points in object space. Then the Fourier transform directly provides spatial frequency. The choice of defining data in time or space depends upon the test procedure.

MTF measurements appear straight forward. Variation in results may be due to nonisoplanatism, electronic nonlinearity, inadequate removal of offset or trends, jitter, noise effects, sample-scene phase effects and improper normalization. The magnitude of these effects depends upon the specific infrared imaging system design and measurement techniques.

8.2.2. ISOPLANATISM

An optical system is isoplanatic if the translation or rotation of an object produces a proportional translation or rotation of the image. An isoplanatic region is that region within the field-of-view where the optical transfer function may be considered shift-invariant within measurement accuracy for all spatial frequencies of interest. Since aberrations do not generally change radically, it is reasonable to treat a small area as isoplanatic or to call this area an isoplanatic patch. Most optical systems are rotationally symmetrical, have minimal aberrations and may be considered isoplanatic. Isoplanatism applies only to the optical subsystem.

8.2.3. SPATIAL SAMPLING

Sampling, due either to the discrete location of the detectors or by the internal A/D converter, poses a special problem in measuring the MTF. Sampling can limit the number of data points that are available resulting in an undersampled LSF, ESF or PSF. Depending upon the sample-scene phase, the resultant MTF will vary. However, two methods are available to construct the full LSF. The first is to move a slit[17,18] or edge[19] in sub-pixel increments across the detector and measure the response for each target location. Similarly, the point spread function can be obtained by moving a small laser beam or spot (flying spot scanner) across a detector element[20]. The second is to create a periodic array of point sources whose locations are at noninteger locations with respect with the projected detector locations[21]. This noninteger arrangement is equivalent to varying the sample-scene phase between the point sources and the detectors[10].

With the moving slit approach, the output of a single detector is monitored as the slit image moves across it. The slit increments are measured (sampled) in object space (e.g., mrad) and the Fourier transform directly transforms the data into the spatial frequency domain with units of cy/mrad. The moving slit approach may accentuate[17] image modifiers such as line-to-line interpolation. These nonlinear processes may or may not be visually obvious when viewing general imagery.

In the periodic array approach[21] (Figure 8-10), each "point" source size and relative location must be known precisely. Each response must be normalized according to its point source intensity which is proportional to the point source size. Each response also must be normalized to each detector's responsivity. Overlaying the various outputs permits complete reconstruction of the PSF

(Figure 8-11). The point source locations are usually measured in angular space and the Fourier transform directly provides the spatial frequency response.

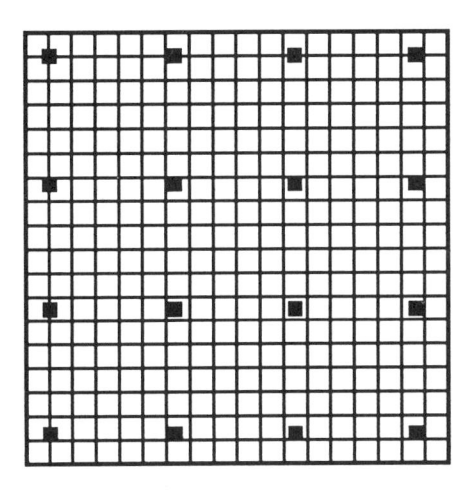

Figure 8-10: Point source array used to reconstruct the PSF for staring arrays. The point source locations are at noninteger locations with respect to the detector locations. The detector array is 18 x 18 and the source array is 4 x 4.

A one dimensional response can be obtained using a slit array (Figure 8-12). As with point sources, the slit widths must be known accurately so that the outputs can be normalized. Both horizontal and vertical MTFs can be obtained by rotating the target 90 degrees. As with the pinhole array target, the slit edges should be aligned parallel to the detector array axes. The pinholes (slits) should be sufficiently separated to insure that the individual PSFs (LSFs) do not overlap. The location of the data points must be carefully defined so that the Fourier transform is done properly. Since most Fourier transform algorithms require equispaced data points, it may be necessary to interpolate between the data points. Creating the equispaced data points through interpolation is often called resampling.

252 TESTING & EVALUATION OF IR IMAGING SYSTEMS

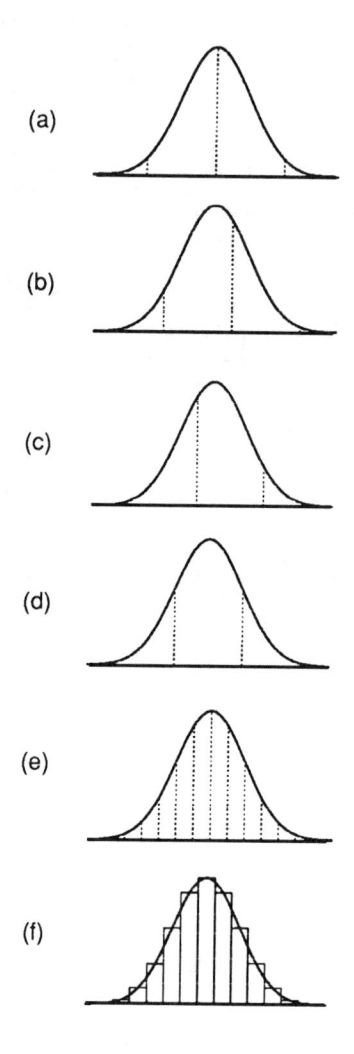

Figure 8-11: PSF reconstructed from the point sources on the 3rd horizontal line in Figure 8-10. The four point sources produce the data points shown in (a) through (d). (e) is the sum and (f) is the effective response. The original PSF is shown on each figure.

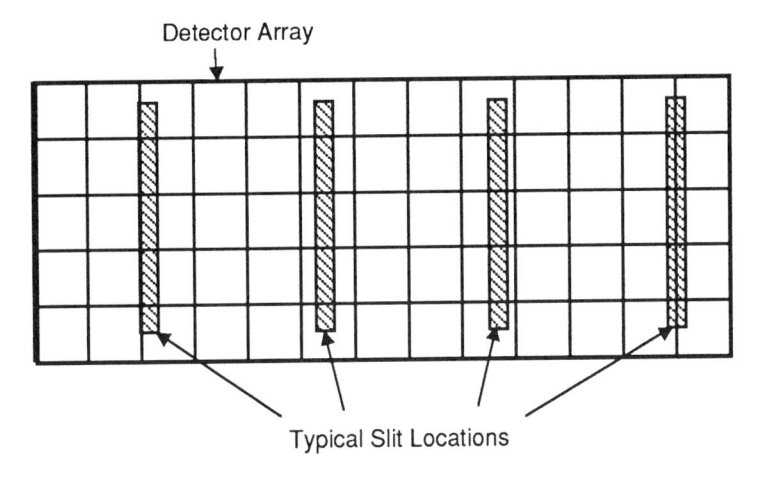

CONTRAST, MODULATION & PHASE TRANSFER FUNCTIONS 253

Figure 8-12: Variable phase slits can be used to reconstruct the LSF for staring arrays.

Varying phases can be obtained by skewing an edge or slit with respect to the detector array axes. The edge spread function can be extended by skewing a horizontal edge such that the first detector in the array is fully covered and the last detector is fully exposed (tangential sampling[22]). This results in a ramp function across the array output (Figure 8-13). Transforming the axes according to the slope of the edge provides the edge response function. Similarly, the edge can be skewed with respect to the vertical axis. Here lines are overlaid to obtain the complete ESF[12]. Either of these approaches may encounter problems with systems that employ line-to-line interpolation schemes or replicate lines. If the scan direction is orthogonal to the TDI direction, the MTF is very sensitive to edge skewness[23].

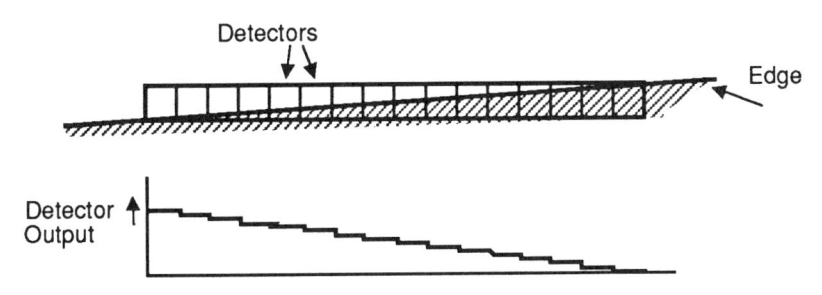

Figure 8-13: Tilted edge to extend the edge spread function and thereby increase the ESF accuracy. (a) Tilted edge and (b) the output of the detectors.

254 TESTING & EVALUATION OF IR IMAGING SYSTEMS

With staring arrays, nonuniform detector responsivity may cause variations in the edge or slit response. Element-to-element uniformity is important. Two-point correction may not be adequate to insure a uniform response for a range of uniform inputs. This nonuniformity must be treated as a separate problem.

The phasing errors discussed so far may not be significant in some infrared imaging systems[24]. Depending upon the design, the optical PSF and detector cross talk may mask phasing effects. Nevertheless, the phasing effects or the lack of it must be verified for the particular infrared system under test.

8.2.4. SYSTEM LINEARITY

It is important to operate in a linear region of the responsivity curve, or more accurately, in a linear region of the slit response function when using a slit. To verify this, the MTF should be obtained for various source intensities. If the system is operating in a linear region, the MTF, within measurement accuracy, will be independent of the source intensity. Figure 8-14 illustrates a responsivity curve that exhibits a nonlinear response when the input ΔT exceeds $0.5\,°C$.

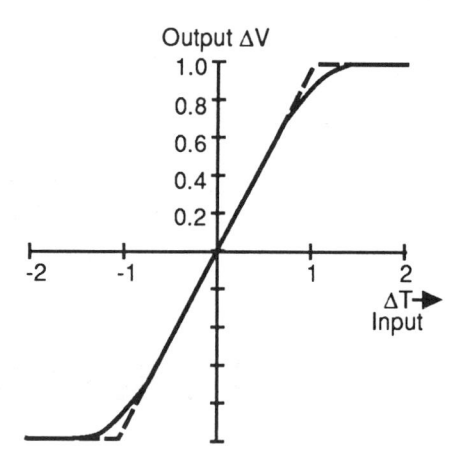

Figure 8-14: Typical responsivity curve illustrating a nonlinear response above an input of $\Delta T = 0.5°C$.

CONTRAST, MODULATION & PHASE TRANSFER FUNCTIONS 255

Assuming an analog scanning system, the output analog video will exhibit the time traces illustrated in Figure 8-15 for different source intensities. For $\Delta T > 0.5\,°C$, the LSF is affected by the system nonlinearity, but this is not obvious from an individual LSF trace. Only when the signal reaches saturation (the maximum available output of unity) does the nonlinear responsivity affect the LSF in a visually obvious way. The resultant MTFs are shown in Figure 8-16. For this example, within measurement accuracy, the MTF is independent of ΔT up to input intensities of $0.6\,°C$.

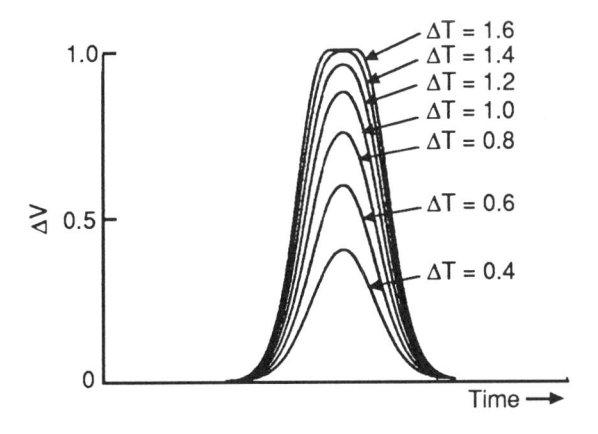

Figure 8-15: LSFs obtained from the responsivity function shown in Figure 8-14. When the ΔT is greater than $0.5\,°C$, the output LSF is distorted. Distortion is visually obvious when ΔT approaches $1.6\,°C$.

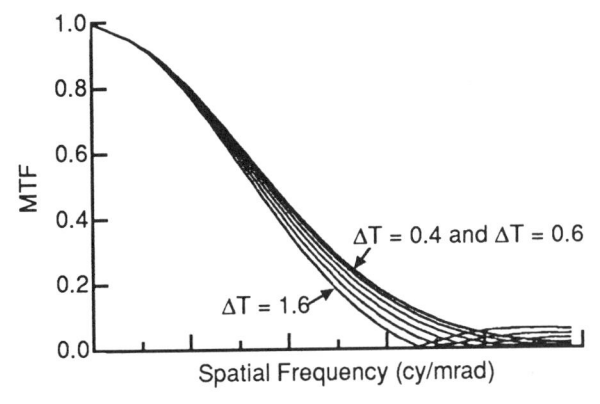

Figure 8-16: MTFs for the LSFs shown in Figure 8-15. When the input ΔT is greater than $0.6°\,C$, the resultant MTF is distorted.

8.2.5. TEST EQUIPMENT DIGITIZATION

The effects of sampling on the computation of the MTF can be dramatic. Two sampling lattices are present: the inherent sampling lattice determined by the detector array or the internal A/D converter and the sampling rate of the test equipment. The test engineer can minimize the infrared imaging system phasing effects through appropriate test methodology and minimize the potential test equipment phasing errors by selecting the appropriate test equipment.

Most infrared imaging systems produce an analog video signal that is compatible with most monitors. This analog signal is then digitized by the test equipment to obtain the data points that will be Fourier transformed. Figure 8-17 illustrates the apparent variation in a digitized LSF due to undersampling by the test equipment. Increasing the sampling rate will minimize these variations and the associated artifacts.

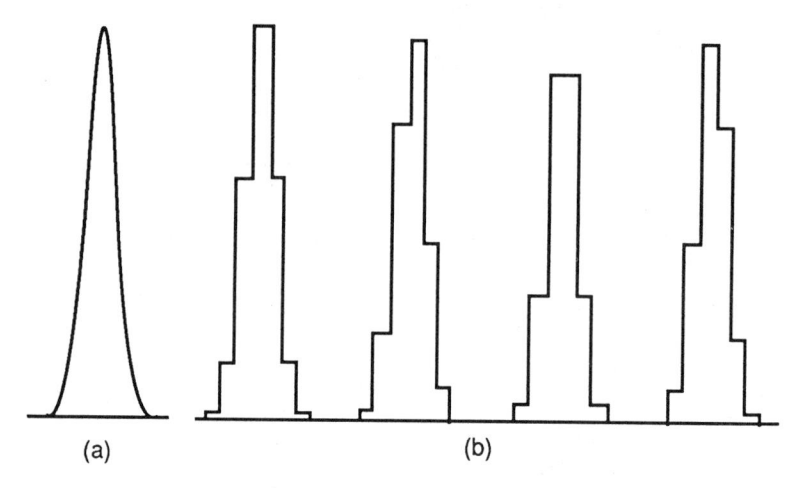

Figure 8-17: Undersampled LSF. (a) Analog signal, and (b) digitized signal that appears different depending upon the phase.

Because the LSF may have long tails, it is difficult to define the width of the LSF. Therefore, a better criterion is selected: the full width of the LSF is measured at half maximum (50%) amplitude (FWHM). When there are at least 4 samples across the LSF at FWHM phasing artifacts are minimized. The requirement of 4 samples per FWHM is approximately equivalent to 10 samples across the entire LSF (Figure 8-18). This is more conservative than the six samples across the edge spread function (full width of the LSF) recommended by Granger[25] and less restrictive than the 8 samples FWHM recommended by

CONTRAST, MODULATION & PHASE TRANSFER FUNCTIONS

Dainty[26] for noiseless systems. When no noise is present, White and Marquis[19] also demonstrated that only 10 samples across the ESF were sufficient to reproduce the MTF. However, as system noise increases, the number of required samples increases.

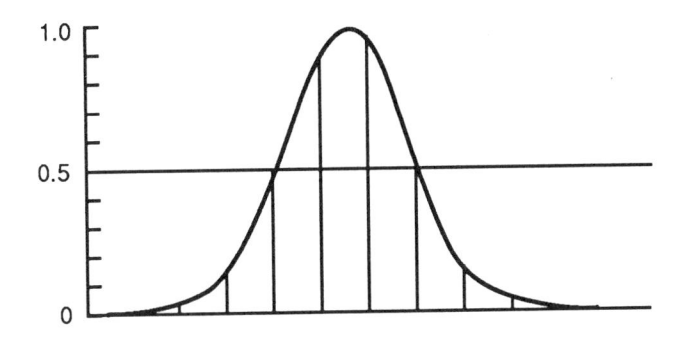

Figure 8-18: Required sampling rate. There should be at least 4 samples FWHM across the LSF to minimize phasing effects. This is equivalent to approximately 10 samples across the entire LSF.

Example 8-2
TEST EQUIPMENT SAMPLING RATE

What should the test equipment sampling rate be to digitize a diffraction limited infrared imaging system that has a circular aperture of 25 cm and a FOV of 20 mrad? The scanning system operates in the 3 to 5 μm region.

The average wavelength is 4 μm. The Airy disk angular subtense is

$$\theta = 2.44 \frac{\lambda_{AVE}}{D} = 2.44 \frac{(4 \times 10^{-6})}{(0.25)} = 39 \times 10^{-6} \ rad$$

This represents $39 \times 10^{-6}/0.020 = 1.95 \times 10^{-3}$ of the FOV.

In the RS 170 time domain, the Airy disk will be 0.195% of the active line rate

$$t = (1.95 \times 10^{-3})(52 \,\mu sec) = 101 \; nsec$$

The FWHM of an Airy disk is approximately 0.42 times the disk diameter and with 4 samples across FWHM, the required sampling rate is

$$f = \frac{4}{(0.42)(0.101 \times 10^{-6})} = 94 \; MHz$$

If the system is truly diffraction limited and the electronic MTF = 1, the test equipment must have a sampling rate of 94 MHz. An even higher sampling rate is desirable but it may be more difficult to purchase equipment with higher sampling rates.

For scanning systems, a frame grabber is not appropriate for horizontal MTF measurements unless it can capture 4 samples per FWHM of the line spread function. A frame grabber designed for RS 170 which captures 640 pixels horizontally has a sampling rate of $640/(52 \,\mu sec) \approx 12.3$ MHz. For staring systems, the frame grabber may be synchronized with the array so that the frame grabber collects data from every available pixel. In the vertical direction a frame grabber typically collects all this data appropriately since the number of vertical lines is usually fixed. For example RS 170 consists of 480 lines and frame grabbers are designed to capture all of the lines. If measuring the monitor with a solid state camera (Figure 4-31: page 119), the FWHM of the LSF should project onto at least 4 camera detector elements.

The 4 samples FWHM requirement also applies to those methods discussed in Section 8.2.3 on reconstructing the LSF or PSF. When using the moving slit approach, the slit should be moved in increments such that there are at least 4 sample FWHM. Similarly when using the periodic array of point sources, the point sources should be placed such that there are at least 4 samples FWHM. If using a scanning photometer (Figure 4-30: page 119) or a fiber optic scanning probe (Figure 4-32: page 120), the probe or slit must be incremented such that at least 4 samples occur at FWHM.

Example 8-3
MOVING SLIT INCREMENTS

The MTF of a staring array will be measured by the moving slit approach. If the system operates in the 8 to 12 μm region and has a clear circular aperture of 4 inches, what should the slit increments be to insure that there are 4 samples FWHM across the LSF? The IFOV is 0.25 mrad.

The best MTF will be due to the diffraction limited optical MTF. All other subsystems will reduce the optical MTF. A diffraction limited, circular lens will provide an Airy disk whose FWHM is approximately:

$$\theta \approx (0.42)\left(2.44\frac{\lambda_{ave}}{D}\right) \approx \frac{\lambda_{ave}}{D} = \frac{10 \times 10^{-6} \, m}{(4 \, in)(.0254 \, m/in)} = 98.4 \, \mu rad$$

The slit should be moved in about 25 μrad increments to achieve 4 samples FWHM. If the slit is place in a collimator whose focal length is 2 meters, the slit should be moved in 50 μm increments or about 0.002 inches. In the presence of noise, more samples are required and the increments must be smaller.

Example 8-4
SLIT ARRAY SPACING

A slit array target (Figure 8-12: page 253) will be used in a 2 meter focal length collimator to test the infrared imaging system described in Example 8-3. What should the slit spacing be?

Each succeeding slit should be spaced by an additional 0.002 inches. With an IFOV of 0.25 mrad, the IFOV represents a distance of 0.02 inches in the 2 meter collimator. If each slit is placed 4 IFOVs apart to avoid LSF overlap, the spacing between slits would be (4·IFOV + 0.002) or 0.082 inches.

260 TESTING & EVALUATION OF IR IMAGING SYSTEMS

8.2.6. BACKGROUND REMOVAL

Before calculating the Fourier transform, the background response or pedestal must be removed. This is done either by making a separate measurement of the pedestal or by assuming the pedestal is uniform. The advantage of measuring the pedestal is that any inhomogeneity in the field-of-view can be removed. The pedestal can be measured either by closing the shutter between the source and the target plate or by setting the source so that the input intensity provides a zero contrast as determined by the responsivity curve. As illustrated in Figure 8-19, if the pedestal is improperly removed, the measured MTF may be too high (too much pedestal removed) or too low (insufficient pedestal removed).

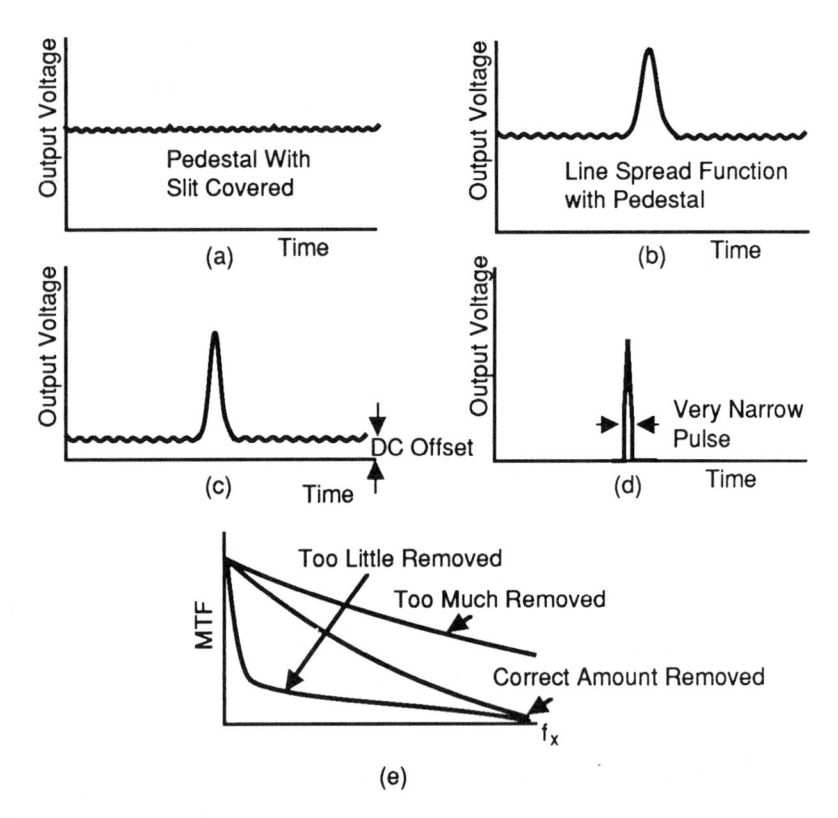

Figure 8-19: Effects of incorrect pedestal removal. (a) Pedestal, (b) LSF with pedestal, (c) too little pedestal removed, (d) too much pedestal removed and (e) resultant MTFs. MTF appears low with too little pedestal removed. With too much pedestal removed, the MTF appears better than expected.

CONTRAST, MODULATION & PHASE TRANSFER FUNCTIONS

Background removal is more difficult when trends, flicker or drift is present. Trends can occur in scanning or staring systems whereas flicker and drift maybe more prevalent in scanning systems. Trends are due to target inhomogeneities, shading, nonuniformity and, for scanning systems, 1/f noise. Drift is a slowly varying temporal change in the background level whereas flicker refers to a level change from frame-to-frame. Trends can be removed by applying a first order least-squares fit to the background data (i.e., a straight line represented by $y = mx + b$). This line is then subtracted from the LSF (Figure 8-20). The corrected response should have zero mean on either side of the LSF response. Since the LSF is a *high* frequency response, the data can be passed through a high pass filter such as the filter used to separate the high and low noise frequency components (Figure 7-11: page 204). If the trend or other low frequency components are not removed, the resultant MTF will have a low frequency peak.

No matter what method is used to remove the background, for comparison, the MTF should be calculated with and without the background removed to verify that the background removal provided the desired results. In addition, the LSF should be plotted before and after background removal to insure that it was not adversely affected. The method used for pedestal and trend removal must be stated with the MTF results.

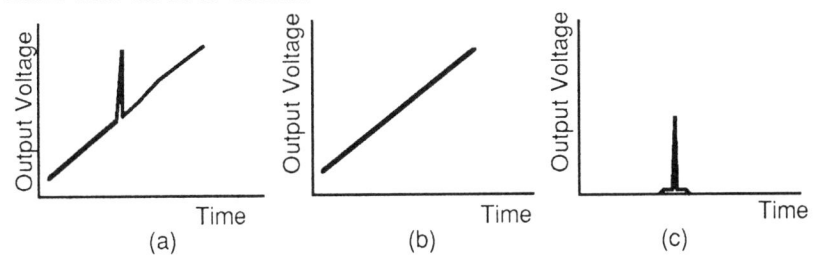

Figure 8-20: Trend removal. The background (b) must be subtracted from the data (a) to obtain the correct LSF (c).

8.2.7. JITTER

When jitter is present (Figure 8-21), the position of the LSF peak varies from trace-to-trace. The jitter may be due to system movement relative to the target during the test (mechanical vibration), inherent jitter in the scanner, or jitter in the system's synchronization signals. If the jitter is due to synchronization problems in the transient recorder, then the potential MTF degradation is due to the measurement technique. Placing the source, target, collimator and infrared imaging system on a vibration-isolated optical table minimizes mechanical vibration.

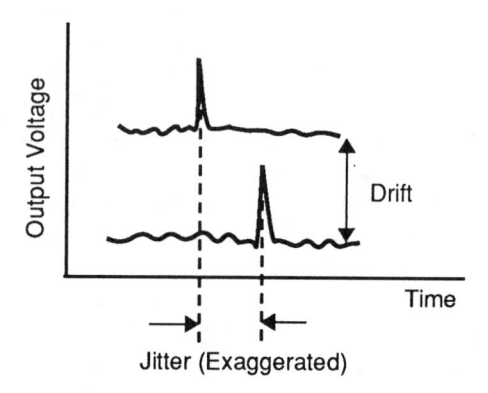

Figure 8-21: Jitter and drift exaggerated. When jitter and drift are present, averaging will broadened the resultant LSF.

If several frames are averaged (time domain averaging) to improve the SNR, jitter in successive frames can cause broadening of the averaged line spread function. This broadened LSF produces a lower MTF. It may be desirable, however, to leave the jitter in the measurements if jitter is an inherent property of the system[27]. That is, the observer sees the entire system response including any inherent jitter. The eye, and somewhat the display, averages system jitter. Therefore, the removal of jitter will increase the MTF but may not represent what the observer sees.

One way to align the LSF traces is to integrate them to obtain edge spread functions. The slope of the edge spread function is then calculated via a least-squares approach (Section 4.7: page 124). Assuming a symmetrical LSF, the location of the 50% point on the edge spread function corresponds to the center of the LSF. This value then can be used to overlay each LSF (Figure 8-22). By overlaying traces, the LSF is effectively oversampled. This increases frequency resolution and increases the accuracy in obtaining the MTF and LSF[12]. The composite LSF usually has irregularly spaced data points. Since most Fourier analyses assume equispaced data points, data interpolation is necessary.

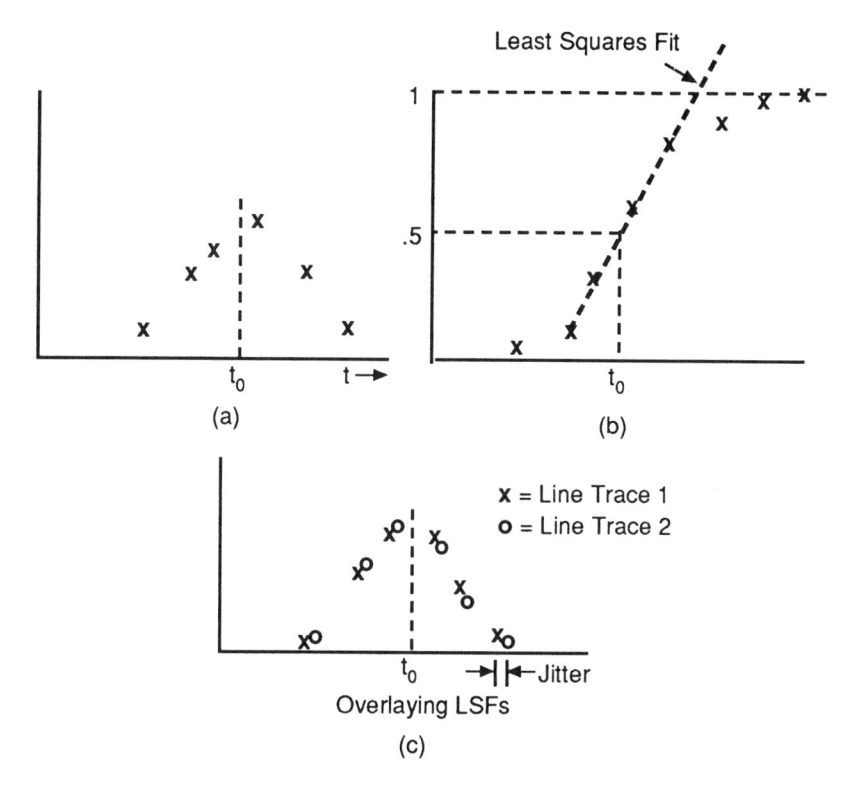

CONTRAST, MODULATION & PHASE TRANSFER FUNCTIONS 263

Figure 8-22: LSFs can be overlaid by using the center of the LSF as a reference point. (a) Typical data set. (b) ESF obtained by integrating the LSF, and (c) overlaying several LSFs. t_o is the center of the symmetrical LSF.

8.2.8. NOISE

White, additive, temporal noise introduces random errors and adds a positive bias to the estimated MTF[28,29]. The MTF in the presence of noise will always be greater than that without noise. A signal-to-noise ratio as high as 500 may be required for accurate, reproducible MTF measurements[29]. White and Marquis[19] also demonstrated that with a SNR of 500, the MTF can be consistently reproduced from a differentiated ESF. As the noise level increases (lower SNR), more samples are required across the ESF. Even with 200 samples across the ESF noise corrupts the MTF (Figure 8-23).

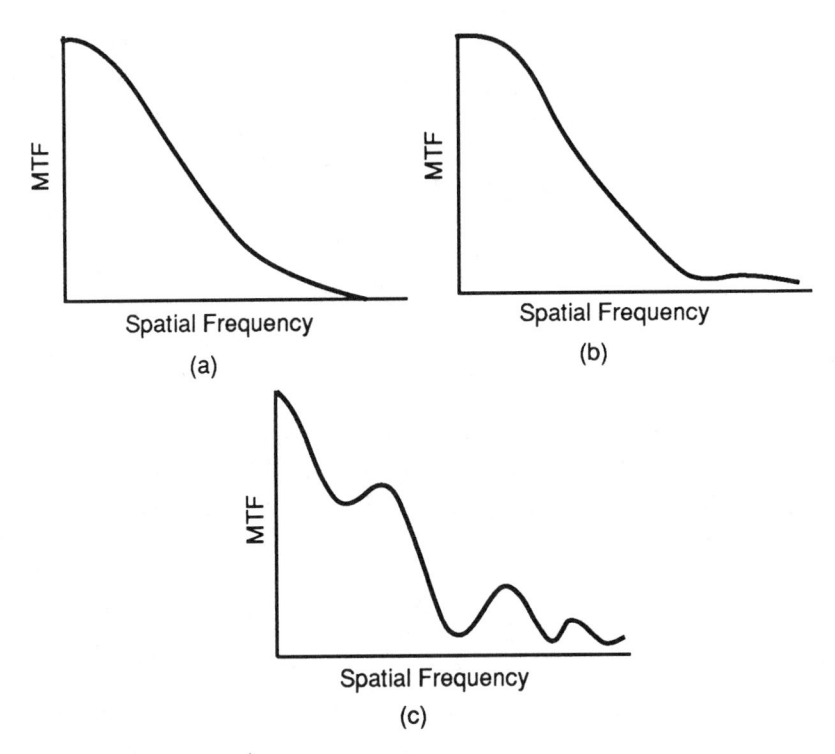

Figure 8-23: MTF degraded by noise. There were 200 samples across the ESF. (a) SNR = 500, (b) SNR = 50, and (c) SNR = 25 (From Reference 19).

CONTRAST, MODULATION & PHASE TRANSFER FUNCTIONS

Averaging in the time domain or in the frequency domain reduces temporal noise. When jitter and drift are present, the individual line spread functions will vary in time. Improper alignment of the individual time traces results in a broaden LSF and subsequent erroneous MTF. Averaging in the frequency domain is discussed in the Section 8.2.10.

If no jitter is present, time averaging offers an easy method to increase the SNR. Averaging can be performed by two methods. The same pixel can be averaged over several frames similar to the way the noise data set N_{VH} was obtained. If using a vertically oriented slit, the pixels can be averaged vertically or horizontally for a horizontal slit. This averaging is usually appropriate for staring arrays since they typically contain minimal jitter. However, the slit must be *critically* aligned to the array if vertical averaging is attempted.

8.2.9. LSF SYMMETRY

Typically, optical systems will produce symmetrical LSF images that are converted into electrical signals by the detectors. For systems that have an analog output, the symmetrical LSF is placed into a serial stream that passes through electronic circuitry. Electrical circuits are causal and will distort the symmetry. When boost circuitry is present, the LSF no longer has symmetry (Figure 8-24). Nonsymmetrical LSFs are more difficult to align when jitter is present.

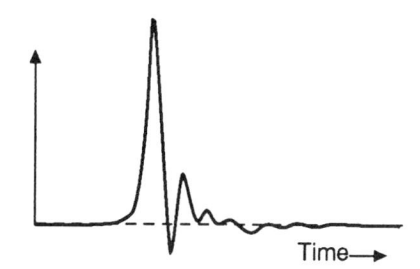

Figure 8-24: Electronic boost and AC coupling can create an asymmetrical LSF. The MTF is shown in Figure 8-28.

8.2.10. FOURIER TRANSFORM

Discrete Fourier transform (DFT) software is readily available. Therefore, the actual transform computation will not be discussed but, rather, data manipulation techniques so that the desired results are achieved. The fast Fourier transform (FFT) is commonly used and is a computationally efficient method of computing the DFT.

When operating on 2N (real-valued) data points, the DFT provides real, $\Re$, and imaginary, $\Im$, components of the complex transfer function, each having 2N points. The MTF from n = 0 to N-1 is duplicated from n = N to 2N-1 and therefore only N points are plotted. After amplitude normalization, the MTF for the nth data point is:

$$MTF(n) = \sqrt{\Re^2(n) + \Im^2(n)} \qquad n = 0, 1, .., (N-1) \qquad 8\text{-}11$$

The spatial frequency associated with the nth data point is related to the sampling rate, f_s, and the number of digitized samples (data points):

$$f_x(n) = \frac{n \, f_s}{2 \, N} \qquad n = 0, 1, ..., (N-1) \qquad 8\text{-}12$$

The frequency resolution, Δf, is $\Delta f = f_s/2N$. The maximum frequency that can be reconstructed is given by the sampling rate Nyquist criterion:

$$f_{max} = \frac{f_s}{2} \qquad 8\text{-}13$$

f_{max} is determined solely by the test equipment or test methodology. Depending upon the test configuration, it may or may not be related to the infrared imaging system Nyquist frequency.

Example 8-5
FREQUENCY RESOLUTION

What is the MTF frequency resolution when measuring a staring array consisting of 640H x 480V detectors? The system output is in the RS 170 format with a line rate of 52 µs. The FOV is 40 mrad by 30 mrad.

If the analog signal is digitized by a transient recorder every 5 nsec, then $f_s = 1/(5 \text{ nsec}) = 20$ MHz. If the LSF and its background are digitized into 1024 samples, then $\Delta f = 97.6$ kHz. The maximum signal that can be recovered by the Fourier transform is $f_s/2 = 10$ MHz. However, the array horizontal Nyquist frequency is $640/(2 \cdot 52 \text{ µs}) = 6.15$ MHz. Oversampling by the test equipment reduces phasing effects.

If a frame grabber is used to measure the vertical MTF and it captures all 480 lines, the vertical sampling interval is 40 mrad/480 = 0.0833 mrad. The Nyquist frequency is 6 cy/mrad. For the vertical direction, the test equipment Nyquist frequency is equal to the array Nyquist frequency.

The frequency resolution is inversely related to the product of the number of data points and the time between samples. Therefore, if a digitizer has a fixed memory size (say, 1024 points), then increasing the sampling rate or equivalently decreasing the time between samples (say, from 10 nsec to 5 nsec) decreases the resultant frequency resolution. Increasing the sample rate does increase the Nyquist frequency, however. Placing more samples on the LSF reduces phasing effects and minimizes potential variability in the computed MTF.

The input to the DFT has a record length of 2N. Padding the ends with zeros of the LSF data set (Figure 8-25) interpolates the computed MTF with a sin(x)/x function and provides a smoother MTF (Figure 8-26). Increasing the frequency resolution does not affect the system's resolution or MTF. It simply increases the accuracy of the computed MTF.

268 *TESTING & EVALUATION OF IR IMAGING SYSTEMS*

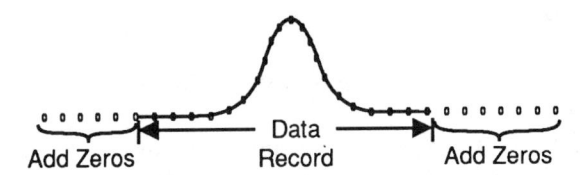

Figure 8-25: Padding zeros to increase the record length. The original data set consisted of 1024 data points. An additional 1024 zeros where added to create a new data record of 2048 points. The LSF should be in the center of the data record.

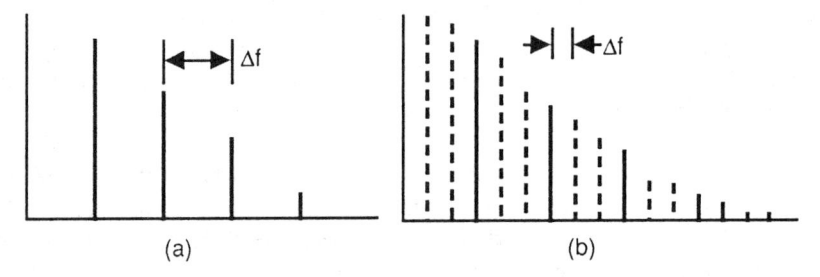

Figure 8-26: Padding zeros increases the apparent MTF resolution by interpolating the MTF with a sin(x)/x function. (a) Original MTF and (b) MTF after interpolation.

Using a finite length data record violates the Fourier transform requirement of continuity for all time. Discontinuities of partial cycles at the beginning or end of the sample interval cause errors. These errors, called leakage, may mask small amplitude frequencies that are present. To minimize leakage, the data is often windowed. The most popular window is the Hanning or raised cosine window[30]. With the Hanning window, the nth data point is multiplied by:

$$w_n = \frac{1}{2}\left[1 - \cos\left(\frac{2n\pi}{2N}\right)\right] \qquad n = 0, 1, ..., (2N-1) \qquad 8\text{-}14$$

The Fourier transform is then performed on the weighted data set. The choice of windows requires knowledge of the data to be collected. Since the LSF is a high frequency response, the window can be approximated by the high pass filter used to separate the noise components (Figure 7-11: page 204). The MTF should be calculated with and without a window and the results compared. However, if the LSF is placed in the center of the data set (i.e., the LSF is centered on data point N), the window should not be required. The window should be used when transforming any data that have nonzero values at the beginning or end of the data record.

Noise introduces a statistical error in the power estimation that is independent of the record length[32]. Increasing the record length does not affect the magnitude of the error. To reduce the error, multiple power spectra must be computed and then averaged at each frequency component to obtain a composite (ensemble average) spectra:

$$P_1(f) = \mathfrak{R}_1^2(f) + \mathfrak{I}_1^2(f)$$
$$P_2(f) = \mathfrak{R}_2^2(f) + \mathfrak{I}_2^2(f)$$
$$\cdots$$
$$P_k(f) = \mathfrak{R}_k^2(f) + \mathfrak{I}_k^2(f)$$

8-15

where k is the number of different data sets from which the spectra are computed. That is, the LSF is obtained k times and the power spectrum is calculated from each transformed LSF. The composite spectra is

$$\overline{P(f)} = \frac{\sum_{i=1}^{k} P_i(f)}{k}$$

8-16

and the average MTF is the square root of the averaged power spectra:

$$\overline{MTF(f)} = \sqrt{\overline{P(f)}}$$

8-17

The methodology is illustrated in Figure 8-27.

270 *TESTING & EVALUATION OF IR IMAGING SYSTEMS*

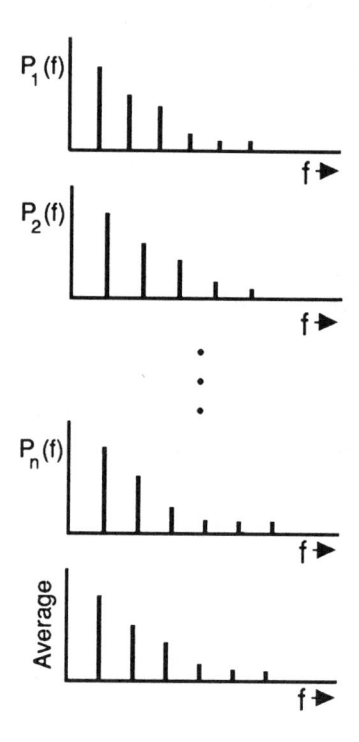

Figure 8-27: Methodology to average power spectral density curves to obtain a composite power spectra. The MTF is the square root of the averaged spectra.

8.2.11. AMPLITUDE NORMALIZATION

For passive, linear-phase-shift systems, the MTF is normalized to unity at zero spatial frequency and it decreases as the frequency increases. Setting the MTF to unity at zero spatial frequency is equivalent to normalizing the area under the LSF to unity. Image enhancement techniques and boost circuitry (active filter circuits) may increase the MTF at some frequencies. With the normalization to unity at zero spatial frequency, the system MTF may be greater than unity at the boost frequency (Figure 8-28). With AC coupled systems, the DC (zero spatial frequency) component is suppressed and this prevents normalization at zero. Here, it is recommended to normalize to unity at a spatial frequency approximately one decade above the AC cut-on (break) frequency or at the first maximum. Often the AC coupling break frequency is below the resolution of the Fourier transform and therefore is not even measured.

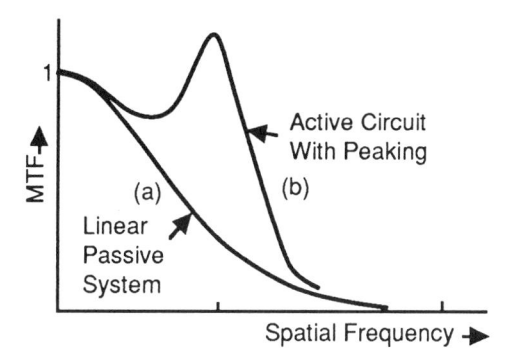

CONTRAST, MODULATION & PHASE TRANSFER FUNCTIONS 271

Figure 8-28: MTF obtained from (a) a symmetrical LSF and (b) the LSF shown in Figure 8-24 (page 265).

8.2.12. FREQUENCY SCALING

The DFT transforms the data into nonspecific frequency units of cycles/sample. The user is left to convert the DFT into the appropriate units. If the data samples are measured in object space (mrad), as would be the case for the moving slit or periodic array of pinholes, the DFT directly provides spatial frequency (cy/mrad). When analog time data are used, the DFT transforms the data into the electrical frequency domain. The conversion of electrical frequency data, f_{Hz} into spatial frequency data, f_x, requires knowledge of the field-of-view and the active time (TV) for one TV line:

$$f_x = \frac{TV}{FOV} f_{Hz} \qquad 8\text{-}18$$

where TV has units of seconds and the FOV is measured in mrad to create cy/mrad. The system line rate may vary from the published values and the system field-of-view may vary up to 5% due to focal length variations. This greatly affects the scale factor and therefore the interpretation of the resultant MTF. Using different scale factors produces MTFs that appear different.

Example 8-6
SCALE FACTOR

If the focal length varied by 5% from system-to-system, what effect will it have on the computed MTF. The system's nominal FOV is 600 mrad and the line rate is 52 μsec.

A LSF was obtained on the analog video and is assumed to be Gaussian in shape with $\sigma = 39 \times 10^{-9}$ sec. The Fourier transform will produce the Gaussian MTF:

$$MTF = e^{-2\pi^2 \sigma^2 f_{Hz}^2}$$

Conversion to spatial frequency provides

$$MTF = e^{-2\pi^2 \sigma^2 \left(\frac{FOV}{TV}\right)^2 f_x^2}$$

Since the detector array size is assumed fixed, a 5% variation in focal length will produce approximately a 5% variation in the FOV (570 to 630 mrad). This variation results in approximately a 10% variation in MTF at mid spatial frequencies. In Figure 8-29 the variations are plotted and the absolute difference in the MTF for a +5% variation in FOV. Methods to measure the FOV are provided in Chapter 9.

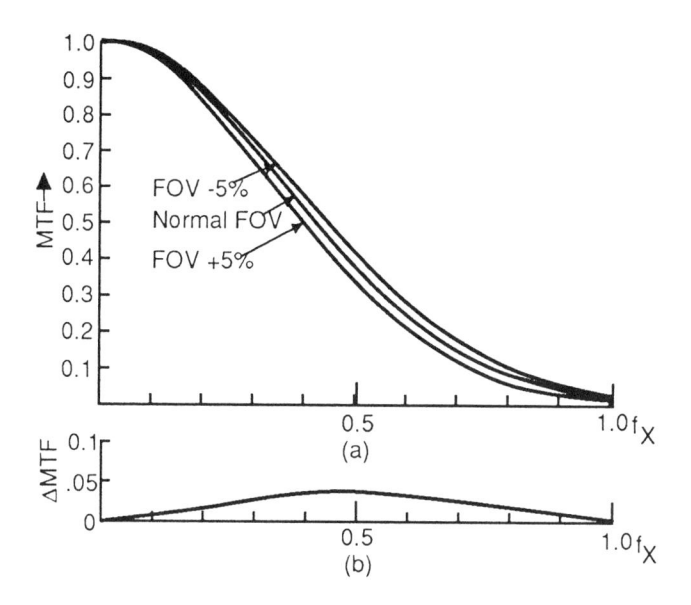

CONTRAST, MODULATION & PHASE TRANSFER FUNCTIONS 273

Figure 8-29: Apparent variations in MTF due to different scale factors. (a) Since the scale factor is proportional to the field-of-view, a 5% variation in the FOV results in a 5% variation in the scale factor and (b) the absolute difference for a +5% change in FOV.

If the infrared imaging system operates in the digital domain and the clock rate is R_{clock} (pixels/sec) and the number of pixels is P per line, then:

$$TV = \frac{P}{R_{clock}} \qquad 8\text{-}19$$

The time-to-frequency scale factor can be *estimated* by measuring the system MTF when the system is viewing a large slit. If the slit angular subtense, α, is large compared to the IFOV, the measured MTF will equal zero at $f_x = k/\alpha$ where k is an integer (Figure 8-30). The zero crossing provides an estimation of the conversion factor from electrical frequency to spatial frequency. The electrical frequency, f_1, at which the MTF equals zero provides $f_x = f_{Hz}/(\alpha f_1)$. Δf limits the accuracy of this approach.

274 TESTING & EVALUATION OF IR IMAGING SYSTEMS

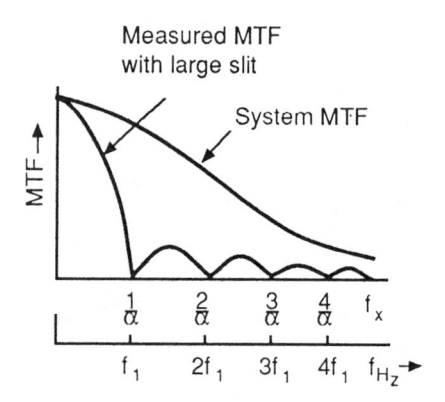

Figure 8-30: Scale calibration with a large slit. The MTF is zero at $f_x = 1/\alpha$ where α is the slit angular subtense.

Example 8-7
SLIT WIDTHS

The infrared imaging system's IFOV is 2 mrad. What size slits are required to measure the MTF and for scale factor calibration? The collimator focal length is 1.5 m.

The MTF requires a slit whose angular subtense is 0.1·IFOV or 0.2 mrad. The required slit width is $(0.2 \times 10^{-3})(1.5 \text{ m}) = 0.3$ mm. For the scale factor, it is desired to have 2 additional zeros across the system MTF curve. The detector first MTF zero occurs at $f_x = 1/\text{IFOV} = 0.5$ cy/mrad. Desired zeros occur at 0.167, 0.333 and 0.5 cy/mrad. The slit angular subtense is associated with the first zero and $\text{IFOV}_{scale} = 1/0.167 = 6$ mrad. The required slit width is $(6 \times 10^{-3})(1.5 \text{ m}) = 9$ mm. If another subsystem component other than the detector dominates the system MTF, then the first zero, or at least where the MTF appears to approach zero is used for the calculation.

CONTRAST, MODULATION & PHASE TRANSFER FUNCTIONS 275

8.2.13. TEST CONFIGURATION MTF

With linear shift-invariant systems the measured MTF is a product of all the subsystem MTFs. These subsystems include the system under test and the test equipment (Figure 8-31). In principle, the introduction of collimators, slits, additional lenses, and cameras can be accurately accounted for in the resultant data. These additional component MTFs can be factored out of the data to leave the MTF of the system only:

$$MTF_{SYSTEM} = \frac{MTF_{MEASURED}}{MTF_{COLLIMATOR} \, MTF_{TARGET} \, MTF_{DATA \, ACQUISTION}} \qquad 8\text{-}20$$

If a slit of angular width α is used, the target MTF is given by $MTF_{TARGET} = \sin(\pi\alpha f_x)/\pi\alpha f_x$. An edge does not require a correction. If the collimator aperture is larger than the system entrance aperture, $MTF_{COLLIMATOR} = 1$. When measuring the analog video the data acquisition MTF should approach unity. If the monitor is measured with a scanning photometer or slit/sensor combination additional MTF corrections maybe necessary. This slit must be considered as another element with its own MTF.

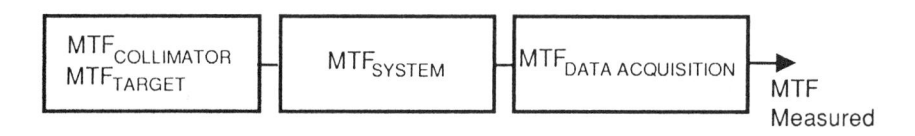

Figure 8-31: The measured MTF depends on the test equipment MTFs.

8.2.14. MTF TEST PROCEDURES

This section summarizes the preceding sections. A generic MTF test configuration was shown in Figure 8-8 (page 248) with the data methodology shown in Figure 8-9 (page 249). The collimator aperture should be larger than the infrared imaging system aperture (Figure 4-20: page 110). No correction is needed for the collimator transmittance since the MTF is a relative measurement.

The DFT provides only a finite number of data points. Rather than represent the MTF as a continuous curve, a better representation would show

276 *TESTING & EVALUATION OF IR IMAGING SYSTEMS*

the discrete frequency components. This highlights the discrete nature of the measurement process and the frequency resolution of the output (Figure 8-32). The Fourier transform does not *know* where the system Nyquist frequency is located and therefore produces an MTF that is limited by the measuring equipment sampling rate. As a result, the system Nyquist frequency should be shown on all graphs. If padded zeros were added, it also should be noted on the graph.

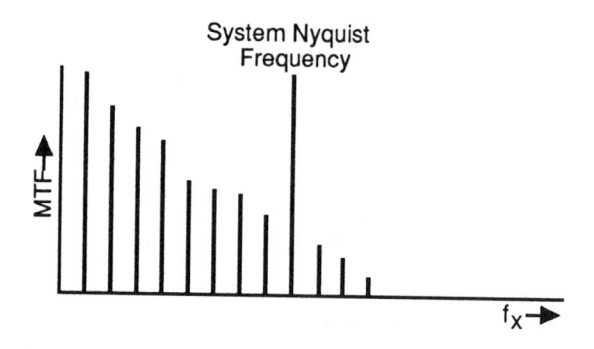

Figure 8-32: Discrete representation of the MTF with the system Nyquist frequency added.

The MTF test set up is given in Table 8-5 and the MTF test procedures are shown in Table 8-6. Possible causes of errors are given in Table 8-7. The large slit used for frequency scaling can also be used for algorithm verification. Here the shape of the curve is important rather than the scale factor. If the slit MTF dominates the measured MTF, the measured MTF should follow $\sin(\pi \alpha f_x)/\pi \alpha f_x$.

Table 8-5
MTF TEST SETUP

1. Establish a test philosophy, criteria for success and write a thorough test plan (Section 1.3: page 12).
2. Verify that the test equipment is in good condition and that the test configuration is appropriate (Figure 8-8: page 248). Ask previous users if any problems were noticed. Place the source, target, collimator and infrared imaging system on a vibration-isolated optical table. Minimize all air currents (Section 4.4: page 112).
3. Verify infrared imaging system is in focus (Section 5.2: page 137).
4. Insure that the infrared imaging system has reached operating equilibrium before proceeding.
5. Set the source intensity at a sufficiently high intensity so that the signal-to-noise ratio is high.
6. Align the target with the array axes.
7. Verify that the system is operating in a linear region. This can be estimated from the responsivity curve (Table 6.3: page 175) and is verified by performing the Fourier transform for different source intensities (Figure 8-15 and Figure 8-16: page 255).

Table 8-6
MTF TEST PROCEDURES

1. Reconstruct the LSF from the various data sets (Section 8.2.3: page 250). Verify that there are at least four samples across the LSF at FWHM (Figure 8-18: page 257).
2. Remove the background pedestal (Section 8.2.6: page 260).
3. Remove jitter (Section 8.2.7: page 261).
4. Perform the Fourier transform and calculate the power spectrum.
5. Normalize the amplitude (Section 8.2.11: page 270).
6. Repeat steps 1-6 as necessary to obtain a statistical representation of the power spectrum. Average all power spectra (Section 8.2.10: page 266), then take the square root to obtain the MTF.
8. Convert the normalized DFT frequencies to spatial frequency. (Section 8.2.12: page 271).
9. Divide the measured MTF by the equipment MTF (Section 8.2.13: page 275).
10. Fully document any test abnormality and all results. The recorded data should include as a minimum: the raw data, window used, source intensity, ambient temperature, method to remove trends, and any other pertinent measuring data. The measured LSF and calculated MTF should be shown both graphically and in tabular form with the system Nyquist frequency clearly indicated.

Table 8-7
REASONS FOR POOR MTF

- Measuring equipment MTFs not removed (Section 8.2.13: page 275).
- System cannot provide optimum focus.
- Amplitude not normalized appropriately (Section 8.2.11: page 270).
- Signal out of linear region (Figure 8-15: page 255).
- Poor trend or pedestal removal (Section 8.2.6: page 260).
- Blackbody temperature or ambient temperature varies during the test.
- Inaccurate frequency conversion factor (Section 8.2.12: page 271).

8.3. PHASE TRANSFER FUNCTION

When operating on 2N (real valued) data points, the DFT provides real, $\Re$, and imaginary, $\Im$, components of the complex transfer function, each having 2N points. The PTF from n = 0 to N-1 is repeated for n = N to 2N-1 and therefore only N points are presented. The PTF for the nth data point is obtained from the complex transfer function:

$$PTF(n) = \tan^{-1}\left(\frac{\Im(n)}{\Re(n)}\right) \qquad 8\text{-}21$$

The PTF is obtained by the same methodology given in Tables 8-5 and 8-6. When a symmetrical LSF is placed at the origin, no phase shift exists whereas if the symmetrical LSF is displaced from the origin, a linear-phase-shifts results (Figure 8-33). Linear-phase-shifts provide no information since the choice of coordinates can eliminate it. Any PTF deviation from linearity is a measure of the LSF asymmetry[33]. Optical systems may have approximately symmetrical LSFs if there are minimum aberrations, but electronic circuitry will generally distort the LSF. AC coupling, boost circuitry and nonlinear image processing algorithms will emphasize the asymmetry (Figure 8-24: page 265).

The linear portion of the PTF can be removed by placing the LSF mathematically at the origin (or at least at the data point nearest the origin). The shift to the origin is equivalent to placing the LSF in the center of the digital data. For example, if the transient recorder collects 2N samples, the LSF should be centered at N+1. To slide the LSF to the origin before performing the Fourier transform usually requires interpolation and thereby slightly corrupts the data. An alternate approach is to calculate the LSF centroid location and use this information to remove the linear-phase-shift. The LSF centroid is located at:

$$m = \frac{\sum_{n=1}^{2N} n\,D(n)}{\sum_{n=1}^{2N} D(n)} \qquad 8\text{-}22$$

where the nth data point of the LSF has a value of D(n).

280 TESTING & EVALUATION OF IR IMAGING SYSTEMS

The centroid is offset from the origin by

$$t_o = \frac{m - (N + 1)}{f_s} \qquad 8\text{-}23$$

The linear-phase-shift,

$$\theta(f) = -2\pi f t_o \qquad 8\text{-}24$$

can be subtracted from the PTF to leave the nonlinear portion:

$$PTF(f)_{NONLINEAR} = \tan^{-1}\left(\frac{\Im(f)}{\Re(f)}\right) - \theta(f) \qquad 8\text{-}25$$

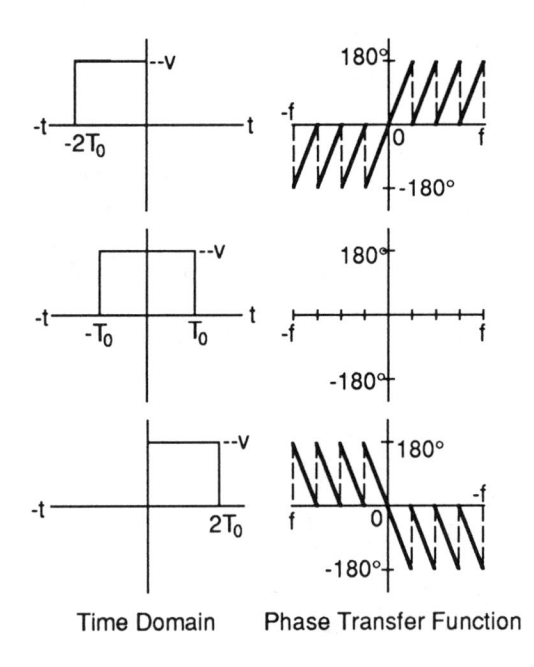

Figure 8-33: Effects of the origin location on the phase transfer function. With a symmetrical input, the PTF is linear for any arbitrary origin.

CONTRAST, MODULATION & PHASE TRANSFER FUNCTIONS 281

Most DFT software provides the real and imaginary components of the complex transfer function. However, it is usually presented by magnitude and phase. For convenience, the phase shift is typically bounded from -180° to +180°. $\theta(f)$ must track with the calculated PTF. When the PTF switches from -180° to +180°, $\theta(f)$ must switch by 360°.

Jitter dramatically affects the centroid location. As a result, the linear-phase-shift must be removed from each independent complex transfer function before averaging. The large slit used to estimate the scale factor can be simultaneously used to verify the phase relationship. For the large slit, the phase changes by 180° after each MTF zero (Figure 8-34).

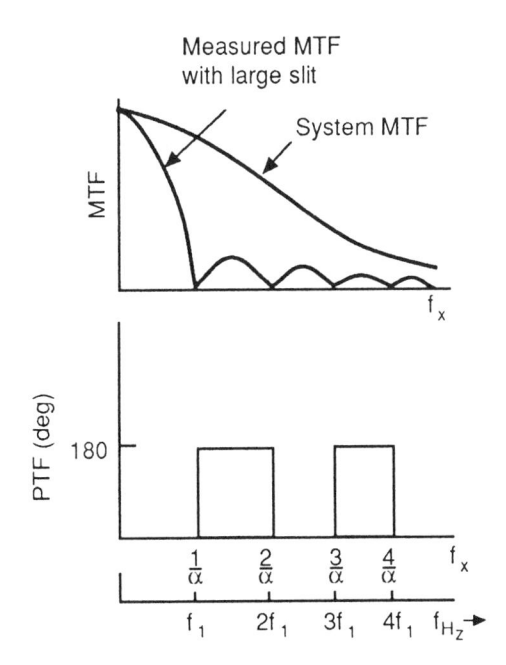

Figure 8-34: PTF calibration using a large slit. The PTF will change by 180° at $f_x = k/\alpha$.

8.4. SUMMARY

The MTF is a fundamental parameter used to characterize an infrared imaging system's ability to reproduce signals as a function of spatial frequency. It guides system design and predicts system performance. The contrast transfer function and slit response function can be calculated from the MTF. Coupled with the three-dimensional noise model and the eye response, the minimum resolvable temperature and minimum detectable temperature can be calculated[34]. By monitoring the MTF or CTF at a specific spatial frequency as the focus lens-assembly is moved, "best" focus can be determined (Section 5.2.3: page 142). Resolution measures that are derived from the MTF or CTF include the instantaneous-field-of-view, effective instantaneous-field-of-view and limiting resolution.

The test engineer must understand the complexity of CTF, MTF and PTF testing so he can choose the appropriate test configuration and data analyses. There are two general approaches for determining the MTF. The direct method; which consists of measuring the response to sinusoidal or bar targets; and the indirect method, in which a one-dimensional MTF is obtained from the Fourier transform of the line spread function. There are benefits and shortcomings to both methods. Sinusoidal patterns are available in the visible but not easy to fabricate for the infrared. One can use square (bar) targets to obtain the contrast transfer function, CTF, and mathematically convert to the sinusoidal response (MTF) using a series approximation.

Theoretically, to obtain the MTF at frequency f_x, an infinite number of square wave responses must be measured. However, the number required is limited by the spatial cutoff frequency, $f_{MTF=0}$, where the MTF approaches zero and remains zero there after. For bar targets whose spatial frequencies are above ⅓ $f_{MTF=0}$, the MTF is equal to $\pi/4$ times the measured CTF. That is, above ⅓ $f_{MTF=0}$, only one target is necessary to provide the MTF. Unless there is evidence otherwise, the optical spatial frequency cutoff should be used for $f_{MTF=0}$. For sampled data systems, the mathematical relationship between CTF and MTF has not yet been developed. Therefore the relationship should be used cautiously for sampled data systems, if at all. For staring arrays and scanning arrays that are sampled, unambiguous CTF values can only be obtained at $f_x = f_N/k$ for k = 1, 2, ... These spatial frequencies provide the in-phase CTF.

MTF, CTF and PTF measurements, while they appear straight forward, may be difficult to perform due to electronic nonlinearity, digitization effects, phasing effects, background removal, jitter, noise, and improper normalization. The magnitude of each of these effects depends upon the infrared imaging system design and measurement techniques. Methods to measure the MTF depend upon both optical and electronic signal considerations. An MTF graph without appropriate commentary is essentially useless. To achieve reproducibility, it is necessary to define the measuring conditions. With different measurement philosophies, it is difficult to ascertain whether different results are due to the difference in measurement technique, test equipment, data analysis or due to differences in the system under test: the desired result. In fact, results from different laboratories can have variations[35;36] as large as ±0.05. A particular laboratory can achieve consistent, small variance results, but with a constant bias from the true MTF. As a result, verification of an MTF specification should be approached with care.

Sometimes the MTF is specified at only one spatial frequency, f_o (pronounced f naught). f_o is $1/(2\alpha)$ where α is the detector IFOV. For staring arrays with 100% fill factor, f_o is equal to f_N. Just specifying the MTF at one or few frequencies does not fully characterize the system and provides little insight into system performance; it does, however, represent convenient test points.

Since systems have chromatic aberrations, the MTF is dependent upon the wavelength of the incident radiation. For broad band spectral response systems, the MTF is a function of the spectral response of the system as well as the spectral characteristics of the source. It is appropriate to call the resultant MTF the polychromatic MTF.

A wealth of information may be found in the shape of the MTF curve. A summary of these effects is given in Table 8-8 and the effects are illustrated in Figure 8-35. Unexpected LSFs or MTFs can occur when there are image processing algorithms or line-to-line interpolation algorithms present.

Table 8-8
MTF VARIATIONS

MTF VARIATIONS	POSSIBLE CAUSES
Suppressed at zero spatial frequency (Figure 8-35a).	AC coupling.
Peak at low frequency (Figure 8-35b).	Trends not removed from the LSF data.
Peak in curve (Figure 8-35c).	Boost circuitry, nonlinear image processing.
Too low (Figure 8-35d).	System out of focus. System in nonlinear region. Jitter broadened LSF (when averaging). Wrong frequency scale factor. Insufficient samples on LSF.
Too high (Figure 8-35e).	Too much background removed. Wrong frequency scale factor. Noise bias. Insufficient samples on LSF.
Low frequency components (Figure 8-35f).	Insufficient background removed.
Not reproducible.	Variations in all of the above.

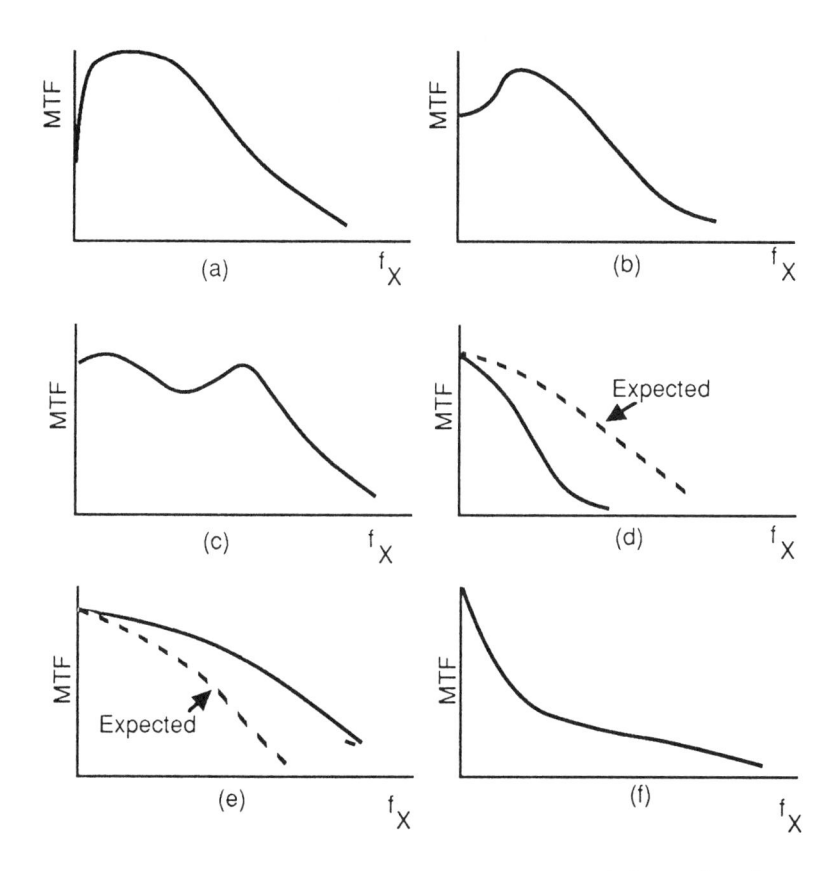

Figure 8-35: Variations in the MTF. See Table 8-8 for explanation.

8.5. REFERENCES

1. E. M. Granger, "Visual Limits to Image Quality", in Digital Image Processing: Critical Review of Technology, A. Tescher, ed.: SPIE Proceedings Vol. 528, pp. 95-102 (1985).
2. J. M. Lloyd, Thermal Imaging, pp. 109-114: Plenum Press, New York (1975).
3. W.J. Smith, "Optical Design", in Electro-Optical Components, W. D. Rogatto, ed., Volume 3 of The Infrared & Electro-Optical Systems Handbook: Environmental Research Institute of Michigan, Ann Arbor, Mich, (1993).
4. K. J. Barnard, G. D. Boreman, A. E. Plogstedt, and B. K. Anderson, "Modulation-transfer Function Measurement of SPRITE Detectors: Sine-wave Response", Applied Optics, Vol. 31 (1), pp. 144-147 (1992).
5. M. Marchywka and D. G. Socker, "Modulation Transfer Function Measurement Technique for Small-pixel Detectors", Applied Optics, Vol. 31(34), pp. 7198-7213 (1992).
6. M. Sensiper, G. D. Boreman, A. D. Durchame, and D. R. Snyder, "Modulation Transfer Function Testing of Detector Arrays Using Narrow-band Laser Speckle", Optical Engineering, Vol. 32(2), pp. 395-400, (1993).

7. A. H. Lettington and Q. H. Hong, "Measurement of the Discrete Modulation Transfer Function", Journal of Modern Optics, Vol. 40(2), pp. 203-212 (1993).
8. J. W. Coltman, "The Specification of Imaging Properties by Response to a Sine Wave Input", Journal of the Optical Society of America, Vol. 44(6), pp. 468-471 (1954).
9. I. Limansky, "A New Resolution Chart for Imaging Systems", The Electronic Engineer, Vol. 27(6), pp. 50-55 (1968).
10. S. K. Park and R. A. Schowengerdt, "Image Sampling, Reconstruction and the Effect of Sample-scene Phasing", Applied Optics, Vol. 21(17), pp. 3142-3151 (1982).
11. T. S. Lomheim, L. W. Schumann, R. M. Shima, J. S. Thompson, and W. F. Woodward, "Electro-Optical Hardware Considerations in Measuring the Imaging Capability of Scanned Time-delay-and-integrate Charge-coupled Imagers", Optical Engineering, Vol. 29(8), pp. 911-927 (1990).
12. S. E. Reichenbach, S. K. Park, and R. Narayanswamy, "Characterizing Digital Image Acquisition Devices", Optical Engineering, Vol. 30(2), pp. 170-177 (1991).
13. W. Wittenstein, J. C. Fontanella, A. R. Newberry, and J. Baars, "The Definition of the OTF and the Measurement of Aliasing for Sampled Imaging Systems", Optica Acta, Vol. 29(1), pp. 41-50 (1982).
14. J. C. Felz, "Development of the Modulation Transfer Function and Contrast Transfer Function for Discrete Systems, Particularly Charge Coupled Devices", Optical Engineering, Vol. 29(8), pp. 893-904 (1990).
15. S. K. Park, R. A. Schowengerdt and M. Kaczynski, "Modulation Transfer Function Analysis for Sampled Image Systems", Applied Optics, Vol. 23(15), pp. 2572-2582 (1984).
16. L. deLuca and G. Cardone, "Modulation Transfer Function Cascade Model for a Sampled IR Imaging System", Applied Optics, Vol. 30(13), pp. 1659-1664 (1991).
17. S. J. Pruchnic, G. P. Mayott, and P. A. Bell, "Measurement of Optical Transfer Function of Discretely Sampled Thermal Imaging Systems" in Infrared Imaging Systems: Design, Analysis, Modeling and Testing III, G. C. Holst, ed.: SPIE Proceedings Vol. 1689, pp. 368-378 (1992).
18. T. L. Williams and N. T Davidson "Measurement of the MTF of IR Staring Imaging Systems", in Infrared Imaging Systems: Design, Analysis, Modeling and Testing III, G. C. Holst, ed.: SPIE Proceedings Vol. 1689, pp. 53-63 (1992).
19. B. White and M. Marquis, "Vertical MTF Measurements", in Infrared Imaging Systems: Design, Analysis, Modeling and Testing IV, G. C. Holst, ed.: SPIE Proceedings Vol. 1969 (1993).
20. S. R. Hawkins, R. P. Farley, G. Gal, A. K. Gresle, S. B. Grossman, and W. G. Opyd, "Focal Plane Spatial Response and Modulation Transfer Function (MTF) Measurements using a Computer-aided Flying Spot Scanner" in Modern Utilization of Infrared Technology VIII, I. Spiro, ed.: SPIE Proceedings Vol. 366, pp. 41-49 (1982).
21. R. F. Rauchmiller and R. A. Schowengerdt, "Measurement of the Landsat Thematic Mapper Modulation Transfer Function using an Array of Point Sources", Optical Engineering, Vol. 27(4), pp. 334-343 (1988).
22. D. J. Bradley, C. J. Braddiley, and P. N. J. Dennis, "The Modulation Transfer Function of Focal Plane Arrays", in Passive Infrared Systems and Technology, H. M. Lamberton, ed.: SPIE Proceedings Vol. 807, pp. 33-41 (1987).
23. H-S. Wong, "Effect of Knife-edge Skew on Modulation Transfer Function Measurements of Charge-coupled Device Imagers Employing a Scanning Knife Edge", Optical Engineering, Vol. 30(9), pp. 1394-1398 (1991).
24. W. W. Frame, "Minimum Resolvable and Minimum Detectable Contrast Prediction for Monochrome Solid-state Imagers", SMPTE Journal, Vol. 96(3), pp. 454-459 (1987).
25. E. M. Granger, "Image Quality Analysis and Testing for Infrared Systems", SPIE Tutorial Short Course, T60, Orlando Fl, (March 1989).

26. J.C. Dainty and R. Shaw, Image Science, pg. 204: Academic Press, New York (1974).
27. H. J. Pinsky, "Determination of FLIR LOS Stabilization Errors" in Infrared Imaging Systems: Design, Analysis and Testing II, G. C. Holst, ed.: SPIE Proceedings Vol. 1488, pp. 334-342 (1991).
28. F. H. Slaymaker, "Noise in MTF Measurements", Applied Optics, Vol. 12(11), pp. 2709-2715 (1973).
29. D. Dutton, "Noise and Other Artifacts in OTF Derived From Image Scanning", Applied Optics, Vol. 14(2), pp. 513-521 (1975).
30. H. P. Stahl, "Infrared MTF Measurements of Optical Systems", Lasers and Optronics, Vol. 10(4), pp. 71-72 (1991).
31. F. J. Harris, "On the Use of Windows for Harmonic Analysis with the Discrete Fourier Transform", Proceedings IEEE, Vol. 66(1), pp. 51-83 (1978).
32. J. S. Bendat and A. G. Piersol, Random Data: Analysis and Measurement Procedures, 2nd edition, pp. 283-286: John Wiley and Sons, New York, (1986).
33. R. E. Jodoin, "Linear Phase Shift Removal in OTF Measurements", Applied Optics, Vol. 25(8), pp. 1261-1262 (1986).
34. L. Scott and J. D'Agostino, "NVEOD FLIR92 Thermal Imaging Systems Performance Model" in Infrared Imaging Systems: Design, Analysis, Modeling and Testing III, G. C. Holst, ed.: SPIE Proceedings Vol. 1689, pp. 194-203 (1992).
35. Z. Bing-Xun and C. Gen-Rui, "Optical Transfer Function (OTF) Measurement Technique Application in China", in Assessment of Imaging Systems, T. L. Williams, ed.: SPIE Proceedings Vol. 274, pp. 60-68 (1981).
36. A. C. Marchant, E. A. Ironside, J. F. Attrude, and T. L. Williams, "The Reproducibility of MTF Measurements", Optica Acta, Vol. 22(4), pp. 249-264 (1975).

EXERCISES

1. CTF measurements on a MWIR infrared system provided the following values. The system's clear aperture is 3.6 cm. What is the calculated MTF?

f_x	1	2	3	4	5	6	7	8
CTF	0.99	0.98	0.91	0.85	0.72	0.40	0.35	0.10

2. The same CTF values given in Exercise 1 were obtained on a MWIR infrared system whose clear aperture is 5 cm. Calculate the CTF. Discuss the difference between the two systems?
3. Discuss the problems associated with measuring the CTF.
4. List five concerns about measuring the MTF.
5. What is the frequency scale factor if the specifications state: (a) the FOV is 40 mrad and the output video is RS170, (b) the output pixel rate is 640 pixels/line and the clock rate is 1 MHz? What is the correct way to obtain the scale factor?
6. Using Example 8-6 as a guide, plot the apparent MTF differences if the FOV varied by $\pm 10\%$. The FOV is 600 mrad.

9

GEOMETRIC TRANSFER FUNCTION

The geometric transfer function is a generic term incorporating all input-to-output transformations that include system geometric performance and replication of target geometry. System geometric performance includes field-of-view, distortion and scan linearity. With a perfect infrared imaging system, the image seen on the monitor will be an exact representation of the object in every geometric respect. Automated vision systems (machine vision) rely on geometric transfer properties in that the output is derived from the target geometry: measuring shape, size, motion, or counting objects. These systems have applications in variety of fields such as aerospace (trackers, correlators and automatic target recognizers), manufacturing (pattern recognition, part location, part inspection, and quality control), guidance of robots, medicine and biology (cell counting and identification).

The optical subsystem can cause distortion in both the horizontal and vertical directions. With scanning systems, distortion also can be caused by scan nonlinearity (distortion in the scan direction) or by improper alignment of the forward and interlace fields (cross scan distortion). Distortion requirements may vary depending upon the region of interest (Figure 9-1). Generally, the best imagery (minimum distortion) is required in the center of the field-of-view ("sweet spot") with allowable imperfections increasing toward the periphery. Unless severe, distortion is generally not noticed in real-world imagery since objects tend to have smooth contours. In the laboratory, distortion becomes obvious when viewing well-defined geometric targets. The scan nonlinearity tests also can be used for aligning the scan mirror and verifying electronic timing. For systems employing TDI, the test methodology also can be used to adjust scan velocity to maximize the MTF.

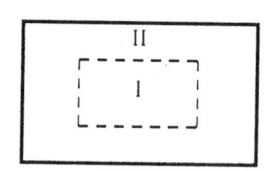

Figure 9-1: Distortion Requirements. Distortion in region I is usually required to be less than in region II.

Since distortion tests are often visual observations, the monitor aspect ratio must be matched to the infrared imaging system's aspect ratio[1]. For example, if the system has an aspect ratio of 1:1 but is formatted into a modified RS 170 video signal, the monitor also should be capable of providing a 1:1 aspect ratio image when driven by that signal.

Because information is processed differently by machine vision systems, observer interpretation of image quality may not be appropriate for machine vision evaluation. Important factors for machine vision include responsivity, noise, MTF, ATF, the speed to process target information, repeatability, accuracy and the ability to distinguish closely spaced objects. At this time, tests to characterize automatic vision systems are formative.

9.1. FIELD-OF-VIEW

The field-of-view (FOV) is the maximum vertical and horizontal angular extent viewed by the infrared imaging system. The field-of-view can be measured by a variety of techniques. It is important to insure that the full field-of-view is presented on the monitor and that the monitor is not a limiting factor (e.g., insure that the monitor is not in an over scan mode).

The FOV can be measured by moving a small target from one edge of the FOV to the other. The total angle through which the target has moved is the FOV. Rotating the infrared system about an axis passing through the entrance nodal point of the system (Figure 9-2) also moves the target across the FOV. By pivoting about the nodal point, there are no vignetting and all the radiation from the collimator that is incident onto the entrance aperture enters the FOV. The test procedure is straight forward:

A cross hair, pinhole target, or other similar target is placed in the focal plane of the collimator. The size of the target is not critical if it is small enough to provide the required measurement accuracy. The infrared system is rotated in azimuth until the target is visually at the edge of the field-of-view and the rotary table angular position is recorded. Then the system is rotated in the opposite direction until the target is visually on the other side of the field-of-view and this angular position is recorded. The horizontal field-of-view is the total angle through which the system has been rotated. The test is repeated for the vertical field-of-view by tilting the system in elevation. If a tilting table is not available, then the infrared system can be rotated 90 degrees (e.g., "turned on its side") with respect to the holder and again moved in azimuth.

290 *TESTING & EVALUATION OF IR IMAGING SYSTEMS*

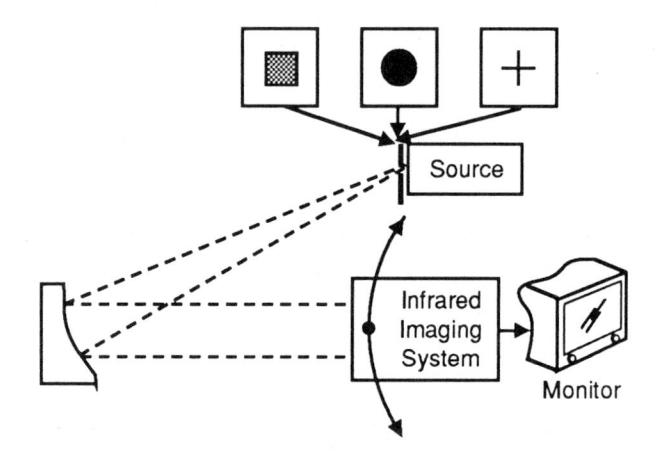

Figure 9-2: Generic test configuration to measure field-of-view with the infrared imaging system mounted on a rotary table. The pivot point must be at the entrance nodal point.

If the imaging system is too large to be placed on a rotary table, a small collimator can be placed on a movable boom (Figure 9-3). The boom pivot point should be at the system's entrance nodal plane. Using the same test methodology, the boom is rotated in azimuth until the target can no longer be visually seen and the angle is recorded. The boom is then rotated in the other direction until the target again just disappears. The horizontal field-of-view is the full angle over which the boom was moved. It is generally impractical to design a boom that can move in elevation. Therefore to measure the vertical field-of-view, the infrared imaging system is rotated 90 degrees with respect to the holder and the boom is moved in azimuth again. It is important that the target moves perpendicularly to the edges of the FOV.

The visual method introduces some ambiguity depending upon the size of the target. Measurement accuracy can be improved by monitoring the output as a function of target angular location. The FOV is the angular width where the system output is 50% of the maximum (Figure 9-4). This method is nearly insensitive to the target size.

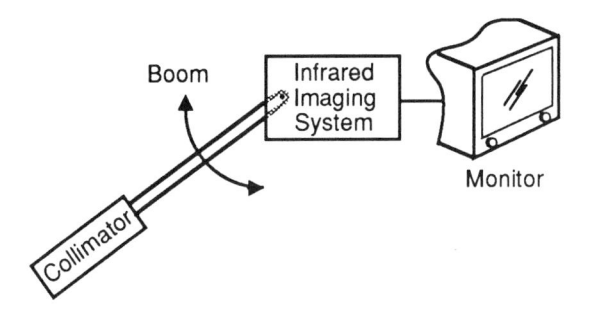

Figure 9-3: Generic test configuration to measure field-of-view using a movable boom. The boom pivot point must be at the system's entrance nodal point.

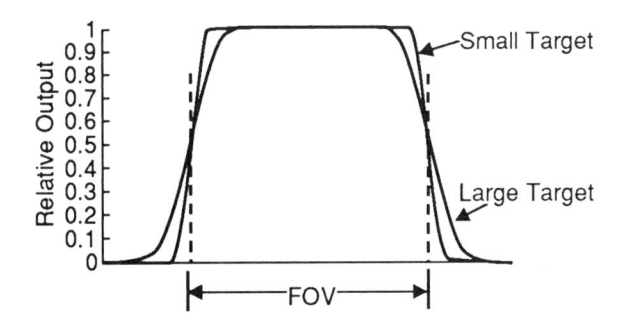

Figure 9-4: Field-of-view definition. A small target provides sharper edges than a large target but the 50% response points remain constant.

If it is not convenient to measure the full field-of-view by the above methods, choose a target as large as possible and note the size of the image on the monitor and compare this to the total size of the displayed FOV (Figure 9-5).

The horizontal and vertical field-of-views are estimated from the ratios:

$$HFOV = \alpha_x \frac{D_X}{d_X}$$

$$VFOV = \alpha_y \frac{D_Y}{d_Y}$$

where α_x and α_y are the horizontal and vertical angular extent of the target, d_x and d_y are the measured image sizes on the monitor, and D_x and D_y are the physical dimensions of the monitor screen. If a frame grabber is used for this method, it is important that the frame grabber output matches the field-of-view (Section 4.6: page 115). Since FOV tests do not require a temperature controlled target, a passive target can be used (Figure 4-11: page 102).

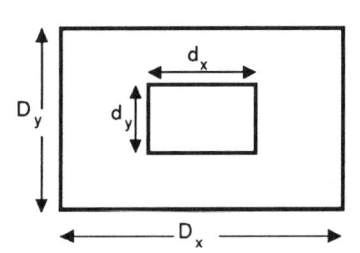

Figure 9-5: Using a rectangular target to measure the field-of-view.

Figure 9-6 illustrates a target suitable for pass/fail testing. The width of bar A should equal the field-of-view tolerance, T. For example, if the required field-of-view is 10° ± 0.1°, then the bar width should be 0.1° with the bar centers separated 10°. The bars labeled B should be separated 10.1°. If both bars A can be seen and the bars B cannot, then the system passes the FOV test. If the both bars A cannot be seen then the FOV is too small. If either bar B can be seen then the FOV is too big. For scanning arrays, variations in the cross scan direction are caused by differences in the focal length and by the scanning mechanism in the scan direction. For staring arrays, differences in the focal length cause FOV variations in both directions.

GEOMETRIC TRANSFER FUNCTION 293

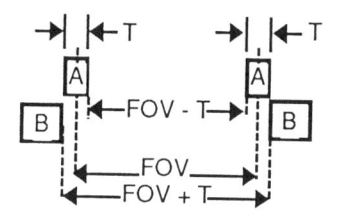

Figure 9-6: Pass/fail field-of-view test target. T is the FOV tolerance.

Example 9-1
FOV ERRORS

What target size should be used to minimize the error in measuring the field-of-view?

Figure 9-7 illustrates the variation in output for two different sized images on the detector array. As the image moves across the detector edge, the output varies from zero to a maximum. The distance over which the gradient takes place is approximately equal to the angular width of the target. The visual method produces an error approximately equal to the spot size on each side of the field-of-view. For example, if the spot size is equal to an IFOV, the maximum error would be 2·IFOV. With a 128 x 128 detector array, the maximum error is approximately $2/128 = 1.56\%$.

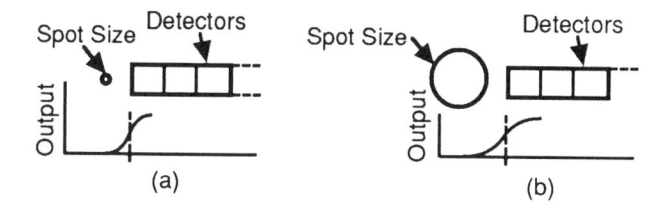

Figure 9-7: Edge response. (a) Small target compared to the IFOV and (b) large target compared to the IFOV.

The geometric spot size is

294 TESTING & EVALUATION OF IR IMAGING SYSTEMS

$$d_{image} = \frac{fl_{SYS}}{fl_{COL}} d_{target}$$

where d_{target} is the physical size of the target and d_{image} is the image of the target on the detector array. As the target size decreases, the image size decreases proportionally. The point spread function or, if using a line, the line spread function limits the minimum spot size. The error becomes negligible when measuring the 50% points (Figure 9-4: page 291).

9.2. GEOMETRIC DISTORTION

Distortion generically describes any displacement of a line or point from its expected location. From now on, distortion is defined as the polar distance between the expected location and its actual location of a point source divided by the vertical field-of-view (Figure 9-8). Types of distortion include barrel distortion that causes rectangles to bulge outward, pincushion distortion that causes rectangles to be pinched inward, and S-distortion that causes straight lines to appear as the letter "S" (Figure 9-9).

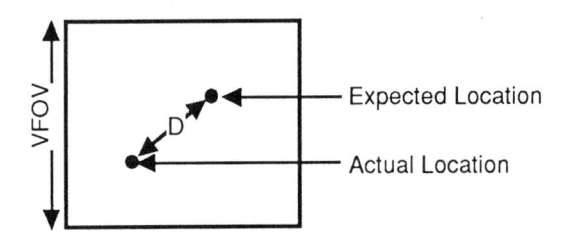

Figure 9-8: Distortion is the polar distance divided by the vertical field-of-view.

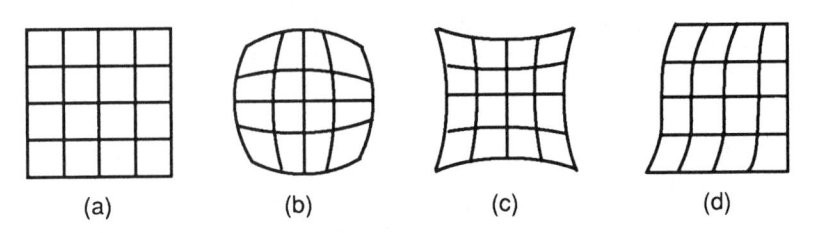

Figure 9-9: Distortion. (a) Input grid. Image of grid with (b) barrel distortion, (c) pin cushion distortion, and (d) S-distortion.

GEOMETRIC TRANSFER FUNCTION 295

The generic test configuration to measure distortion is shown in Figure 9-10. Targets can either be an array of pinholes or a rectangle. The pinhole radii can be equal to the amount of allowed distortion. For example, the hole radius may be 10% of the VFOV corresponding to 10% distortion. Ideally, the pinhole target should completely fill the field-of-view. If collimator constraints prevent filling the entire system FOV, then a target that fills 25% of the field-of-view is recommended. The FOV center must contain one pinhole as a reference point.

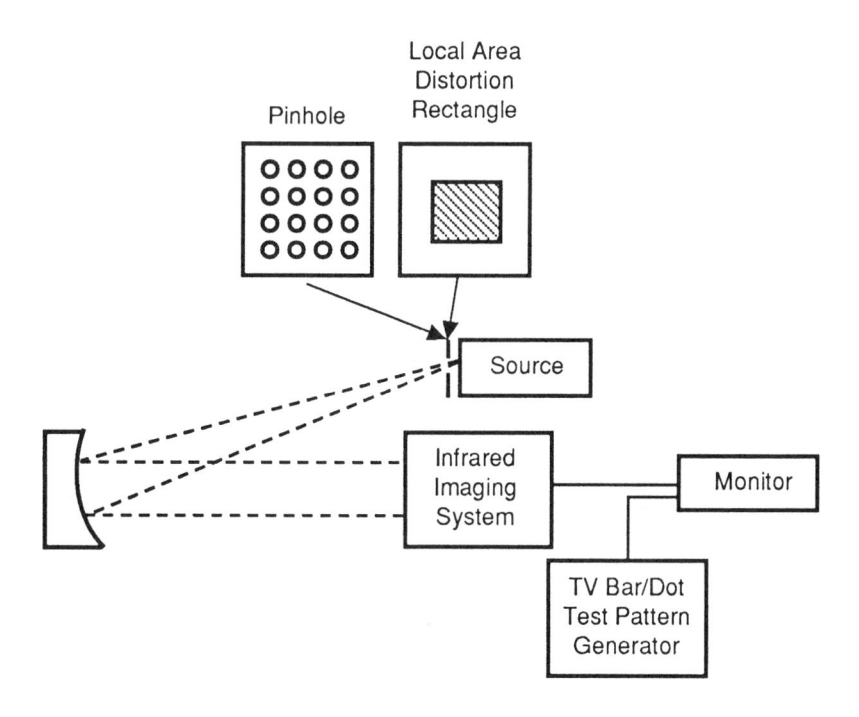

Figure 9-10: Generic test configuration to measure distortion.

To measure distortion, a TV bar/dot test pattern is superimposed onto the pinhole image. The bar/dot pattern generator must be adjustable so it can match the infrared imaging system's aspect ratio. The superimposed pattern is adjusted such that it coincides with the pinhole in the center of the FOV and with the expected location of the remaining pinholes. Distortion can be visually estimated by comparing the pinhole image diameter to the location of the dot pattern

296 TESTING & EVALUATION OF IR IMAGING SYSTEMS

(Figure 9-11). If the dot test patterns falls within each pinhole image, then the system passes the distortion test: the distortion is less than the specification. For quantifiable results, a frame grabber can be used if it provides sufficient resolution and has the appropriate aspect ratio. When using the frame grabber, the centroid location of each dot produced by the bar/dot generator is calculated. Then the centroid location of each pinhole image is calculated. Distortion at each point is the maximum polar distance between the two centroids divided by the VFOV.

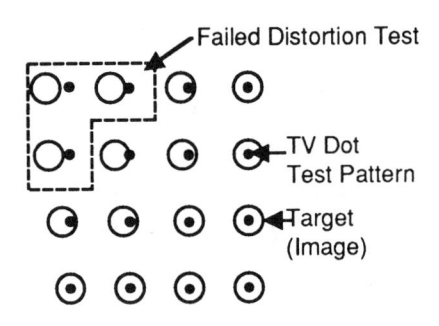

Figure 9-11: Testing distortion with pinhole array test target. The target diameter is equal to the allowed distortion.

If it is impossible to produce a target large enough to fill at least one quarter of the FOV, then a local area distortion measurement provides an alternative. A rectangular target can be placed in the center of the field-of-view. The width, S_{x1}, and height, S_{y1}, are measured. The diagonals, D_{1a} and D_{1b} are calculated (Figure 9-12). A frame grabber is useful for this application. The target is then moved to another location and the image width, S_{x2}, and the height, S_{y2}, are again measured and the diagonals, D_{2a} and D_{2b}, are calculated. Distortion is:

$$Distortion = \left| \frac{D_{1a} - D_{1b}}{VFOV} \right|$$

or

$$= \left| \frac{D_{2a} - D_{2b}}{VFOV} \right|$$

which ever is greater. The diagonals have been referred back to object space. Local area distortion does not produce the accuracy of the pinhole method.

GEOMETRIC TRANSFER FUNCTION

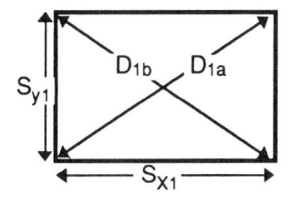

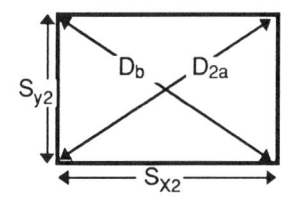

Figure 9-12: Local area distortion test. A rectangular target is placed in various locations throughout the FOV. The diagonals are calculated for each location.

The geometric distortion test procedure using the pinhole target is given in Table 9-1. System artifacts that interfere with distortion measurements include vertical line-to-line interpolation, image processing algorithms, phasing effects, chromatic aberrations and diffraction. These artifacts blur lines. An inherent limitation is the IFOV. Although sub-pixel centroid location measurements are possible, it is prudent to limit distortion measurements to $\pm$ IFOV/2.

Table 9-1
PINHOLE DISTORTION TEST PROCEDURE

1. Establish a test philosophy, criterion for success and write a thorough test plan (section 1.3: page 12).
2. Verify that the test equipment is in good condition and the test configuration is appropriate (Figure 9-10: page 295).
3. Verify that the imaging system is in focus (Section 5.2: page 137).
4. Verify the spectral response of the system and its relationship to the source, collimator spectral transmittance, and atmospheric spectral transmittance (Sections 4.3: page 105, and Section 4.4 page 105).
5. Verify that the system is operating in a linear region (Section 6.1).
6. Align the target with one pinhole in the center of the FOV.
7. Superimpose a TV dot or bar test pattern onto the video signal.
8. Adjust the superimposed pattern such that it coincides with the pinhole in the center of the FOV and with the expected location of the remaining pinholes.
9. As a pass/fail test, if a superimposed dot falls outside the pinhole image, the system fails the distortion test.
10. For quantifiable results, the pinhole and dot centroids are calculated. Distortion is the maximum difference divided by the VFOV.

298 TESTING & EVALUATION OF IR IMAGING SYSTEMS

9.3. SCAN NONLINEARITY

Scan nonlinearity includes both interlace errors (vertical nonlinearity) and nonlinear scanning (horizontal nonlinearity). With improper alignment of the interlace field, the detectors do not scan the appropriate position in object space (Figure 9-13). Scan nonlinearity may not be obvious with rectangular shaped targets whose axes are parallel to the detector axes (e.g., the target orientation required for the tests given in Chapters 4-8) but becomes obvious with diagonal lines. The diagonal line target hereafter will be called the interlace target.

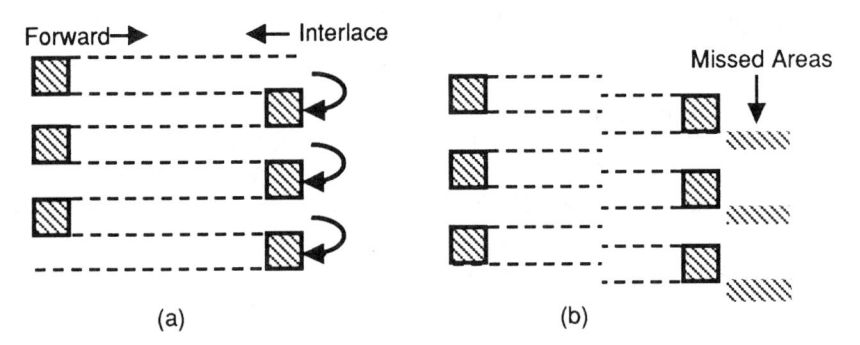

Figure 9-13: Vertical distortion due to incorrect scan mirror alignment with bidirectional scanning (a) correct alignment of forward and interlace scanned field, and (b) incorrect alignment.

Figure 9-14 illustrates the ideal output and the interlace target image when there is a strapped detector element* or timing errors. The image edges may be softened by line-to-line interpolation or other image processing algorithms. The image seen will depend upon the interlace ratio and scanning scheme.

The generic test configuration is shown in Figure 9-15. Ideally the interlace target should fill the entire field-of-view. Since the deviation from linearity is of interest, the target width is not critical. The target intensity is not critical but should be within the system's linear response range. Before evaluating the image, focus the infrared imaging system. The test is simply to view the interlace target image and compare it to the images shown in Figure 9-14. The interlace target also can be used to adjust the scan mirror or timing to achieve the desired image.

* *When an array has a defective element, the output of an adjoining element is "strapped" to the defective element output. The image will then have two identical lines.*

GEOMETRIC TRANSFER FUNCTION 299

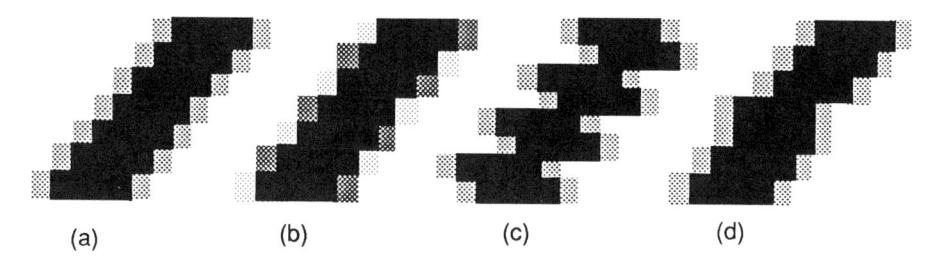

Figure 9-14: Typical outputs when viewing the interlace target with a system that has a 2:1 interlace ratio. (a) Expected output, (b) misaligned interlace, (c) reversed fields or timing errors and (d) strapped detector.

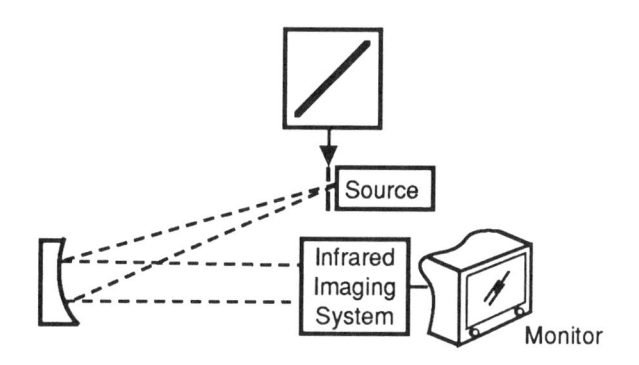

Figure 9-15: Generic test configuration to measure scan linearity.

Low frequency variations in the scanner velocity can provide horizontal distortion in the image. This distortion appears as condensed or rarefied bar targets. Figure 9-16 illustrates how a four-bar target may appear field-to-field in the presence of the low frequency sinusoidal oscillation. The interlace field may be superimposed onto the main field or may be offset depending upon the phase between the low frequency oscillation and the scanner frequency. If the interlace image is offset, the bars will have serrated edges that may be smoothed by line-to-line interpolation. It is impossible to measure the oscillation frequency or amplitude by just examining bar targets because it changes from frame-to-frame.

300 TESTING & EVALUATION OF IR IMAGING SYSTEMS

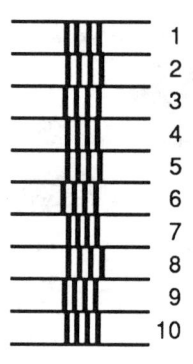

```
━━━┃┃┃━━━ 1
━━━┃┃┃━━━ 2
━━━┃┃┃━━━ 3
━━━┃┃┃━━━ 4
━━━┃┃┃━━━ 5
━━━┃┃┃━━━ 6
━━━┃┃┃━━━ 7
━━━┃┃┃━━━ 8
━━━┃┃┃━━━ 9
━━━┃┃┃━━━ 10
```

Figure 9-16: Rarefaction and condensation of a four-bar target due to variations in the scan velocity for 10 different fields.

Figure 9-17 illustrates how the interlace target may appear when a fixed low frequency oscillation is present (assuming no line-to-line interpolation and no interlace). The amplitude has been exaggerated for illustrative purposes. The superimposed oscillation frequency is:

$$F = \frac{v}{T_S}$$

where T_s is the period of oscillation measured in object space and the scan velocity, v, is

$$v = \frac{HFOV}{\eta \, F_T}$$

For example if HFOV = 15 mrad, the scan efficiency (See Example 2-1: page 29), η, is 0.75 and the field time, F_T, is 1/60 sec, then the scan velocity is 1200 mrad/sec. The oscillation period is obtained from the FOV and the fraction of that value that provides the full period. In Figure 9-17, the oscillation repeats itself every 10.2 mrad corresponding to a frequency of 1200/10.2 or 118 Hz. Random vibration will not produce a clear oscillatory pattern.

GEOMETRIC TRANSFER FUNCTION 301

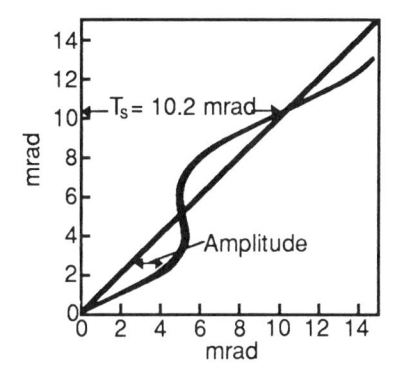

Figure 9-17: Variation in the interlace target image when low frequency oscillations are present.

With systems employing TDI, the MTF is affected by the scan velocity. For maximum MTF, the scan velocity must be matched to the delay element. It was demonstrated in Section 2.3 (page 36) that periodic bar patterns whose spatial frequencies approach Nyquist frequency will produce beat frequencies. Scan velocity mismatch changes the beat frequency[2]. By adjusting the scan velocity to create the required beat frequency pattern, the proper scan velocity is obtained.

Scan nonlinearity also can cause pixel misregistration from line-to-line. This effect is noticeable on pure serial systems and line scanners. For line scanners that have a wide field-of-view, scan nonlinearity can cause large rarefaction or compression of the image. If the pixels are not aligned, a straight line will appear either serrated or wavy. The relative location of each pixel can be specified by the angular position of the center of each pixel.

302 TESTING & EVALUATION OF IR IMAGING SYSTEMS

For example, each pixel center shall be within ½ IFOV of

$$\theta = \frac{\left(n - \frac{1}{2}\right)}{N} \cdot FOV$$

where n is the pixel number from the start of the line, N is the total number of pixels on a line. n - ½ is the center of pixel n.

9.4. MACHINE VISION PERFORMANCE

Automated vision systems measure image size, image location, measure movement, and count objects. The ability to locate, size or count depends on distortion, system MTF, noise and the specific imaging processing algorithm (software) used. When noise is present, accuracy decreases as the target contrast (intensity) decreases. Appropriate performance tests estimate the accuracy and repeatability of machine vision systems. Depending upon the system application, different targets may be used. Benchmarks have been recommended that measure the speed to perform specific functions but test targets and test methodologies are formative at this time. Errors in machine vision sizing and location of targets are caused by the artifacts induced by the infrared system and the specific image processing algorithm used.

9.4.1. MEASURING LOCATION AND COUNTING OBJECTS

The American National Standards Institute (ANSI) target[3] (Figure 9-18) was recommended for performance testing of machine vision systems that measure the distance between objects and count objects. The target consists of eight randomly spaced equidiameter circles. The circles are at (5,-14), (21,4), (-10,-25), (-18,4), (1,18), (-16,-11), (1,1), (-14,22) with a circle diameter of 10. Different sized targets can be created by simply multiplying all values by a constant. For example, a target 3 times larger would have target locations at (15,-42), (63,12), (-30,-75), (-54,12), (3,54), (-48,-33), (3,3), (-52,66) with a circle diameter of 30. The appropriate size is selected according to the system requirements and the collimator focal length.

The target is mounted on a translation/rotation stage. The operator verifies that the machine vision system correctly locates the 8 targets and correctly verifies the size. The target is then rotated or translated by a known increment and the test is repeated. The machine vision system must correctly measure the rotational/translational distance for each target. This procedure is repeated as often as necessary to determine the automatic vision system's accuracy and repeatability (Section 4.7: page 124). The ANSI target does not simulate any specific real world situation.

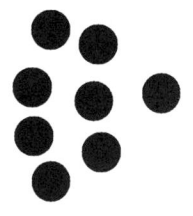

Figure 9-18: ANSI target used to measure relative movement and position.

For some applications (notably IRST systems), it is important to be able to distinguish closely spaced objects (CSO). Figure 9-19 represents a series of CSOs. The test is simply to determine how closely spaced two objects can be and still be recognized as two objects. The relative size and distance between the circles can be adjusted according to the system requirements.

Figure 9-19: Closely spaced object target.

The Standard Correlation Target Set[4], SCTS®, combines the ANSI target with the CSO target to create a realistic target set. It consists of varying sized circles at random locations (Figure 9-20). As the complexity of the targets increases, the range of diameters increases. The simplest SCTS target consists of 20 equidiameter targets. It is similar to the ANSI target but contains more targets. The next SCTS target contains 40 and the last contains 60 targets. The fixed diameter circles can be used to verify the counting ability of automated vision systems. Complex targets simulate actual situations (e.g., varying cell sizes in biology or multi-targets in IRST applications). Overlapping targets simulate overlapping cells (clusters) or overlapping targets (closely spaced objects) in the IRST application. The calibration set can be tailored (size and location) to the exact application.

304 *TESTING & EVALUATION OF IR IMAGING SYSTEMS*

The test procedure for the SCTS is identical with that used for the ANSI target. A single calibration target is presented to the automated vision system at a random location and random orientation. The automated vision system must correctly count, measure distance or size the images. The target is then moved to a new location and orientation. The process is repeated as often as necessary to determine the system's accuracy and repeatability.

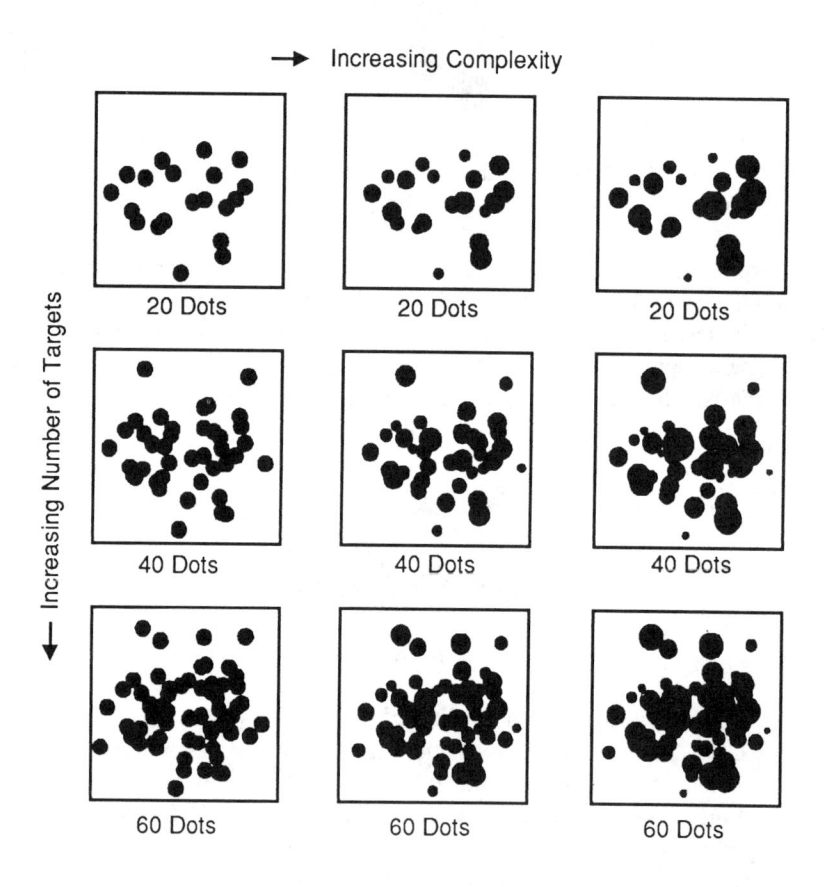

Figure 9-20: Standard Correlation Targets. These targets are a combination of the ANSI and CSO targets. Size, relative position and algorithm counting ability can be evaluated.

rotational/translational distance for each target. This procedure is repeated as often as necessary to determine the automatic vision system's accuracy and repeatability (Section 4.7: page 124). The ANSI target does not simulate any specific real world situation.

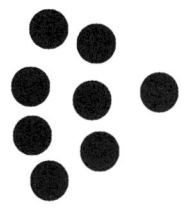

Figure 9-18: ANSI target used to measure relative movement and position.

For some applications (notably IRST systems), it is important to be able to distinguish closely spaced objects (CSO). Figure 9-19 represents a series of CSOs. The test is simply to determine how closely spaced two objects can be and still be recognized as two objects. The relative size and distance between the circles can be adjusted according to the system requirements.

Figure 9-19: Closely spaced object target.

The Standard Correlation Target Set[4], SCTS®, combines the ANSI target with the CSO target to create a realistic target set. It consists of varying sized circles at random locations (Figure 9-20). As the complexity of the targets increases, the range of diameters increases. The simplest SCTS target consists of 20 equidiameter targets. It is similar to the ANSI target but contains more targets. The next SCTS target contains 40 and the last contains 60 targets. The fixed diameter circles can be used to verify the counting ability of automated vision systems. Complex targets simulate actual situations (e.g., varying cell sizes in biology or multi-targets in IRST applications). Overlapping targets simulate overlapping cells (clusters) or overlapping targets (closely spaced objects) in the IRST application. The calibration set can be tailored (size and location) to the exact application.

304 TESTING & EVALUATION OF IR IMAGING SYSTEMS

The test procedure for the SCTS is identical with that used for the ANSI target. A single calibration target is presented to the automated vision system at a random location and random orientation. The automated vision system must correctly count, measure distance or size the images. The target is then moved to a new location and orientation. The process is repeated as often as necessary to determine the system's accuracy and repeatability.

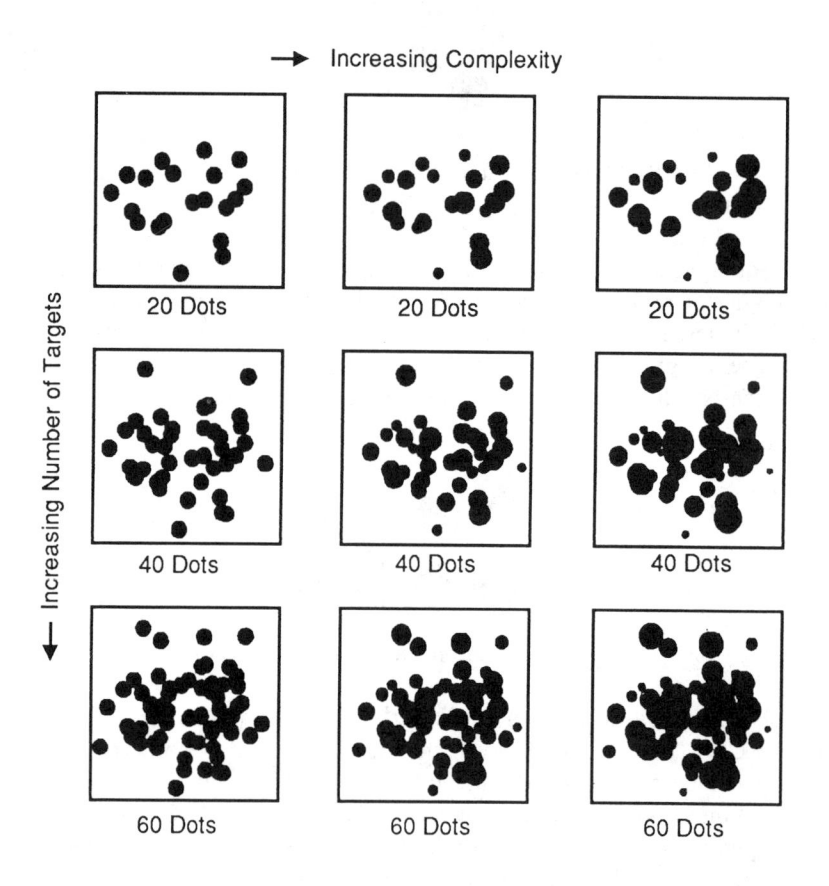

Figure 9-20: Standard Correlation Targets. These targets are a combination of the ANSI and CSO targets. Size, relative position and algorithm counting ability can be evaluated.

9.4.2. ALGORITHM EFFICIENCY

The Abingdon Cross benchmark[5] test is a measure of algorithm efficiency. The task is to find the medial axis of the symmetrical cross (Figure 9-21) as quickly as possible. The original benchmark cross was designed to test image processing algorithms and therefore the noise level and cross intensity were computer generated[6,7]. By equating the area and intensity of the cross to the area and intensity of the object to be detected during actual usage, the time to find the cross is representative of how long it would take to detect and identify the desired object. This provides the correlation between actual use and laboratory testing. When placed in a collimator, the total solid angle subtended by the cross area is $(9x/fl_{COL})^2$. This is equated to the solid angle subtended by the real object or (target area)/(range)^2. The test results are affected by all the artifacts created by the infrared system: system noise, aberrations, distortion, and digitization effects.

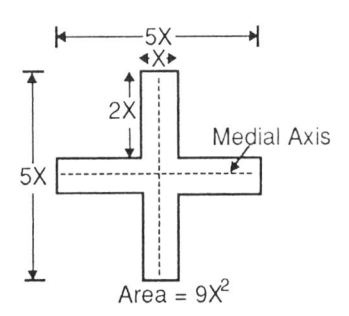

Figure 9-21: Abingdon cross. The time required to find the medial axis is a measure of algorithm efficiency.

9.5. SUMMARY

The geometric transfer function is a generic term incorporating all input-to-output transformations concerned with the preservation of images and shapes. The geometric transfer function includes field-of-view, distortion, and scan nonlinearity. With scanning systems, distortion tests also can be used to adjust the scan mirror to maximize the MTF. Since phasing effects are well understood, phase variations can be used to adjust the scan velocity in systems that employ TDI detectors.

306 TESTING & EVALUATION OF IR IMAGING SYSTEMS

Machine vision requirements are different from systems designed for visual observation and therefore many tests used for visual testing are inappropriate for machine vision testing. The Standard Correlation Target Set evaluates the ability to distinguish closely spaced objects, count objects, size objects and measure relative position. For calibration of trackers or other automated systems that measure motion, the SCTS target can be placed on a moving table or the infrared system can be rotated in a manner similar to that used for measuring the field-of-view. Procedures for testing trackers have not been defined at this time. However, the SCTS can be used to measure tracking tenacity as a function of target intensity and angular velocity.

Typical specifications are listed in Table 9-2. Machine vision metrics have not been established at this time. The machine vision tests illustrated in this chapter are only guidelines. As the technology matures, new test requirements will emerge and concurrently new targets and test methodologies will be established.

Table 9-2
TYPICAL SPECIFICATIONS

FIELD-OF-VIEW:
- The field-of-view shall be $2.5° \pm 0.05°$.

DISTORTION:
- The distortion shall not be greater than 0.5% in the central 50% of the field-of-view.
- The distortion shall not be greater then 1% over the entire field-of-view.

SCAN LINEARITY (serial scan or line scanners):
- Each pixel shall be within ½ IFOV of the equivalent pixel on the adjoining line.

MACHINE VISION:
- The algorithm shall be able to locate the Abingdon cross medial axis within 3 msec.
- The system shall correctly locate and count all 60 objects on the simple SCTS target for 10 different random orientations.

9.6. REFERENCES

1. N. Sampat, "The RS-170 Video Standard and Scientific Imaging: The Problems", Advanced Imaging, pp. 40-43, February 1991.
2. T. S. Lomheim, L. W. Schumann, R. M. Shima, J. S. Thompson, and W. F. Woodward, "Electro-Optical Hardware Considerations in Measuring the Imaging Capability of Scanned Time-delay-and-integrate Charge-coupled Imagers", Optical Engineering, Vol. 29(8), pp. 911-927 (1990).
3. "American National Standard for Automated Vision Systems - Performance Test Measurement of Relative Position of Target Features in Two-Dimensional Space", Report ANSI/AVA A15.05/1-1989, American National Standard Institute, 1430 Broadway, New York, NY 10018 (1989).
4. SCTS® is available from Santa Barbara Infrared Inc., 312A North Nopal Street, Santa Barbara, Calif 93103-3225
5. K. Preston, Jr., "Benchmark for Image Processing", Advanced Imaging, pp. 30-38, May 1990.
6. K. Preston, Jr., "Benchmark Results - The Abingdon Cross", in Evaluation of Multicomputers for Image Processing, I. Uhr et. al., eds., pp. 34-54: Academic Press, Cambridge, Mass (1986).
7. K. Preston, Jr., "Sci/Industrial Image Processing: New System Benchmark Results", Advanced Imaging, pg. 46, September 1992.

EXERCISES

1. To measure the FOV, the pinhole target must move perpendicular to the FOV being measured. Calculate the error as a function of angle from perpendicularity.
2. Using the rectangle method to determine the FOV, estimate the errors when the image on a monitor is measured with a ruler.
3. Using the rectangle method to determine the FOV, estimate the errors when the image is captured by a frame grabber with 480 x 640 pixel resolution.
4. If the detector array consists of 128 x 128 elements, what is the minimum measurable distortion?
5. Sketch the interlace target image when high frequency (jitter) movement is present.
6. Can a periodic square wave target (e.g., a MRT target) be used for local area distortion measurements? What is the accuracy of this approach?
7. An infrared imaging system can be focussed at 10 feet. Design a pass/fail target that can be used to verify that the field-of-view is $3° \pm 0.05°$.
8. For the SCTS target containing 40 equi-sized targets, estimate the number of individual targets that may be identified.
9. Of the resolution measures listed in Table 5-2 (page 146), which ones may be appropriate for machine vision systems?

10

OBSERVER INTERPRETATION OF IMAGE QUALITY

MINIMUM RESOLVABLE TEMPERATURE and MINIMUM DETECTABLE TEMPERATURE

Observer detection of standard targets is an industry-wide method of estimating image quality. The minimum resolvable temperature (MRT) and the minimum detectable temperature (MDT) are subjective measures that depend upon the infrared imaging system's resolution and sensitivity. MRT is a measure of the ability to resolve detail and is inversely related to the MTF, whereas the MDT is a measure to detect something. Very small targets cannot be resolved and therefore MDT is inversely proportional to the ATF. The MTF and ATF are the system's response to high-contrast noiseless targets whereas the MRT and MDT deal with an observer's ability to perceive low-contrast targets embedded in noise.

MRT and MDT are not absolute values but are perceivable temperature differentials relative to a given background. Sometimes they are called the minimum resolvable temperature difference (MRTD) and minimum detectable temperature difference (MDTD). The terms *difference* or *differential* are often omitted since it is understood that it is a differential measurement.

These tests depend upon decisions made by an observer. The results vary with training, motivation and visual capacity of the observer as well as the environmental setting. Because of the considerable inter- and intra-observer variability, several observers are required. The underlying distribution of observer responses must be known so that the individual responses can be appropriately averaged together. To obtain any degree of consistency, it is assumed that the observers are *qualified*.

Consistency is monitored by the semiautomatic test procedure. During this test, the observer tracks his own threshold. Automatic tests or objective tests are desirable because of the insufficient number of trained personnel and because the subjective test may take several hours to perform. This chapter addresses the procedures of all three tests: subjective (manual), semiautomatic and objective (automatic).

10.1. OBSERVER VARIABILITY

The process of seeing is somewhat a learned ability. It is a perceptual one, accomplished by the brain, affected by and incorporating other sensory systems such as emotions, learning and memory. The relationships are many and not well understood. Seeing varies between individuals and temporally within an individual. The use of any metric (such as detection, recognition, or identification) must be treated statistically rather than as an absolute value. Because each person is different, individual characteristics can affect detection threshold. Age, experience, personality, mood, and intelligence are just a few. During an evaluation, many changes take place that could influence the outcome of a test. People learn new methods, have new experiences, change moods and become fatigued.

Two different observers can obtain different detection thresholds (Figure 10-1) with either response being equally likely. The responses may be both above the average, below the average or one above and one below. Further, the same person may obtain different results depending upon the time of day.

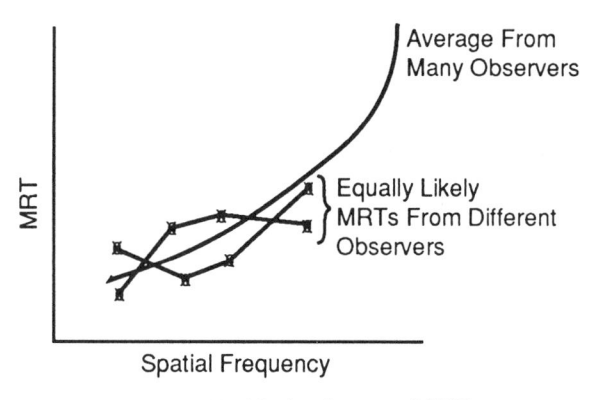

Figure 10-1: Equally likely observer MRT responses.

Visual performance can fluctuate considerably over a period ranging from a few minutes to several weeks[1,2]. The psychophysical frequency of seeing curve includes these fluctuations. At any particular moment in time, an individual's response may be significantly different from the population average. Because so many changes can occur and since they may not be easily controlled, the consistency from one session to another can be severely compromised. MRT variability as high as 50% are often cited from laboratory-to-laboratory with 20% variability reported at one laboratory[3]. This variability is the test precision (σ_p) discussed in Section 4.7 (page 124). The full range of values ($\pm 3\sigma$) is much higher. Differences in detection criteria, fatigued observers, differences in test procedure and differences in test equipment may be contributors to the variability.

It is unreasonable to expect that all observers should obtain precisely the same detection threshold value. But if an observer does worse than expected, he is considered a *poor* observer and his result is often deleted from the analysis. It seems that the results from only *good* observers (those who obtain low threshold values) are reported. This practice inappropriately reports low observer detection values. Brown[4] stated "It seems a fairly standard industry practice that the way in which you improve MRTD is to tune the test technician; not change the device." This practice is avoided by using the appropriate statistical analysis and realizing that observer variability exits. Good engineering practices require well-defined specifications, test procedures, data analysis methodology and well motivated, experienced observers.

10.1.1. FREQUENCY OF SEEING RESPONSE

Statistically speaking, at threshold the target is detected 50% of the time by one observer or, equivalently, detected by 50% of the population. Detection is sometimes erroneously treated as an absolute value in which all observers cannot see the target below threshold and all can see the target above threshold (i.e., a step response). It is the variation of thresholds (frequency of seeing response) that accounts for the variation in detection values. As the width of the frequency of seeing curve (Figure 10-2) increases, the variation in individual responses increases. Observers are considered consistent if their threshold value does not change significantly over time and if their frequency of seeing curve is narrow or equivalently, has a small σ.

To average observers' responses appropriately, it is necessary to understand the underlying frequency of seeing distribution. Holst and Pickard[5] reported an experiment in which 76 observers detected standard four-bar MRT targets of various spatial frequencies embedded in noise. The targets were computer generated and broad band white noise was added to the video signal.

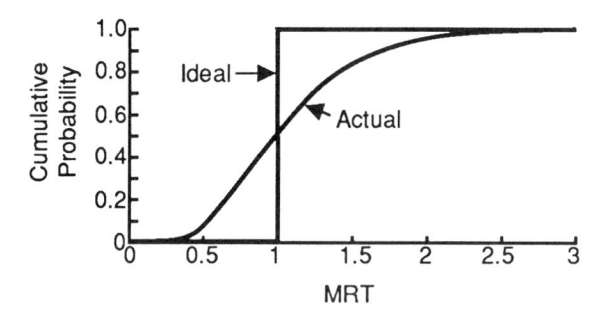

Figure 10-2: Frequency of seeing distributions. Ideally the probability of seeing an MRT target is a step function. The actual response is log-normally distributed. When plotted on a linear scale it appears skewed.

The 2700 detection responses could be mathematically described by a log-normal distribution:

$$P(x) = \frac{1}{\sqrt{2\pi}\log(\sigma)} e^{-\frac{1}{2}\left[\frac{\log(x)-\log(\mu)}{\log(\sigma)}\right]^2} \quad 10\text{-}1$$

where $\sigma = 1.58$ and μ is the population average. With a frequency of seeing curve that is log-normal distributed*, the estimated mean, m, is the geometric average of the responses. Therefore, the average MRT for N observers is

$$MRT_{ave} = \left[\prod_{i=1}^{N} MRT_i\right]^{\frac{1}{N}} \quad 10\text{-}2$$

where MRT_i is the individual detection threshold. For example, if the MRT values for three observers were 0.5, 0.6 and 1 °C, the average for these observers is $[(0.5)(0.6)(1)]^{1/3} = 0.67$ °C. Although not experimentally verified, it is reasonable to assume that MDT thresholds follow the same log-normal distribution.

*A log-normal distribution appears as the usually Gaussian shaped distribution when plotted on a logarithmic axis. Over a limited region, it can be approximated with a linear Gaussian distribution with the same mean. The standard deviations are not related.

312 *TESTING & EVALUATION OF IR IMAGING SYSTEMS*

10.1.2. VISUAL ANGLE

The eye's detection capability depends upon the visual angle subtended by the target size and the distance from the monitor to the observer. As shown in Figure 10-3, in the absence of noise, the eye's contrast threshold is characteristically J-shape. The eye is most sensitive to periodic targets whose spatial frequencies[6] between 3 and 10 cycles/deg.

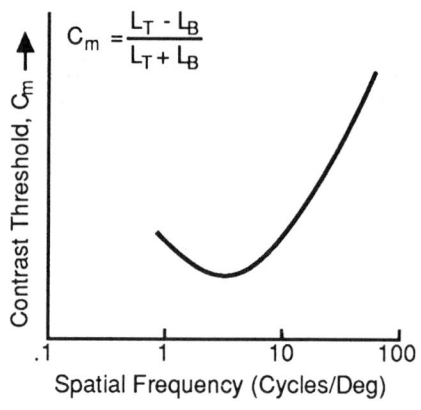

$$C_m = \frac{L_T - L_B}{L_T + L_B}$$

Figure 10-3: Representative observer contrast threshold curve. L_T and L_B are the target and background luminances respectively.

MRT should follow, in part, the shape of the eye's threshold curve. The eye appears to be AC coupled and it becomes more difficult to resolve very large targets. This reduced detection capability should appear as an increased MRT for very low frequencies. From a practical point of view, the largest target that can be seen is determined by the experimental configuration, the monitor size and the distance the observer is from the monitor. Thus, the very low end of this curve is not usually measured.

Two measurements are possible: (1) allowing the observer to move his head and (2) the head is fixed in space. Since the eye's detection capability depends upon the angular subtense of the target relative to the eye's location, head movement may provide different results than if the head is fixed at the same location. In the laboratory, the distance to the monitor is not usually specified nor limited in any way. To maximize detection capability (stay on the minimum of the contrast threshold curve), an observer subconsciously moves toward the monitor to perceive small targets and further away to see larger targets. With head movement, the observer will approximately move to the distance that places the target on the minimum of his contrast threshold curve. With a minimum contrast between 3 and 10 cy/deg, the distance selected will vary from observer-to-observer (Figure 10-4). The distance suggested can exceed the size of the laboratory when viewing large targets. But unless otherwise instructed, an observer probably will not be more than 4 to 8 times the picture height from the monitor which is a comfortable viewing distance (*comfort zone*) for living-room TVs. If the observer moves to the distances suggested in Figure 10-4, the MRT will asymptote to zero as the spatial frequency approaches zero[7]. On the other hand, if the observer stays within his comfort zone, the MRT asymptotes to 0.3 to 0.7 times the NEDT. While it is possible to perceive low spatial frequency targets at a very long distance, these responses are not representative of actual system usage. These long distances may be considered just an exercise in visual testing rather than infrared imaging system characterization. If the infrared imaging system is designed for a specific application in which the observer sits a fixed distance away from the monitor, then it is appropriate to restrain head movement to those distances that are commensurate with the operational environment.

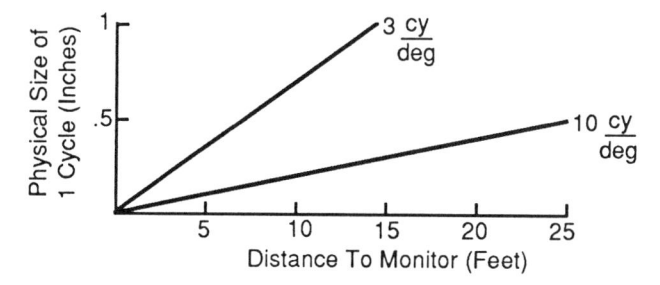

Figure 10-4: Optimal viewing distances for 2 different observer spatial frequencies as a function of one cycle on the monitor.

314 TESTING & EVALUATION OF IR IMAGING SYSTEMS

Example 10-1
OPTIMAL VIEWING DISTANCE

What is the optimal viewing distance for an infrared imaging system that has a 30 mrad x 30 mrad FOV and uses a 14 inch x 14 inch monitor?

The size of one cycle on the monitor is:

$$d_{cycle} = \frac{MONITOR\ SIZE}{f_x \cdot HFOV} = \frac{0.0389}{f_x}\ feet$$

where f_x has units of cy/mrad. The distance from the monitor is:

$$R = \frac{d_{cycle}}{2\tan\left(\frac{\theta}{2}\right)}$$

where θ is the angle subtended by one cycle. If the observer strives to maximize his detection threshold at 3 cy/mrad then:

$$R = \frac{6.686}{f_x}$$

If he desires 10 cy/mrad then:

$$R = \frac{22.88}{f_x}$$

The range of viewing distances is given in Table 10-1. The comfortable viewing distance, which is 4 to 8 times the monitor size, is 4.7 to 9.3 feet. While staying within his comfort zone, the observer will not optimize his detection threshold for the low spatial frequency targets. As the observer moves toward the monitor to resolve higher frequency targets, he will begin to discern the raster pattern. At some point he will feel that the raster interferes with his detection process. This represents a lower limit on his distance from the monitor. For this distance, smaller targets will no longer be on his minimum contrast detection curve and therefore may be more difficult to perceive.

Table 10-1
OPTIMAL VIEWING DISTANCES
(Based upon 30 mrad FOV and a 14" monitor)

SPATIAL FREQUENCY f_x (cy/mrad)	OPTIMAL DISTANCE (feet)
0.5	13.4 - 45.8
1.0	6.7 - 22.8
2	3.3 - 11.4
4	1.6 - 5.7
6	1.1 - 3.8
8	0.8 - 2.9
10	0.7 - 2.2

Example 10-2
LAB - FIELD CORRELATION

In actual (field) operation, the observer will be 24 inches from a 6 inch monitor. In the laboratory, a 14 inch monitor will be used. How far away should the test engineer be from the laboratory monitor to simulate actual usage?

$$\theta_{FIELD} = 2\tan^{-1}\left(\frac{MONITOR\ SIZE}{2 \cdot DISTANCE}\right) = 2\tan^{-1}\left(\frac{6}{2 \cdot 24}\right) = 0.25\ rad$$

$$D_{lab} = \frac{MONITOR\ SIZE}{2 \cdot \tan\left(\frac{\theta_{FIELD}}{2}\right)} = \frac{14}{2\tan(0.125)} = 56\ inches$$

The laboratory observer must be 56 inches from the laboratory monitor to subtend the same angle as the observer in the field.

316 *TESTING & EVALUATION OF IR IMAGING SYSTEMS*

10.1.3. NOISY IMAGES

Although the contrast threshold is reported as J-shaped, the actual shape depends upon the noise power spectral density[8,9,10]. If the noise is restricted to certain spatial frequencies, then the detection of targets of comparable spatial frequencies becomes more difficult. The observer's ability to see a specific spatial frequency target depends upon the noise content in the neighborhood of that spatial frequency (Figure 10-5). Low spatial noise frequency components will interfere with detecting low frequency targets (large objects). Mid spatial frequency noise increases the contrast threshold curve at mid frequencies and so on.

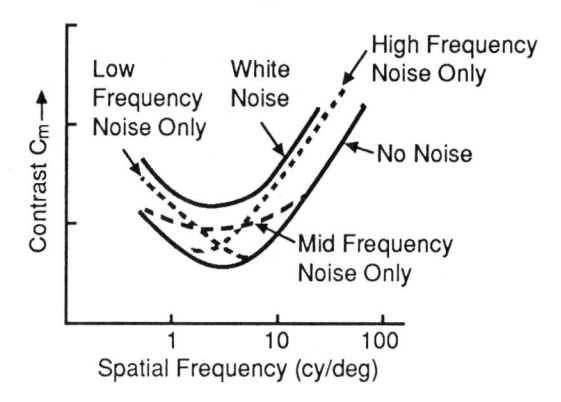

Figure 10-5: Effects of spectral noise on the contrast threshold. The MRT should follow the shape of the contrast threshold curve.

At long distances, the pixel angular subtense will less that what the eye can resolve. Therefore high frequency noise will not be perceived and the background will blend into a neutral gray. This permits the MRT to approach zero as the distance increases[7]. This presents an image that is contrary to that required by the MRT test: detection of targets embedded in noise. However, if excessive low frequency noise is present (nonuniformity), detection becomes more difficult and at long distances, the MRT approaches a finite value[7,11].

The spectral properties of noise cannot be directly assessed visually but only be inferred by the resultant detection responses. With spectral noise, it may be possible to pass some specifications and fail others. If the test procedures described in this chapter produce unexpected results, it is prudent to measure the noise power spectral density for each component of the three-dimensional noise model.

10.1.4. OBSERVER QUALIFICATION

A *qualified* observer is difficult to define. The qualified observer should be consistent, have a low detection threshold and have good visual acuity. If an individual performed the tests often then he is experienced. NVESD[*] has introduced[12] the "Thermal Image Training Package" which consists of a set of software programs designed to teach military observers to recognize and identify thermal images of military vehicles. While this software package was specifically designed for the military observer, it can form the basis for training individuals in the operation and detection of images without employing a thermal imaging system. The Thermal Image Model[13] developed for the US Army Tank Automotive Command provides realistic imagery that also can be used to train observers. This model also can simulate fixed pattern noise and phasing effects.

New observers should compare their detection thresholds to those obtained by existing qualified observers. The semiautomatic test procedure (Section 10.2.2.) provides a measure of observer variability. As the individual's variability is reduced, he becomes a consistent observer. But this does not mean that he will have a lower detection threshold value. It simply means that he will perform consistently and that his particular frequency of seeing curve is narrow (small σ). This training may take up to six months[3]. There is no industry wide standard by which to rate observers nor any national school to train observers.

There is a myriad of methods available to test visual capacity. The observer should have corrected 20/30 or better acuity with corrected astigmatism. Visual acuity tests only provide information about how well an observer can resolve high contrast targets. On the other hand, MRT and MDT are measures of threshold detection of low contrast targets embedded in noise. The Vision Contrast Test System, VCTS[14] determines contrast sensitivity. Since detection threshold is inversely proportional to the contrast sensitivity, individuals with high contrast sensitivity can obtain lower detection threshold values than those who have lower contrast sensitivity. Figure 10-6 illustrates a typical Vision Contrast Test System chart and Figure 10-7 shows the evaluation form. Contrast sensitivity is inversely related to the contrast threshold (Figure 10-3). The chart is only for initial screening since it does not include the effect of noise.

[*]*Over the years, the US ARMY proponent agency for analyzing, testing and evaluating infrared imaging systems has changed its name: Night Vision Laboratory, NVL; Center for Night Vision and Electro Optics, CNVEO; Night Vision and Electro Optical Laboratory, NVEOL; Night Vision and Electro Optical Directorate, NVEOD; and (as of this writing) Night Vision and Electronic Sensor Directorate, NVESD. It will probably change again!*

318 *TESTING & EVALUATION OF IR IMAGING SYSTEMS*

Finally, sustained effort is required from well-trained staff to maintain consistency within a given laboratory. The observer must be healthy and alert. An operator who is neither tired nor under stress probably could achieve consistency but this is not guaranteed.

VISION CONTRAST TEST SYSTEM

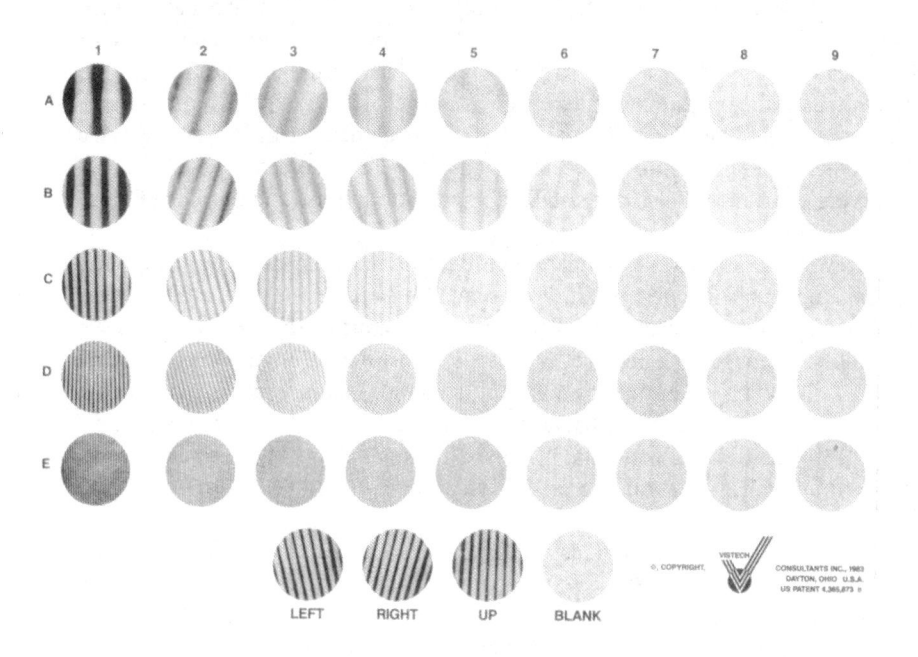

Figure 10-6: VCTS test chart. For each spatial frequency target (A, B, ..., E) the observer reports the target that he can just perceive (1, 2, ..., 9). Contrast decreases as the target number increases.

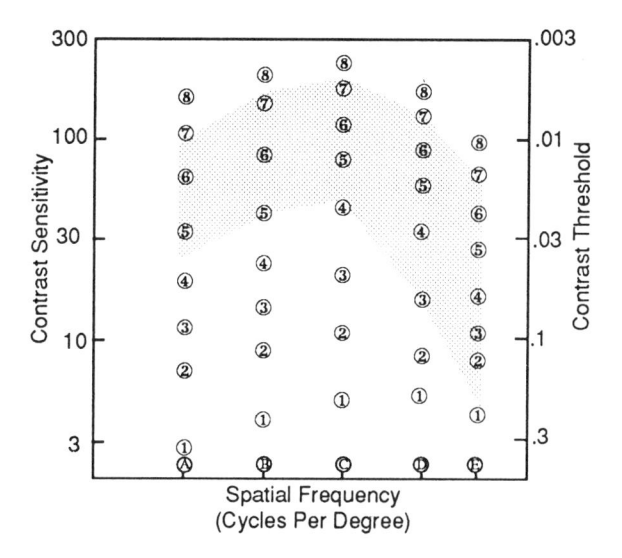

Figure 10-7: Typical observer response. Good observers will possess high contrast sensitivity. The gray band represents the range of values responses expected from the general population.

10.1.5. MEASURED VALUES VERSUS SPECIFICATIONS

Because of the large observer variability, the measured MRT can vary from test-to-test and from observer-to-observer. The MRT is not one number but a range of numbers due to the observer variability. These numbers can vary from below the specification to above the specification (Figure 10-1: page 309).

Appropriate statistical analysis can separate system response (the desired outcome) from observer variability. The statistical approach provides two alternate decisions: (1) the measured MRT is greater than the specification and (2) there is no reason to believe that the measured MRT is greater than specification. The standard Z statistic[15] is used to select one of the alternate decisions. Since the frequency of seeing curve is log-normally distributed, the Z statistic also must be expressed in logarithms:

320 TESTING & EVALUATION OF IR IMAGING SYSTEMS

$$\log(MRT_{MAX}) = \log(MRT_{SPEC}) + Z \frac{\log(\sigma)}{\sqrt{N}} \qquad 10\text{-}3$$

where MRT_{MAX} is the maximum allowable value of averaged MRT responses based upon N observers, MRT_{SPEC} is the specification value and σ is the standard deviation obtained from the frequency of seeing curve. σ is assumed to be 1.58. If the measured averaged MRT is less than MRT_{MAX} then the system passed the specification. That is, there is no reason to believe that the measured MRT is statistically different from specification *although the measured value may be greater than the specification*. Equation 10-3 can be rewritten as

$$MRT_{MAX} = K \cdot MRT_{SPEC} \qquad 10\text{-}4$$

where

$$K = 10^{Z \frac{\log(\sigma)}{\sqrt{N}}} \qquad 10\text{-}5$$

K is a multiplicative factor that dramatically increases as the number of observers decreases. Z is determined from a confidence level. When the measured MRT is less than that given by Equation 10-4, it can be said that "at, say, the 90% confidence level, the system met specification." Table 10-2 illustrates the multiplicative factor for three levels of confidence.

Table 10-2
MRT MULTIPLICATIVE FACTORS

LEVEL OF CONFIDENCE	Z	K						
		N = 1	N = 2	N = 3	N = 4	N = 6	N = 100
90%	1.29	1.80	1.52	1.41	1.34	1.27	1.06
95%	1.65	2.13	1.71	1.55	1.46	1.36	1.07
99%	2.33	2.90	2.12	1.85	1.70	1.55	1.11

For example, suppose the MRT specification is 0.1°C, and the geometrically averaged MRT of 3 observers is less than 0.155°C. Then it can be said that at a 95% confidence level the system passed the specification. This apparently large spread is due to the population's broad (large σ) frequency of

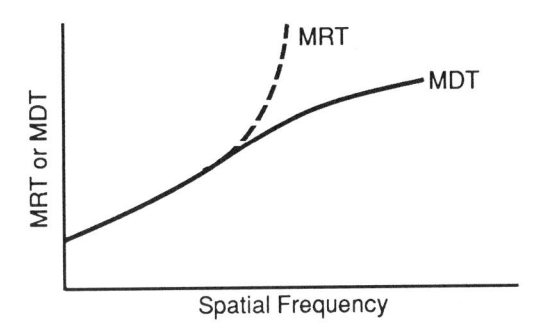

Figure 10-12: MRT and MDT as a function of spatial frequency.

MRT measurements are performed with the target axes both parallel and perpendicular to the scan direction or detector array axes. Because of different sampling in the vertical and horizontal directions and because noise factors may be different in each direction, the horizontal and vertical MRTs may be different. The noise is measured via the three-dimensional noise model. σ_{TH} affects MRT_H and σ_{HV} affects MRT_V. MRT can be measured with the target axes at 45° but this data is not widely used due to the greater uncertainty in calculating the observer's response. Observer threshold approaches a minimum for vertically and horizontally oriented targets. For patterns oriented at 45 degrees the threshold increases[16] by 15 to 25%. As a result, patterns oriented 45° provide worse threshold values than those oriented vertically or horizontally. This effect is due to the visual path way response and is not at all related to phasing effects or raster pattern effects. Phasing effects further degrade the image and further worsen the threshold value.

The observer is given an unlimited viewing time to detect (MDT test) or resolve (MRT test) the target. So that the tests more closely simulate field conditions, a limited-time test may be appropriate. An individual's accuracy increases when he has more viewing time to determine his threshold. Therefore it is expected that the limited viewing time test would provide worse (higher) values[17]. The limited viewing time test requires more observers to ensure that enough data points have been obtained for adequate statistical analyses.

Infrared imaging systems are subject to sampling effects. The MRT and MDT do not have a unique value for each spatial frequency but have a range of values depending upon the location of the target with respect to the sampling lattice. As a result it has become widely accepted to "peak up" the targets. That is, to adjust the targets to achieve the best visibility. For the MRT test, it is important that the observer count the number of bars to insure that the required number is present.

In Section 2.3 (page 36), it was shown that distortion in pulse width and amplitude occurs during the digitization process. If the Nyquist frequency is f_N and the target frequency is f_x, then when f_x/f_N is greater than about 0.9, the target phase can be adjusted so that the image appears to be a 4-bar target. When f_x/f_N is less than about 0.6, the output modulation nearly replicates the input modulation but there is some slight variation in pulse width and amplitude. In the region where f_x/f_N is approximately between 0.6 and 0.9, no matter what phase is chosen, adjacent bar modulation is always less than the input modulation (Figure 2-22: page 40). In this region, bar targets will never *look* correct. One or two bars may be either much wider than the others or one or two bars may be of lower intensity than the others. As a result the MRT may be higher in this region. For staring arrays, the MRT is well behaved (in the sense that the system precisely replicates the input) when the target spatial frequency is f_N/k where k is an integer.

Webb[18] measured the in-phase and out-of-phase MRT of a Mitsubishi IR-5120A Pt:Si staring array. By convention, the MTF is portrayed to abruptly drop to zero at the Nyquist frequency since higher frequency signals cannot be faithfully reproduced. Equivalently, the MRT is shown to approach infinity at the Nyquist frequency. The anticipated MRT is shown with a solid line in Figure 10-13. In the region of where f_x/f_N is between 0.6 and 0.9, the four-bar pattern becomes more difficult to perceive resulting in elevated results. Above $f_x/f_N = 0.9$, the MRT is again well behaved if the appropriate phase is selected. At Nyquist frequency, the out-of-phase response is zero whereas the in-phase response is finite. Webb also demonstrated that it is possible to perceive four-bar targets above Nyquist frequency if the targets are in-phase (see Figure 2-23: page 41). Out-of-phase bars are not discernible when $f_x/f_N > 1$.

These results are not often seen due to observer variability and lack of targets in the region of interest. If Webb had not collected the data at $f_x/f_N = 0.92$ and $f_x/f_N = 1.04$, the resultant MRT[19] would have appeared "well behaved" (Figure 10-14). This illustrates the need to measure the MRT at a variety of spatial frequencies. It is *not* appropriate to make measurements at a few selected spatial frequencies and then to draw a line between the data points.

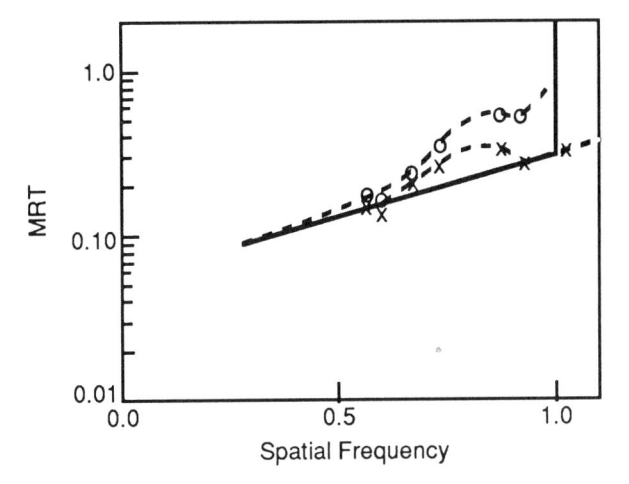

Figure 10-13: MRT of a Mitsubishi IR-5120A Pt:Si staring array (From reference 18). The MRT is shown to approach infinity at the Nyquist frequency. The anticipated response is shown as a solid line. The dashed line for $f_x/f_N > 1.0$ is the expected response for an equivalent scanning system. Both in-phase (crosses) and out-of-phase (circles) data are shown.

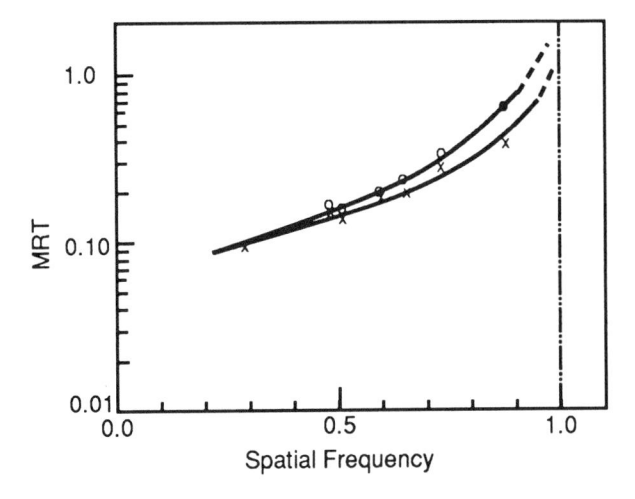

Figure 10-14: Same data as Figure 10-13 except that the high frequency values have been omitted (From reference 19). A smooth curve has been drawn through the data points.

328 *TESTING & EVALUATION OF IR IMAGING SYSTEMS*

Infrared imaging systems respond to flux and not temperature. As discussed in Section 3.1.5 (page 73), as the background (ambient) temperature drifts, the flux difference between the target and the background change for a fixed thermometric difference. Using Figure 3-12 (page 75) as an example, a 1° C drift in the ambient temperature can appear as an apparent variation of $\Delta T = 0.02$ degrees in the 8 to 12 μm region. This value will vary according to the specific system design. MRT testing take several hours depending upon the number of observers and the number of different spatial frequencies selected. Ambient temperature conditions can usually change over this long period and therefore the MRT measurements are particularly susceptible to ambient temperature changes. For reproducible results, extreme care must be exercised in controlling the ambient temperature. For low ΔT measurements, the test configuration must be baffled so that room air currents do not affect the target temperature. The reflective target configuration (Figure 4-8: page 100) provides a method to control the background temperature. MRT and MDT tests should be performed at the same background temperature as the NEDT.

For staring arrays, the amount of noise present depends upon how often the nonuniformity correction takes place and the gain/level reference points. Without continual NUC, 1/f noise can increase the fixed pattern noise. As the ambient temperature changes, the amount of noise changes depending upon the relationship between the temperature references and the background temperature (Figure 2-28: page 47).

10.2.1. SUBJECTIVE TEST METHODOLOGY

The observer is usually allowed unlimited viewing time. He can continually adjust the system (gain and level) and monitor (contrast and brightness) to optimize the image for his detection criterion. Usually the observer adjusts the monitor to a low brightness level and a high contrast level to make the image noisy. The monitor brightness and contrast control settings selected provide insight into the man-machine operation for just detectable thresholds. Observers generally choose similar gain and level settings. Large deviations from the average settings suggest that a different detection criterion may have been used. Those individuals who select unusual settings may require further instructions or training. The ambient lighting should approximately match the monitor luminance. The observer must be allowed sufficient time to dark adapt to the reduced ambient lighting before proceeding with the test. It is important that the observer not be influenced by the environment. This includes, for example, extraneous light sources, noise (e.g., air conditioning, machinery, or fans) and other people in the room.

OBSERVER INTERPRETATION of IMAGE QUALITY 321

seeing curve. The large variability is illustrated[5] in Figure 10-8. With this large variability, these tests should not be used to distinguish subtle changes such as focus errors (Figure 5-5: page 140). It cannot be emphasized strongly enough that this large variability is for the population as a whole. Any one individual or group of individuals may have much smaller variability.

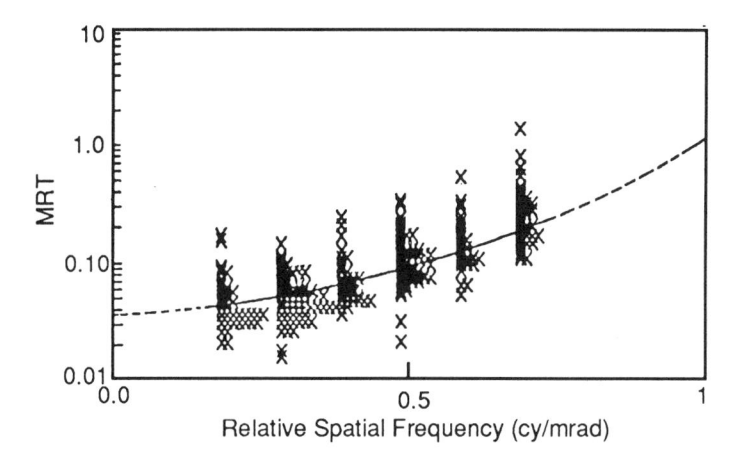

Figure 10-8: Distribution of individual MRT responses (From reference 5) for a scanning infrared imaging system. Data was collected at 0.2, 0.3, ..., 0.7 cy/mrad and is presented in a histogram format. The range of values (minimum to maximum) was approximately a factor of 10 for all spatial frequencies. (By courtesy of Martin Marietta).

☞

Example 10-3
MRT SPECIFICATION

The MRT specification is 0.1 °C. Six observers obtained thresholds of 0.08, 0.08, 0.1, 0.2, 0.2, and 0.35 °C. Did the system pass the specification?

$$MRT_{ave} = \left[\prod_{i=1}^{6} MRT_i\right]^{\frac{1}{6}}$$

322 TESTING & EVALUATION OF IR IMAGING SYSTEMS

$$MRT_{ave} = [(0.08)(0.08)(0.1)(0.2)(0.2)(0.35)]^{\frac{1}{6}} = 0.144$$

For 6 observers, K = 1.36 at a 95% confidence level and K = 1.55 at the 99% confidence level. Then MRT_{MAX} is 0.136°C and 0.155°C at the 95% and 99% confidence levels respectively. At the 95% confidence level, the specification was not met. However, there is a (100 - 95) = 5% chance that an error was made. That is, a good system was rejected. On the other hand, at the 99% confidence level, the specification was met. But the 99% value (0.155°C) is rather large and precision is traded for confidence.

This apparent dilemma is solved by repeated testing. With a good system, the average of six observers will be less than 0.136°C 95% of the time for each independent test. If the system fails the test, repeat the test. As shown in Table 10-3, 5% of the good systems will fail the test the first time. If the failed units are retested, then again 5% of these will fail. At the end of the second test, only 0.25% of the total units will have failed. Table 10-4 illustrates the pass/fail percentages when the 90% level is chosen where MRT_{MAX} = 0.127°C for 6 observers. If the system is out of specification, it should fail the first and second retests[*].

Table 10-3
FRACTION OF UNITS TESTED WITH 95% CONFIDENCE LEVEL

FRACTION TESTED	TOTAL FRACTION THAT PASSED	TOTAL FRACTION THAT FAIL
1	0.95	0.05
1st retest 0.05	0.9975	0.0025
2nd retest 0.0025	0.99988	1.21×10^{-4}

[*]The purist may also want to consider the probability of passing a unit that is defective. In statistics, this is called the "error of the second kind" and it is the probability of false alarm used in detection theory.

Table 10-4
FRACTION OF UNITS TESTED WITH 90% CONFIDENCE LEVEL

FRACTION TESTED	TOTAL FRACTION THAT PASSED	TOTAL FRACTION THAT FAIL
1	0.90	0.10
1ˢᵗ retest 0.10	0.99	0.01
2ⁿᵈ retest 0.01	0.9999	0.001

10.2. MRT and MDT TESTS

The MRT curve is usually examined in three places (Figure 10-9). First the high spatial frequency asymptote is estimated. This is simply an indication of the maximum spatial frequency resolvable through the infrared imaging system. For staring arrays and undersampled systems, the highest spatial frequency that can be faithfully reproduced is limited by the Nyquist frequency.

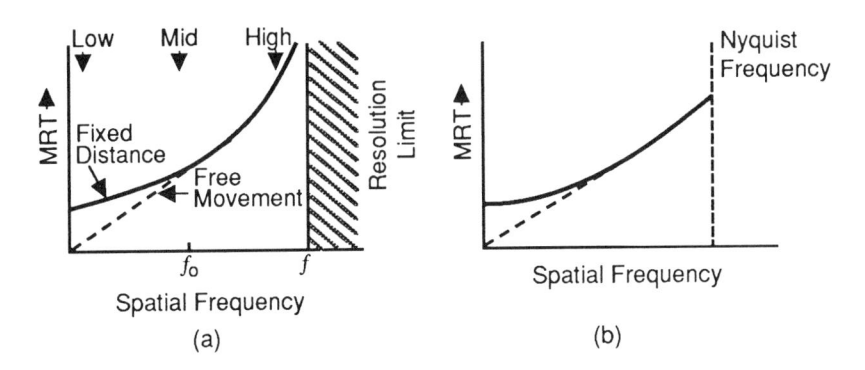

Figure 10-9: Typical MRT responses. (a) Appropriately sampled scanning systems and (b) staring array. While maintaining the head at a fixed location, the MRT will asymptote to 0.3 to 0.7 times the NEDT. With head movement, the MRT may asymptote to zero. The sampled data system response is limited by the system Nyquist frequency.

324 TESTING & EVALUATION OF IR IMAGING SYSTEMS

A second point is f_o (pronounced f-naught) which is $f_o = 1/(2 \cdot \text{IFOV})$. The MRT at f_o is an average type of sensitivity number and is often used to compare different systems. For staring arrays with 100% fill factor, f_o is the highest spatial frequency that can be reproduced and it is equal to the array Nyquist frequency. Finally the low spatial frequency value is examined. This is strongly dependent upon the viewing distance. When minimal head movement is allowed, The low frequency response asymptotes to 0.3 to 0.7 times the NEDT. When the observer is totally free to move, the low frequency asymptotes to zero in the absence of excessive low frequency spatial noise.

MDT has no limit. Small objects can be seen as long as they have sufficient intensity (Figure 10-10). MDT, also called hot spot detection, is usually plotted as a function of the target's angular subtense. For comparison to the MRT, the MDT target can be considered as one-half of a cycle (Figure 10-11). With this approach, the MDT can be plotted as a function of a fictitious spatial frequency (Figure 10-12). At low and mid spatial frequencies, the eye acts as an edge detector and the MDT and MRT tend to have the same values.

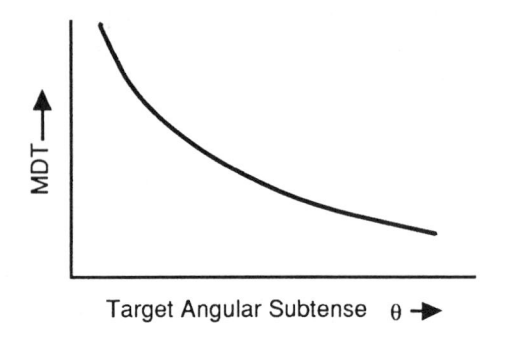

Figure 10-10: Typical MDT response.

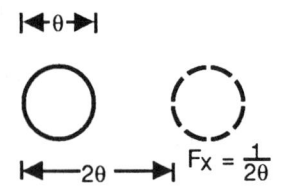

Figure 10-11: Fictitious spatial frequency associated with a MDT target.

OBSERVER INTERPRETATION of IMAGE QUALITY 329

For MRT and MDT testing, there may exist an offset between the actual temperature differential and the reported temperature differential. This occurs when the SiTF does not pass through the origin. This offset is removed from the final MRT or MDT value by obtaining the MRT or MDT for both a positive and negative contrast target and averaging the results together. The amount of offset depends upon the target-background characteristics and test set design. The offset may be different for each separate target used. It does not appear reasonable to assume that there is a universal offset for the test configuration.

The generic test configuration is shown in Figure 10-15. Since MRT and MDT are detection criteria for noisy imagery, the infrared imaging system's gain must be sufficiently high so that the image is noisy. Since high spatial frequency response is of interest, it is necessary to mount the source, targets, collimator and system on a vibration-isolated optical table. The entire MRT versus spatial frequency curve should be measured. Otherwise it is reasonable to say that the system cannot image targets with spatial frequencies greater than the maximum spatial frequency measured. Because a discrete target set is used, the location where the MRT asymptotes to infinity may not be measured. The curve may asymptote between the last resolvable target and the next available target. It is therefore imperative that the first target that is not resolvable is recorded as "CNR" (cannot resolve). No entry on the data sheet implies that the target was not used. As a conservative approach, the observer should distinguish the entire bar from top to bottom.

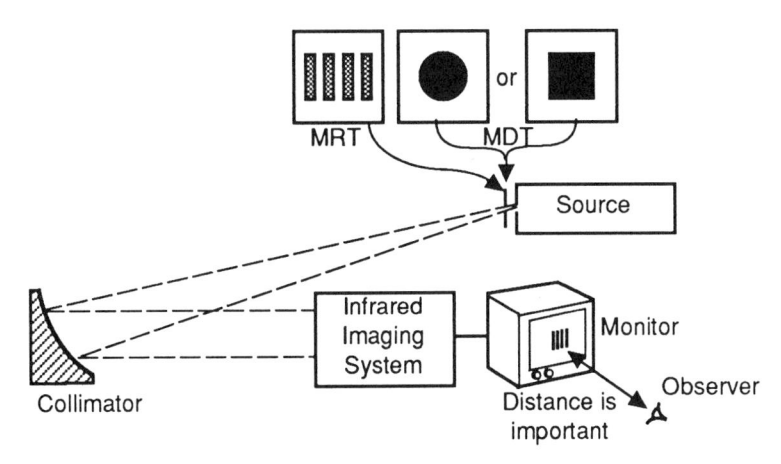

Figure 10-15: Generic MRT and MDT test configuration. The entire set up should be placed on a vibration-isolated optical table.

330 TESTING & EVALUATION OF IR IMAGING SYSTEMS

For MRT tests, the targets should range from low spatial frequencies to just past the system cutoff. Targets must span the entire spatial frequency response. Although the original NVL model[20] calculated the MDT based upon square targets, round targets are often used for MDT testing. The MDT target's angular subtense should vary from about 0.1·IFOV to 5·IFOV.

Example 10-4
MRT TARGET SELECTION

An infrared imaging system has a spatial frequency cutoff at 9 cy/mrad. What size targets are required to measure the MRT at 2, 4, ..., 8 cy/mrad. A 140 inch focal length off-axis collimator is available.

With a 7:1 aspect ratio target, the bar length is 7 times the bar width. Each cycle of the target subtends:

$$\theta = \frac{d}{focal\ length}\ mrad$$

where d is the width of one cycle (one bar plus one space). The spatial frequency is:

$$f_x = \frac{1}{1000\ \theta}\ cy/mrad$$

The bar widths, which are d/2, are given in Table 10-5. If the smallest target needed cannot be easily manufactured, then a longer focal length collimator must be used.

Table 10-5
MRT TARGET SIZES FOR A 140" COLLIMATOR

f_x (cy/mrad)	1 CYCLE (inches)	BAR WIDTH (inches)	BAR HEIGHT (inches)
2	0.0700	0.0350	0.2450
4	0.0350	0.0175	0.1125
6	0.0334	0.0117	0.0119
8	0.0176	0.0088	0.0616

OBSERVER INTERPRETATION of IMAGE QUALITY

Three possible monitor configurations are possible: (1) the tests are performed using the dedicated infrared imaging system monitor or an equivalent monitor, (2) the tests are performed with a very high quality monitor and the results are a measure of the system's response up to the analog video signal, or (3) a high quality monitor is used but the analog video is passed through a circuit that approximates the system's actual monitor performance (Figure 4-29: page 118). The monitor aspect ratio should be matched to the infrared imaging system's aspect ratio. For example, if the system has an aspect ratio of 1:1 but is formatted into a modified RS 170 video signal, the monitor also should be capable of providing a 1:1 aspect ratio image when driven by that signal.

The generalized test setup is given in Table 10-6. The MRT and MDT test procedures are given in Table 10-7. Data analysis techniques and test documentation are listed in Table 10-8. Possible causes for variations in test results are listed in Table 10-9.

Table 10-6
SUBJECTIVE TEST SETUP

1. Establish a test philosophy, criterion for success and write a thorough test plan (Section 1.3: page 12).
2. Determine if head movement is permitted (Section 10.1.2: page 312). If not, select a suitable head restraint.
3. Select qualified observers who are not fatigued nor have any other immediate physical, social, emotional, psychological problems that may affect his threshold detection (Section 10.1.4: page 317).
4. Verify that the test equipment is in good condition and that the test configuration is appropriate (Figure 10-15: page 329). Ask previous users if any problems were noticed.
5. Verify that the infrared imaging system is in focus (Section 5.2: page 137).
6. Verify the spectral response of the system and its relationship to the source characteristics, collimator spectral transmittance and atmospheric spectral transmittance (Section 4.3: page 105, and Section 4.4: page 112).
7. Insure that the infrared imaging system has reached operating equilibrium before proceeding.
8. Set the ambient lighting approximately at the same level as the monitor. Allow the observer to dark adapt to the illumination level.

Table 10-7
MRT/MDT TEST PROCEDURE

1. Select a four-bar 7:1 aspect ratio target of the appropriate spatial frequency for the MRT test or an appropriate round or square target for the MDT test.
2. For the MRT test, position the target with the bars oriented vertically to obtain the horizontal MRT or horizontally to measure the vertical MRT.
3. Adjust target phase for maximum visibility.
4. For the MRT test, verify four bars are visible by counting them.
5. Establish a positive subthreshold temperature differential and slowly increment the blackbody temperature differential.
6. Allow the observer to adjust system and monitor controls continuously to optimize the image.
7. Record the temperature differential at which the observer can resolve all four bars 50% of the time for the MRT or detect the target for the MDT.
8. Establish a negative subthreshold temperature differential and slowly decrement the blackbody temperature differential.
9. Record the temperature differential at which the observer can resolve all four bars 50% of the time for the MRT or detect the target for the MDT.
10. MRT and MDT are the average of the absolute values of positive and negative temperature differential recordings.
11. Repeat for other spatial frequency targets.
12. If a target cannot be resolved, record "CNR."

OBSERVER INTERPRETATION of IMAGE QUALITY

Table 10-8
DATA ANALYSIS and TEST DOCUMENTATION

1. Multiply all observations by the spectrally weighted collimator and atmospheric transmittances to determine the effective source temperature differential at the entrance pupil of the infrared imaging system (Section 3.2: page 79).
2. Perform the tests with at least three observers.
3. Geometrically average the individual observer responses.
4. Repeat tests for different target locations and orientations.
5. Fully document any test abnormality and document all results. All data for each observer and the average values should be recorded. Present the data both in tabular and graphical form. Record the head to monitor distance for each target. Record the monitor brightness and contrast control settings that were optimized by the observer. Record the ambient temperature. As appropriate, record the system Nyquist frequency.

Table 10-9
POSSIBLE CAUSES for POOR or NONREPRODUCIBLE RESULTS

- System out of focus (Section 5-2: page 137).
- Unexpected spectral noise components (Figure 10-5: page 316).
- Large observer variability (Section 10.1: page 309).
- Phasing effects (Figure 10-13: page 327).
- Noise characteristics changing with time.
- Ambient temperature not specified (Figure 3-14: page 76).
- Ambient temperature fluctuating (Section 3.1.5: page 73).
- Gain/level normalization variations (Figure 2-28: page 47).

10.2.2. SEMIAUTOMATIC TEST METHODOLOGY

In the subjective (manual) method, the observer defines threshold only once. A single event is insufficient to define a 50% probability point. To minimize variability and to determine rapidly an observer's threshold, a semiautomatic approach can be used[21]. The primary advantage of this "up and down" technique is that it automatically concentrates testing around threshold. The observer tracks his threshold for several minutes and the average value is easily seen.

Figure 10-16 illustrates the semiautomatic test configuration. The observer decreases the temperature differential when he just perceives the target and increases the temperature differential when the target just disappears. The temperature differential is monitored continuously and graphed on a strip chart recorder. Statistical analysis of the input temperature differential excursions yields both the mean and an estimate of the standard deviation[22]. For MRT and MDT, it is sufficient to determine the average value. With the semiautomatic method an individual's threshold is better defined. The method does not change the inter-observer variability and the log-normal distribution with $\sigma = 1.58$ is still appropriate when describing the responses from many individuals.

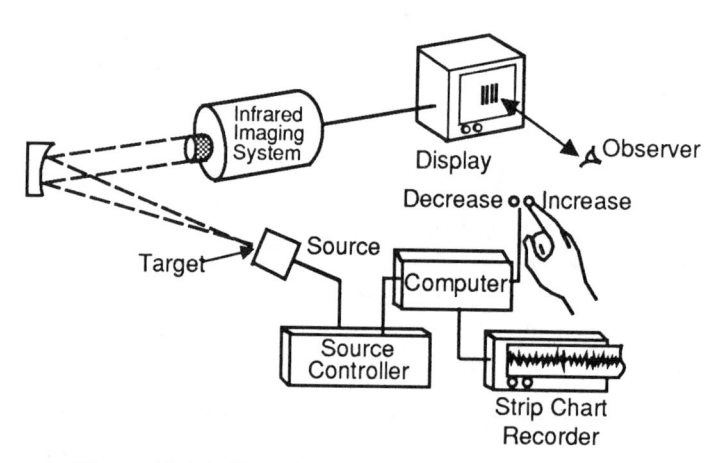

Figure 10-16: Generic semiautomatic test configuration.

The continuous trace provides some very important information about the observer. First, the observer variability is rapidly assessed (Figure 10-17). Observers with small excursions about the mean are generally considered more consistent. Second, there is usually a learning curve associated with reaching minimum detection (Figure 10-18). Observers cannot instantaneously reach threshold but require some time to adjust to the task. It is this learning curve that causes the MRT test to take so much time. Finally, distractions are obvious (Figure 10-19). Even with a distraction, the observer's threshold can still be determined.

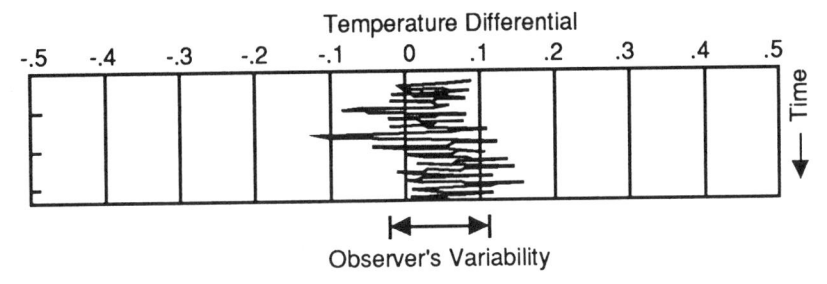

Figure 10-17: Observer variability (From reference 21). (By courtesy of Martin Marietta).

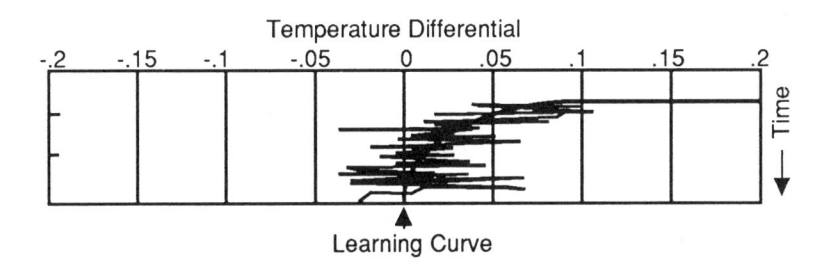

Figure 10-18: Typical learning curve seen on all tests. (From reference 21). (By courtesy of Martin Marietta).

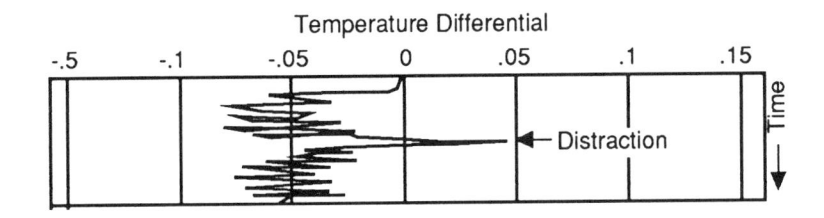

Figure 10-19: Typical distraction (From reference 21). (By courtesy of Martin Marietta).

336 TESTING & EVALUATION OF IR IMAGING SYSTEMS

For successful implementation of the semiautomatic method, the temperature differential increments used for the source must be matched to the eye's probability of seeing curve. If the increments are too small, the observer loses interest because the test takes too long. If the increments are too large, threshold values are passed too quickly and again the observers become disenchanted. The eye responds logarithmically to input stimuli but differential temperature controllers usually respond linearly. For probability of seeing values between 20% and 80%, the eye's response can be approximated by a linear probability curve. During the preliminary design of the semiautomatic procedure, it was experimentally determined that observers were comfortable with differential temperature increments equal to $\sigma/2$. Thus, the entire probability-of-seeing curve can be traced out with approximately 12 equi-spaced σ increments. When scaled into linear space, the increments become dependent upon the expected MRT or MDT value with the increment approximately equal to 0.29 times the expected MRT or MDT value. This illustrated in Figure 10-20. For example, if the expected MRT is 1° C, the temperature controller increments should be approximately 0.29° C. It may be necessary to adjust the incremental value for each observer.

This method requires a computer controlled source that has a fast settling time, fast slew rate and is critically damped. If the source controller does not exhibit these features, large overshoots in the temperature differential can render the test useless. Similarly, with some very slow time constant sources the test takes too long and the observer loses interest. The test procedures are identical with those presented in Table 10-7.

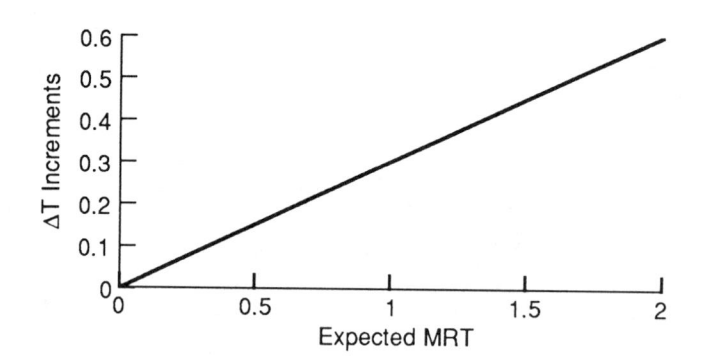

Figure 10-20: ΔT increments required for the semiautomatic test.

10.2.3. OBJECTIVE TEST METHODOLOGY

Because of the large variability in observer responses and the long time it takes to perform a subjective test, recent efforts have concentrated on developing an objective or automated MRT test. The ability to quantify the eye-brain detection process has been perhaps the greatest obstacle to perfecting an objective MRT measurement technique. Cuthbertson et. al.[23] suggested that an objective MRT can be calculated from the measured MTF and NEDT. De Jong and Bakker[24] calculated the MRT using a similar technique but with an improved eye model. The advantage of this approach is that any noise power spectral density spectral feature that would increase the MRT is considered.

Edwards[25] and later Gunderson[26] used a frame grabber to provide a histogram of target and background pixel values. Their underlying assumption is that there is a unique threshold between target and background pixels that is related to the observed MRT. Quantization effects of the frame grabber and the phasing effects associated with insufficient number of samples on the target places a limiting factor on their technique. As shown in Figure 4-34 (page 122), probably about 8 samples per highest frequency are required to reproduce the signal intensity and pulse width faithfully. This requirement affects the highest spatial frequency that can be tested.

Example 10-5
STARING ARRAY AUTOMATIC MRT TESTING
(Pixel histogram approach)

What is the highest MRT spatial frequency that can be measured unambiguously for a staring array consisting of 128 x 128 detectors? The system's FOV is 30 mrad.

To avoid phasing effects between the detector array and the target location, the requirement of 8 samples per highest frequency limits the automatic MRT technique to 0.25 times the Nyquist frequency. The array Nyquist frequency is $128/(2 \cdot 30) = 2.13$ cy/mrad and the highest MRT spatial frequency that can be measured accurately is 0.53 cy/mrad.

Example 10-6
ANALOG VIDEO AUTOMATIC MRT TESTING
(Pixel histogram approach)

What is the highest MRT spatial frequency that can be measured on the analog video when the signal is acquired by a frame grabber? The frame grabber digitizes the analog video into 640 samples horizontally. The system's HFOV is 30 mrad.

The frame grabber Nyquist frequency is 640/(2·30) = 10.67 cy/mrad. By requiring 8 samples per frequency to avoid phasing effects, the highest MRT spatial frequency that can be accurately measured is 2.67 cy/mrad. The highest spatial frequency may be limited by the array (Example 10-5) or by the measurement technique.

The eye provides significant temporal and spatial filtering. For low spatial frequency targets, the eye can perceive images that have a signal-to-noise ratio much less than unity. A single video analog trace often shows a poor signal-to-noise ratio which may appear surprising since the perceived image is good. Thus many frames must be averaged to achieve a video signal that appears representative of the perceived image. This means that to separate target pixels from background pixels, many frames of data must be averaged to reduce the noise so that the target pixels can be measured.

The limited success of predicting laboratory MRT values and comparing subjective to objective results is due, in part, to the variability of observer response, using too few observers, inappropriate use of the arithmetic average, experimental difficulties in obtaining the MTF, incorrect eye modeling and the failure to account for head movement during actual testing.

By allowing the observer to adjust his viewing distance to the monitor, he apparently optimizes several interrelated detection criteria that include striving for equal clarity of all four bars and maximizing his perceived signal-to-noise ratio. This apparently results in an equal detection capability for all spatial frequencies such that eye's contrast sensitivity approaches a constant.

When the observer's distance does not exceed his comfort zone, the predicted MRT[27] is approximately:

$$MRT(f_x) \approx K_2 \frac{NEDT}{MTF(f_x)} \qquad 10\text{-}6$$

This simple relationship appears to be valid for mid and high spatial frequencies where moderate head movement is permitted. It assumes that the monitor does not degrade the system MTF. Thus the measured MTF obtained from the analog video signal before the monitor is considered identical with the MTF if it were measured on the monitor. That is, the monitor MTF is approximately unity over the spatial frequencies of interest. The NEDT (Table 7-4: page 208) and MTF (Table 8-5: page 277) are obtained experimentally.

The MRT values in Figure 10-8 (page 321) were geometrically averaged and a proportionality constant of 0.7 was obtained for a digitally scan-converted system (Figure 10-21). Proportionality constants ranging from 0.3 to 0.7 have been reported for different systems and the "constant" may be a function of spatial frequency.

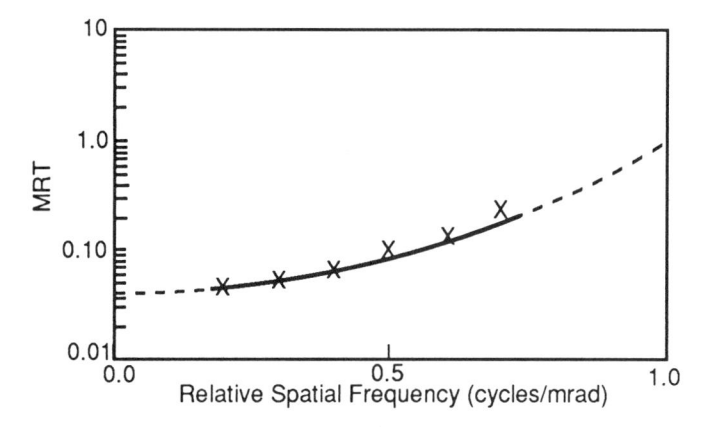

Figure 10-21: Geometrically averaged MRT observations compared to $MRT(f_x) = 0.7\ NEDT/MTF(f_x)$ (From reference 27). (By courtesy of Martin Marietta).

FLIR92 incorporates all the noise sources[28] contained in the three-dimensional noise model. For the horizontal MRTs, the proportionality constant is

$$K_{2H} = \frac{\left[\sigma_{TVH}^2 + k_1 \sigma_{VH}^2 + k_2 \sigma_{TH}^2 + k_3 \sigma_H^2\right]^{\frac{1}{2}}}{\sigma_{TVH}}$$

and for the vertical MRT, the factor is

$$K_{2V} = \frac{\left[\sigma_{TVH}^2 + k_4 \sigma_{VH}^2 + k_5 \sigma_{TH}^2 + k_6 \sigma_V^2\right]^{\frac{1}{2}}}{\sigma_{TVH}}$$

where $k_1, ..., k_6$ are eye spatial and temporal integration factors that may be spatial frequency dependent. Thus, systems with different noise sources will have different proportionality factors.

While the automatic MRT test relieves the test engineer from the MRT task, the test engineer cannot be totally bypassed. It is impossible to predict the effects of spectral noise on image quality. Therefore, an observer is always required to provide the final decision whether the noise is bothersome and whether image quality is acceptable. The objective test may be appropriate during production runs after the proportionality constant has been experimentally obtained. It is not recommended for prototype testing.

10.3. SUMMARY

Every time an infrared imaging system is turned on, the observer subconsciously makes a judgment about image quality according to his internal rating scale. In Section 1.2.2 (page 9), the Cooper-Harper methodology was exploited to standardize the rating scale. MRT and MDT are standard tests that provide a subjective measure of image quality. They provide only one datum in the complex assessment of image quality. As a laboratory test, the results may or may not directly correlate with actual field performance. Field test methodologies must consider platform vibration, ambient lighting and observer distractions.

The average value of individual thresholds is obtained by calculating the geometric average. Because of large observer variability, just one detection threshold is more representative of the observer than the system under test. A single MRT value provides little insight into the system performance unless the observer is well trained, qualified and consistent. These observer attributes have been vaguely defined. They may only be ascertained after the observer has performed the test often and his results have been monitored over a long time. The semiautomatic method provides insight into observer consistency. The semiautomatic approach is better from a statistical sense than the subjective method.

Test results suggest that an observer's detection threshold follow a lognormal distribution. The frequency of seeing curve has a standard deviation of $\sigma = 1.58$ and it can be used to estimate the confidence in the data collected. The statistical approach separates unacceptable systems from observer variability effects. It is sometimes suggested that a target with the specified MRT be presented to the observer. If he can see it then the system passes the specification. If he cannot, the system fails. With large observer variability this does not appear to be a viable approach and therefore not prudent to assign a single value for pass/fail.

Since the eye has a very definite contrast sensitivity response, the distance to the monitor dramatically affects the observer's detection threshold. The observer tends to move to his comfort zone (4 to 8 times the monitor height). However, this limits his detection capability of low spatial frequency targets. This may be just an exercise in visual testing because the intended use of the system rarely requires the observer to be more than a few feet from the monitor. It may be appropriate to constrain head movement to the distance anticipated during actual use.

By placing the test configuration on a vibration-isolated optical table, the best MRT will be obtained. The spatial frequency cutoff for undersampled systems is limited by the sampling frequency. Vibration can mimic microscan and can effectively increase the cutoff frequency above the Nyquist frequency limit. The amount of vibration present is test configuration specific. It depends upon the table used, building construction and the location of vibration sources such as hallways, machinery, roads, traffic, and rail roads.

The automatic test procedure given is a guideline for establishing an objective test procedure. Any automatic method selected must be verified with perhaps hundreds of subjective tests for validation purposes. The equation, MRT = K·NEDT/MTF, appears appropriate for digitally scan-converted thermal imaging systems when the proportionality constant is 0.7. The proportionality constant appears appropriate for mid spatial frequencies. Other systems may require different factors.

The test engineer must be thoroughly knowledgeable in the system operation as well as the specific test requirements and pitfalls. Due to phasing effects, there is not a unique MRT value for each spatial frequency. The MRT will exhibit increased values in the region where f_x/f_N is between 0.6 and 0.9. The targets should be "peaked up" to maximize visibility. Typical specifications are given in Table 10-10.

Table 10-10
TYPICAL SPECIFICATIONS

- The MRT shall not be greater than 0.5 °C (average of three observations) at 5 cy/mrad when the ambient temperature is at 20°C. The visual angle subtended by the observer shall be 25° ±0.2°.
- The MDT shall not be greater than 0.3 °C (average of 3 observations) for at a target whose angular subtense is 3 mrad when the ambient temperature is at 20°C. The observer may adjust the gain and level with no restriction on head location. The monitor shall be a high quality monitor whose bandwidth is greater than the infrared imaging system's bandwidth.

10.4. REFERENCES

1. I. Overington, Vision and Acquisition, pp. 32-47: Pentech Press, London (1976).
2. I. Overington, "Image Quality and Observer Performance", in Image Quality, J. Cheatham, ed., SPIE Proceedings Vol. 310, pp. 2-9 (1981).
3. C. W. Hoover Jr. and C. M. Webb, "What is an MRT? And How Do I Get One", in Infrared Imaging Systems: Design, Analysis, Modeling and Testing II, G. C. Holst, ed.: SPIE Proceedings Vol. 1488, pp. 280-288 (1991).
4. P. S. Brown, "Strategies for Testing Electro-Optical Devices", in AUTOTESCON 1987 Proceedings of the International Automatic Testing Conference, pp. 59-63 (1988).
5. G. C. Holst and J. W. Pickard, "Analysis of Observer Minimum Resolvable Temperature Responses" in Imaging Infrared: Scene Simulation, Modeling, and Real Time Image Tracking, A. J. Huber, M. J. Triplett, and J. R. Wolverton, eds.: SPIE Proceedings Vol. 1110, pp. 252-257 (1989).
6. B. O. Hultgren, "Subjective Quality Factor Revisited", in Human Vision and Electronic Imaging: Models, Methods and Applications, B. E. Rogowitz and J. P. Allebach, eds.: SPIE Proceedings Vol. 1249, pp. 12-22 (1990).
7. J. M. Mooney, "Effect of Spatial Noise on the Minimum Resolvable Temperature of a Staring Array", Applied Optics, Vol. 30(23), pp. 3324-3332 (1991).

8. S. Daly, "Application of a Noise Adaptive Contrast Sensitivity Function in Image Data Compression", Optical Engineering, Vol. 29(8), pp. 977-987 (1990).
9. H. Pollehn and H. Roehrig, "Effect of Noise on the Modulation Transfer Function of the Visual Channel", Journal of the Optical Society of America, Vol. 60, pp. 842-848 (1970).
10. A. Van Meeteren and J. M. Valeton, "Effects of Pictorial Noise Interfering With Visual Detection", JOSA A, Vol. 5(3), pp. 438-444 (1988).
11. C. M. Webb and G. C. Holst, "Observer Variables in Minimum Resolvable Temperature Difference", in Infrared Imaging Systems: Design, Analysis, Modeling and Testing III, G. C. Holst, ed.: SPIE Proceedings Vol. 1689, pp. 356-367 (1992).
12. R. LaFollette and J. Horger, "Thermal Training for Military Observers", in Infrared Imaging Systems: Design, Analysis, Modeling and Testing II, G. C. Holst, ed.: SPIE Proceeding Vol. 1488, pp. 289-299, (1991).
13. T. H. Cook, C. S. Hall, F. G. Smith, and T. J. Rogne, "Simulation of Sampling Effects in FPAs", in Infrared Imaging Systems: Design, Analysis, Modeling and Testing II, G. C. Holst, ed.: SPIE Proceedings Vol. 1488, pp. 214-225 (1991).
14. The Vision Contrast Testing System is manufactured by Vistech Consultants, Inc., 4162 Little York Road, Dayton Ohio 45414-2566
15. W. J. Dixon and F. J. Massey, Introduction to Statistical Analysis, pp. 79- °0: McGraw-Hill, New York (1957).
16. F.W. Campbell, J. J. Kulikowski, and J. Levinson, "The Effect of Orientation on the Visual Resolution of Gratings", Journal of Physiology, Vol. 187, p 427-436 (1966).
17. J. T. Wood, W. J. Bentz, T. Pohle, and K. Hepner, "Specification of Thermal Imagers", Optical Engineering, Vol. 15(6), pp. 531-536 (1976).
18. C. M. Webb, "Results of Laboratory Evaluation of Staring Arrays", in Infrared Imaging Systems: Design, Analysis, Modeling and Testing, G. C. Holst, ed.: SPIE Proceedings Vol. 1309, pp. 271-278 (1990).
19. G. C. Holst, "Effects of Phasing on MRT Target Visibility", in Infrared Imaging Systems: Design, Analysis, Modeling and Testing II, G. C. Holst, ed.: SPIE Proceedings Vol. 1488, pp. 90-98 (1991).
20. J. Ratches, W. R. Lawson, L. P. Obert, R. J. Bergemann, T. W. Cassidy, and J. M. Swenson, Night Vision Laboratory Static Performance Model for Thermal Viewing Systems, US Army Electronics Command Report ECOM Report 7043, Ft. Monmouth, NJ (1975).
21. G. C. Holst, "Semi-automatic MRT Technique", in Infrared Technology XII, I. J. Spiro and R. Mollicone, eds.: SPIE Proceedings Vol. 685, pp. 2-5 (1986).
22. W. J. Dixon and F. J. Massey, Introduction to Statistical Analysis, pp. 318- 327: McGraw-Hill, New York (1957).
23. G. M. Cuthbertson, L. G. Shrake, and N. J. Short, "A Technique for the Objective Measurements of MRTD", in Infrared Technology and Applications, L. Baker and J. Masson, eds.: SPIE Proceedings Vol. 590, pp. 179-192 (1985).
24. A. N. de Jong and S. J. M. Bakker, "Fast and Objective MRTD Measurement", in Infrared Systems-Design and Testing, P. R. Hall and J. S. Seeley, eds.: SPIE Proceedings Vol. 916, pp. 127-143 (1988).
25. G. W. Edwards, "Objective Measurements of Minimum Resolvable Temperature Difference (MRTD) for Thermal Imagers", in Image Assessment: Infrared and Visible, T. L. Williams, ed.: SPIE Proceedings Vol. 467, pp. 47-54 (1983).
26. J. A. Gunderson, "Results of Objective Automatic MRT Testing of Thermal Imagers Using a Proposed New Figure of Merit", in Automatic Testing of Electro-optical Systems, J. Nestler and P. I. Richardson, eds.: SPIE Proceedings Vol. 941, pp. 14-17 (1988).
27. G. C. Holst, "Minimum Resolvable Temperature Predictions, Test Methodology and Data Analysis", in Infrared Technology XV, I. Spiro, ed.: SPIE Proceedings Vol. 1157, pp. 208-218 (1989).

28. L. Scott and J. D'Agostino, "NVEOD FLIR92 Thermal Imaging Systems Performance Model" in <u>Infrared Imaging Systems: Design, Analysis, Modeling and Testing III</u>, G. C. Holst, ed.: SPIE Proceedings Vol. 1689, pp. 194-203 (1992).

EXERCISES

1. Find the average value for the following MRT observations: 0.5, 0.5, 0.6, 0.6, 0.7° C.
2. Find the average value for the following MRT observations: 0.5, 0.5, 0.6, 0.6, 1.3° C.
3. Find the average value for the following MRT observations: 0.6, 0.6, 0.7° C.
4. Find the average value for the following MRT observations: 0.6, 0.6, 1.3° C.
5. Discuss the effect of the 0.7 and 1.3°C observations on the results obtained in Exercises 1 through 5.
6. List 10 psychophysical features that may affect detection threshold.
7. During actual usage, a 18-inch monitor will be viewed at a distance of 48 inches. The test engineer has only a 6-inch monitor. How far away should he be to simulate field usage?
8. If viewing a 4 cy/deg target, what is the observer spatial frequency if the test engineer moves 2 inches closer to the 6" monitor described in Exercise 7. Compare this result when the user moves 2 inches closer to the 18-inch monitor. What can be said about variation in head movement?

INDEX

A/D converter 23, 32, 44, 120, 135, 161, 185, 227, 244, 247, 250, 256
Aberrations 23, 68, 106, 250
Abingdon Cross benchmark 305
AC coupling 32, 63, 103, 105, 160, 212, 265
Accuracy 124, 303
Aerial reconnaissance 152
AGC 32, 105
Airy disk 144, 257
Algorithm efficiency 305
Aliasing 138, 223, 240
Amplitude normalization 270
ANSI target 302
Aperiodic transfer function (ATF) 69, 145, 176
Area weighted average resolution (AWAR) 151
Artifacts 23
Aspect ratio 51, 84, 116
Atmospheric transmittance 4, 17, 61, 112
Atmospheric turbulence 115, 181, 248
Automated MRT 337
Automated vision system 288, 302

Background limited performance (BLIP) 193
Baffles 110
Bar/dot pattern 295
Beat frequency 38, 301
Bias 124
Black hot 54
Blackbody 59, 90
Blur diameter 68, 144
Blur efficiency 71, 176
Boost 25, 44, 265

Calibration points 45, 213
Centroid location 296
Closely spaced objects (CSO) 144, 303
Cold spike 25
Collimator transmittance 63, 108
Collimators 105
Comfort zone 313
Common module 26, 29, 30, 41, 48, 65, 205

Complex target 144
Confidence 320
Contrast sensitivity 317
Contrast threshold 312
Contrast transfer function (CTF) 52, 141, 238
Cooler 30
Cooper-Harper 9, 115, 218, 340
Cosine$^N\theta$ 23, 63, 108, 165

Data analysis 14
DC coupled 160, 212
DC restoration 33, 162
Depth of focus 137
Detection 6
Detector pitch 42, 244
Diagonal line target 298
Diffraction 68, 144
Digitization 21, 36, 247, 256, 326
Distortion 288, 294
Documentation 15
Drift 261
Droop 33, 170
Dynamic range 32, 44, 118, 119, 184, 217, 225

Edge detection 144
Edge spread function (ESF) 249
Effective instantaneous field of view (EIFOV) 147
Emittance 61, 91, 237
Ensquared power 71, 141, 145, 176
EO mux 240
Ergodic 196
Extended source 57, 63, 92

F/# 62, 106
Far infrared (FIR) 4
Fast Fourier transform (FFT) 266
Field of view (FOV) 289
Field testing 16
Fill factor 36
Fixed pattern noise (FPN) 30, 45, 79, 199, 203, 211, 317
Flicker 261
Flood illumination 57, 164, 178, 204, 225

346 TESTING & EVALUATION OF IR IMAGING SYSTEMS

Flying spot scanner 250
Focus 134, 321
Fourier transform 223, 237, 247, 266
Frame grabber 116
Frequency of seeing 310
Frequency resolution 266
Frequency scaling 271

Gain/level correction 23, 30, 45, 93, 168, 200, 203, 213
Gamma 21, 48
Gaussian 126, 193, 311
Gaussian MTF 272
Geometric distortion 26, 294
Geometric transfer function 288
Ghosting 28
Glare 25
Gray level 44, 98, 135
Ground resolved distance (GRD) 134, 152

Hot spot detection 176, 324

Image evaluation 2, 8
Image processing 44, 305
Image quality 1, 5, 9, 44, 51, 137, 144, 235, 289, 308, 340
Imaging resolution 149, 177
In-phase CTF (IPCTF) 52, 244
Infrared search and track (IRST) 2, 92, 145, 189, 303
Input-to-output 5
Interlace target 298
Isoplanatic 247, 250

Jaggies 37
Jitter 261

Kirchoff's law 91
Knife edge response 249

Lambertian source 61
Laser speckle 237
Leakage 268
Least significant bit (LSB) 44
Least squares 127
Light emitting diodes (LED) 42, 240
Line spread function (LSF) 141, 177, 247
Line-to-line interpolation 297
Linearity 32, 45, 49, 74, 169, 184
Log-normal distribution 311
Long wavelength infrared (LWIR) 4

Machine vision 2, 17, 21, 288, 302
Magnification 62
Mean-variance technique 225
Measurement resolution 149, 177
Microphonics 197, 223
Microscan 341
Mid-wavelength IR (MWIR) 4
Minimum detectable temperature (MDT) 2, 9, 156, 234, 308
Minimum resolvable temperature (MRT) 6, 9, 14, 39, 74, 140, 156, 234, 308
Modulation transfer function (MTF) 141, 142, 234, 247
Moire pattern 37, 138, 155
Monitor 21, 48, 51, 83, 115, 135, 137, 312
Monitor luminance 119

Narcissus 25, 200, 217
Nichrome wire 102
Nodal point 289
Noise 264
Noise equivalent bandwidth 222
Noise equivalent differential temperature (NEDT) 74, 170, 199, 201, 205
Noise equivalent flux density (NEFD) 219
Noise equivalent irradiance (NEI) 192
Noise power spectral density (NPSD) 222, 316
Noise statistics 193
Nonlinearity 74, 87
Nonuniformity 29, 34, 199, 214, 223, 261, 316
Nonuniformity correction (NUC) 45
Normalization 79
Nyquist frequency 36, 123, 141, 223, 234, 244, 266, 323, 326

Objective MRT 337
Objective test 308
Observer qualification 317
Observer variability 137, 140, 309, 317
Off-axis parabolic 106
Optical transfer function (OTF) 234
Optical transmittance 61, 81

Pass/fail criterion 14, 292, 341
Passive target 101
Pedestal 260
Phase transfer function (PTF) 234, 279

INDEX 347

Phasing effects 103, 135, 155, 243, 256, 317, 337
Physical measures 8
Picture element (pixel) 149
Pinhole target 295
Planck's blackbody law 59
Plastic target 99
Point source 66, 69, 91, 219
Point source detection 145, 176
Point spread function (PSF) 176, 247
Point visibility factor (PVF) 71, 176, 219
Poisson statistics 193, 226
Precision 124

Quantitative test 137
Quantization 44, 120

Radiant exitance 59
Radiant sterance 58
Radiometry 57
Raster pattern 138, 314
Raster scan 26
Rating scale 9
Rayleigh criterion 144
Reconstruction 51
Record length 267
Rectangular target 295
Reflective target 99
Repeatability 124, 303
Resolution 6, 34, 48, 51, 134, 144, 282
Resolution element (resel) 149
Responsivity 32, 65, 160
Responsivity uniformity 167
Ringing 170
RS 170 48, 116, 258

Sample-scene phase 243, 250
Sampling 36, 38, 116, 121, 247, 250, 325
Saturation 32, 160, 184, 225
Sayce target 138
Scan efficiency 29, 300
Scan noise 25, 200
Scan nonlinearity 298
Scan velocity 301
Scanners 26
Scene generators 98
Semiautomatic MRT 334
Sensitivity 6, 192
Shading 21, 23, 108, 200, 214, 261
Short wavelength infrared (SWIR) 4

Signal transfer function (SiTF) 57, 66, 73, 109, 160, 192, 207, 213
Slit response function (SRF) 149, 176, 234
Sobel operator 144
Sparrow criterion 144
Spatial frequency 82
Spatial resolution 135
SPRITE 26
Square-wave response 238
Standard Correlation Target Set (SCTS) 303
Star target 137
Statistical analysis 124
Step response 249
Strapping 32
Student t-test 173
Subjective 136
Subjective evaluation 9
Subjective test 308
Sweep frequency 137
Sweet spot 214, 288

T/# 64
Tangential sampling 253
Target 93
Target projector 90
Target signature 3, 17
Target simulator 90
Target transfer function (TTF) 71, 176
Target wheels 101
TDI 26, 301
Temperature references 45
Test equipment 12, 13, 120
Test philosophy 12
Test plan 12, 13
Test procedures 12
Thermal derivative 73, 208
Thermoelectric 92
Three-dimensional noise model 192, 195, 316, 325
Threshold 309, 310
Tonal transfer curve (TTC) 54
Trend analyses 15
Trends 170, 261, 210
Tri-bar target 139

Uncooled detectors 170
Uniformity 167

Very long wave infrared (VLWIR) 4
Vibration 181, 341
Vignetting 25, 108
Visual angle 312

Wedge target 137
White hot 54

Young's fringes 237

Z statistic 319